Darwin's Hunch

Darwin's Hunch

*Science, Race and the Search
for Human Origins*

Christa Kuljian

www.christakuljian.com

 https://www.facebook.com/DarwinsHunch/

 @ChristaKuljian

First published by Jacana Media (Pty) Ltd in 2016
Second impression 2017

10 Orange Street
Sunnyside
Auckland Park 2092
South Africa
+2711 628 3200
www.jacana.co.za

ISBN 978-1-4314-2425-2

Cover design by publicide
Cover image: Bernhard Zipfel, University Curator of Fossil
    and Rock Collections, Evolutionary Studies Institute,
    University of the Witwatersrand
Set in Sabon 9.5/13.5pt
Job no. 003085

Also available as an e-book:
d-PDF ISBN 978-1-4314-2489-4
ePUB ISBN 978-1-4314-2490-0
mobi file ISBN 978-1-4314-2491-7

See a complete list of Jacana titles at www.jacana.co.za

For my parents, Robert and Jewel Kuljian,
who made it possible for me to study
the history of science

Science, since people must do it, is a socially embedded activity. It progresses by hunch, vision and intuition. Much of its change through time does not record a closer approach to absolute truth, but the alteration of cultural contexts that influence it so strongly.

– Stephen Jay Gould, *The Mismeasure of Man*, 1981

We are all connected,
the living and the dead.

We arrive in life and then it
walks away from us;
leaving our bones behind
and our minds are as wide as the universe.

Human evolution began in this corner of the earth,
our ancestors left their dust as chromosomes
in each of us, they made a home,
and hominids stood up as humans and walked
languages fell out of our mouths and talked, walked their way
into and out of landscapes and mindscapes …

We are all connected,
the living and the dead,
and our minds
are as wide
as the universe.

– From "Anthem" in *The Everyday Wife*
  by Phillippa Yaa de Villiers

Contents

Note on Language

Nomenclature is often difficult and complex. Many terms that were used historically, such as "Bushmen", "Bantu", "Negro" and "Coloured", cause great offence. The derogatory colonial term "Bushmen" was used to describe various groups of hunter-gatherers and "Hottentots" was used for pastoralists. Today, the terms KhoeSan, Khoisan or San are most commonly used. However, the term "Bushman" is still used by some people to identify themselves and the term "Hottentot" is considered an insult. I have chosen to use inverted commas to indicate that I am using a word as it was used in its historical context. As the terms used changed over time, the text will often indicate when the terms changed and why.

Timeline of Selected Fossil Excavations & Key Events

1856   Neanderthal (*Homo neanderthalensis*), Neander Valley, Germany

1868   Cro-Magnon (*Homo sapiens*), Les Eyzies, France

1871   Charles Darwin publishes *The Descent of Man*

1891   Java Man (*Homo erectus*), Trinil, Java, Indonesia

1912   Piltdown Man is announced, Sussex, England

1923   Raymond Dart arrives in South Africa

1924   Taung Child (*Australopithecus africanus*), Taung, South Africa

1928   Peking Man (*Homo erectus*), Zhoukoudien, China

1933   Kanam skull, Olduvai, Tanzania

1936   Raymond Dart leads Wits expedition to the Kalahari

1936   *Australopithecus africanus*, Sterkfontein, South Africa

1938   *Paranthropus robustus*, Kromdraai, South Africa

1939   /Keri-/Keri dies in Oudtshoorn

1945   Phillip Tobias leads group of Wits students to the Makapansgat Valley

1947   Mrs Ples (Sts5) (*Australopithecus africanus*), Sterkfontein, South Africa

1948   National Party comes into power and begins to implement apartheid

1950   First UNESCO Statement on Race

1951   Hertha De Villiers begins working with Raymond Dart at Wits

1953   Piltdown Man is announced as a hoax

1959   Phillip Tobias takes over from Raymond Dart as the head of department of anatomy, Wits University

1959	Zinj (*Paranthropus boisei*), OH 5, Tanzania
1959	Phillip Tobias begins working with Mary and Louis Leakey
1960	Johnny's Child (*Homo habilis*), OH 7, Tanzania
1961	Tobias publishes *The Meaning of Race*; also leads exhumation of Cornelius Kok II grave at Campbell
1965	Bob Brain begins work at Swartkrans
1966	Tobias re-opens Sterkfontein
1972	KNM 1470 (*Homo rudolfensis*), Kenya
1974	Lucy (*Australopithecus afarensis*), Hadar, Ethiopia
1975	Alan Morris arrives at Wits as PhD student with Phillip Tobias
1984	'Ancestors' exhibit at the American Museum of Natural History
1989	Lee Berger arrives at Wits as PhD student with Phillip Tobias
1989	Himla Soodyall begins studying mitochondrial DNA
1994	PAST (Palaeo-Anthropology Scientific Trust) established
1994/8	Little Foot (*Australopithecus*), Sterkfontein, South Africa
1996	Wits event when Tobias returns Cornelius Kok II skeleton to Adam Kok V
1998	Ron Clarke announces Little Foot skeleton found at Sterkfontein and acknowledges important role of Stephen Motsumi and Nkwane Molefe.
1999	Sterkfontein Valley is named a UNESCO World Heritage Site
2005	Maropeng Visitors Centre opens; and Origins Centre at Wits opens
2008	*Australopithecus sediba*, Malapa, South Africa
2010	Lee Berger and his son Matthew publicly announce *Australopithecus sediba* at Maropeng
2013	Excavations begin at Rising Star Cave
2015	Lee Berger and his team publicly announce *Homo naledi*

Introduction

In 1973, when I was 11 years old, I watched a 13-part documentary series produced by the BBC and written and presented by Jacob Bronowski called *The Ascent of Man*. His title was a play on Charles Darwin's *The Descent of Man*, published in 1871. As a child, I forgave Bronowski for saying "man" instead of "humans", something I would not forgive today. In each episode, he travelled the world to explain the growth of science and art as important factors in human development. There were episodes about stone tools, migration and agriculture, and the use of fire. But in the first episode, he asked: "Where should we begin?" His answer was: "With the creation of man himself." He began the story in the Great Rift Valley of northern Kenya and south-west Ethiopia. With wonder in his voice, Bronowski walked through the layers of volcanic ash and stone from four million years ago that had buckled and sloped over on its side. "The record of time that is usually buried underfoot," he said, had been tipped over so that it was visible to us. "It is almost certain now that man first evolved in Africa." He sat down on a rock and showed his audience a historic skull, found not in the Rift Valley, but in a place called Taung in 1924 made famous by an anatomist named Raymond Dart and called *Australopithecus*. "It's not a name I like," said Bronowski. "It just means southern ape." But it was the Taung child skull from South Africa that had motivated Bronowski to investigate human history in the first place. "I do not know how the Taung baby began life," he said, "but to me it still remains the primordial infant from which the whole adventure of man began."[1]

Of all the TV that I watched as a child, *The Ascent of Man* ranks up there with Alex Haley's *Roots* as the series that had the deepest impact on me and stayed lodged in my mind. I think it played no small part in my decision seven years later to focus on the history of science at Harvard University. The introductory course on the social history of science convinced me. What interested me most was the concept that science does not exist in a vacuum.

Science is moulded by its social and political context. The British scientists in the 1800s were shaped by the Victorian Age. Science in the United States after World War II was shaped by the Cold War, McCarthyism and Sputnik. Racism and sexism in science were themes that came up again and again. One of my professors, palaeontologist and writer Stephen Jay Gould, published *The Mismeasure of Man* in 1981 where he critiqued craniometry (the measuring of skulls) and racial thinking in the eighteenth and nineteenth centuries. He explained the roots of biological determinism, where scientists argued that biology determines culture and intelligence. He showed how these racist views would return to public debate again and again because scientists assumed that there would be biological reasons to explain white superiority. Gould argued that scientific thinking had been wrongly shaped by the false social assumptions of the time.

After taking a course on biology and women's issues with Professor Ruth Hubbard, I became interested in a similar tendency for male scientists to use biology to affirm patriarchy. Hubbard critiqued the biological theories of women's inequality. I wrote my senior thesis about how the scientific and social myths of the early twentieth century limited the choices open to women. In the late 1800s, the respected physician at Harvard Dr Edward H Clarke argued that women should not pursue academic studies because it would negatively affect their biology and their unborn children. In one essay, I wrote: "What an ingenious way for the elite male community to keep women in 'their place'."

I did not follow an academic career in the history of science. In 1989 I began working for the Charles Stewart Mott Foundation. After several trips to South Africa, I was asked by the foundation to open an office in Johannesburg in 1992. My work at Mott focused on providing support to paralegal and advice offices that advocated for the legal rights of people living in rural communities and townships. The Mott Foundation staff in the US and South Africa began a comparative conversation about racism. It was in that context that the publication of *The Bell Curve* by a conservative political scientist, Charles Murray, brought race and intelligence back onto the national agenda in the US. A debate exploded and Stephen Jay Gould was one of *The Bell Curve*'s biggest critics. As Gould had predicted, the argument that racial differences result in differing levels of intelligence would continue to rear its head whenever ruling elites were fearful that disadvantaged groups were promoting social change. Gould brought out a revised edition of *The Mismeasure of Man* in 1996 and wrote: "What argument against social change could be more chillingly effective than the claim that established orders, with some groups on top and others at the bottom, exist as an accurate reflection of the innate and unchangeable intellectual capacities of people so ranked?"[2]

At the same time as the debate about *The Bell Curve* flared in the US, in South Africa apartheid had ended and the country was finally moving away from classifying and ranking human beings according to race. The first democratic elections were held in 1994 and a new constitution was adopted in 1996. The history of science came back into my life by way of the work that my husband, Roger Jardine, was doing. He had studied physics, and given his position as national co-ordinator of science and technology for the African National Congress (ANC), in the period leading up to the 1994 elections he took a leading role in developing science and technology policy for a democratic South Africa. In early 1995, Roger joined Mandela's government as the director general of the department of arts, culture, science and technology (DACST). He worked with Minister Ben Ngubane, Deputy Minister Bridget Mabandla, and Chief Director Themba Wakashe, as well as Phillip Tobias, on the efforts to return Sarah Baartman's remains to South Africa. Along with Deputy President Thabo Mbeki, DACST was supportive of Ron Clarke, and the announcement of Little Foot at Sterkfontein. Accompanying Roger, and the deputy director-general Rob Adam, I had the opportunity to visit Sterkfontein to see Little Foot *in situ* on 23 December 1998.

It wasn't until the mid-2000s that my career shifted to focus on writing and I published my first book, *Sanctuary*, in 2013. For several years, I had been thinking about the possibility of writing about the search for human origins in South Africa. In July 2013, the idea was cemented when I read an article in *The Star* about Desmond Tutu's visit to Sterkfontein. Tutu looked out over the landscape and said: "It is very humbling to think that we emerged only 200 000 years ago, but that this universe is 14 billion years old. Isn't it fantastic that there is a mind that oversaw this incredible journey so that eventually we have emerged, that God should have taken so much trouble." And then he let out his characteristic infectious laugh.[3]

I put together a book proposal that was guided by questions that I believe my professors Stephen Jay Gould and Ruth Hubbard might have asked. How has the changing social and political context shaped the search for human origins? What impact did colonialism and empire have on the views of scientists studying human evolution in the early twentieth century? What influence did apartheid have on the search? How have the changing scientific views about race, and racism, affected the efforts to understand human evolution? How have the beliefs of individual scientists, and the times in which they lived, shaped the narrative of human origins? How has that narrative changed over time? And finally, what does it all mean for our understanding of what it is to be human?

The project was under way. In November 2013, I saw another article, this time in the *Mail & Guardian*. It was no more than half a page and entitled "Skinny Scientists at Palaeo Rock Face".[4] Professor Lee Berger, a palaeoanthropologist at Wits University, had recently recruited six women scientists to climb down into a cave called Rising Star at an undisclosed location in the Cradle of Humankind. It struck me that my book would not only be about history, but also about current unfolding events. Why has the search for human origins continued to be of huge interest around the world and in South Africa with every new "discovery" claiming headlines?

I went back to an old copy of *The Ascent of Man*, which had been published as a book as well as produced as a television series. There, on page 28, was a photograph of the Taung child skull.[5] In the years since many authors and researchers have contributed countless books and articles on human origins. By reviewing their important work, and by taking another look at Raymond Dart and his conclusions about the Taung skull close to a century ago, and working our way up to the present, what would we learn? I was interested to find out.

Part One
Searching for Difference

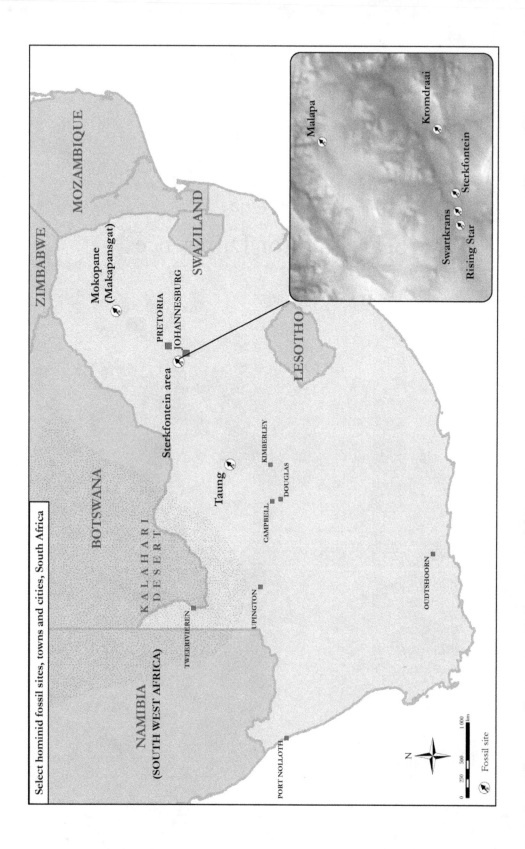

Select hominid fossil sites, towns and cities, South Africa

1

Darwin's Hunch

In sixteenth century Europe, close to 500 years ago, the Christian concept of a Great Chain of Being was widely popular. Many people believed that there was a divine hierarchy that looked much like a ladder, with rocks and minerals at the bottom, proceeding upwards to plants and then animals, with human beings on the top rung, just below God. The Great Chain of Being applied to a hierarchy amongst humans as well, with the top rung assigned to monarchs and kings, which were above the noblemen and the clergy, who in turn were above the commoners. Those on the top claimed that the hierarchy helped to maintain order and everyone was expected to remain in their place.[6]

In the mid-1700s, taking all of the organisms from this ladder, Carl Linnaeus, a Swedish botanist, worked for decades to develop the modern system for organising all living things on earth. Linnaeus assigned all living organisms into three kingdoms – animal, vegetable and mineral. Each kingdom contained categories of order, class, genus and species and he gave each animal and plant two names. For example, *Homo sapiens* indicated that the genus was *Homo* and the species was *sapiens*. Linnaeus, also known as Carl von Linne, made a bold move by placing *Homo sapiens* within the order of primates, which is part of the class of mammals, within the kingdom of animals. His system categorised humans in close relation to apes and chimpanzees. Humans were no longer above animals on the ladder of the Great Chain of Being, but were classified as just another of nature's creations, alongside all other living things. In his long master work, *Systema Naturae*, however, Linnaeus did list *Homo sapiens* first.[7]

Linnaeus divided *Homo sapiens* into four varieties defined largely by geography, skin colour and temperament. He described the four varieties as coming from four geographic regions: *Americanus*, *Europeus*, *Asiaticus* and *Afer* (African). The four groups were not particularly hierarchical. Although he did not list the geographic areas in rank order with Europeans first, he described the behaviour of each group in a way that was clearly influenced by his own opinion and that of his contemporaries, that Europeans were superior: *Europeus* was "governed by laws", *Americanus* governed "by custom", *Asiaticus* governed "by opinions", and *Afer* governed "by impulse". The concept of a hierarchy amongst humans had not faded away with Linnaeus. In fact, in one fell swoop, Linnaeus made clear his hierarchical thinking by creating a fifth group called *Homo monstrosus*, which included "monstrous or abnormal" people. On the basis of his perception of their genitals, Linnaeus classified the Khoisan and "Hottentots" of southern Africa in this category.

With this one deed of giving a different name, he sent a dehumanising and painful ripple-effect across the centuries. Linnaeus illustrated how the act of creating nomenclature, the act of naming, holds enormous power.[8] The process of naming, and often re-naming – for human beings and for hominid fossils – would continue to have great power. And the matter of who does the naming also indicates who holds that power.

It was an admirer of Linnaeus, German physician JF Blumenbach, who helped place Linnaeus's geographical categories of humans even more squarely into a racial hierarchy. Initially, Blumenbach recognised the four initial grouping of human beings that Linnaeus had presented, but by the end of the 1700s he concluded that there were in fact five. The reason for this change was that Blumenbach introduced the term "Caucasian", which he used to describe people from around the Caucasus Mountains, where he thought that the most ideal and most beautiful human beings lived. He concluded that they must be the original humans from which others had descended. Blumenbach drew a model with a Caucasian skull at the centre. To the left, he drew a Native American and Asian skull as diverging from the Caucasians. He drew a Malay skull and an African skull diverging to the right. He invented the fifth racial category – Malay – in order to keep the lineup symmetrical on both sides. This transition, which placed the Caucasians at the centre, was a pivotal change. Blumenbach's model became the accepted one. Although it may not have been his intention, his model imprinted the ranking of human variation and race on Western science. Scientists in Europe at the time were quick to accept Blumenbach's model because it was in line with the growing assumption that Europeans were superior to other peoples around the world.[9]

Another influence on thinking, which developed through the 1800s, was a debate between scientists who believed in monogenism – that all human beings came from one single source – and those that believed in polygenism, the belief that each major race group had been created separately. Over the centuries, the

Biblical story of Adam and Eve had led to the belief that there was one single origin for all humankind. But polygenism was beginning to gain popularity and to be promoted by scientists such as Louis Agassiz in the United States and Georges Cuvier, the famous anatomist in France. Cuvier believed that Adam and Eve were European. In his view, the other races of the world were created by a major catastrophe on earth that had occurred about 5 000 years before and which had resulted in survivors living and evolving in isolated areas of the globe. Robert Knox, a Scottish anatomist and zoologist, also believed in polygenism. In his treatise *The Races of Man*, published in 1850, he wrote "… racial natures … were unchanging over thousands of years, and were so different that they should be called different species."[10]

It was Charles Darwin in *Origin of Species* in 1859, with its promotion of evolution by natural selection, who cast doubt on polygenism and made a strong case for one unified source for human beings. Twelve years later, in 1871, Darwin published *The Descent of Man* and made a major statement suggesting that humans had evolved in Africa:

> "In each great region of the world, the living mammals are closely related to the extinct species of the same region. It is therefore probable that Africa was formerly inhabited by extinct apes closely allied to the gorilla and chimpanzee; and as these two species are now man's nearest allies, it is somewhat more probable that our early progenitors lived on the African continent than elsewhere."[11]

Darwin's theory that humans had originated in Africa was not widely accepted at the time and was derided for close to 150 years. For decades, scientists argued that humans must have evolved in Europe or Asia instead. As late as the 1960s, scientists such as Carleton Coon were still arguing against the theory, and even when geneticists supported Darwin's case with new research in the late 1980s, there was continued scorn for the idea. Today, in 2016, most scientists agree that Darwin was correct on many levels, but for well over a century, there was plenty of debate.

Despite Darwin's deduction that humans evolved in Africa, explorers and scientists began the search for fossil evidence in Europe and Asia. One of the first ancient hominid fossils that was found in support of the theory of human evolution was a skull in the Neander Valley in Germany in 1856.[12] The skull had thick, heavy bones with large ridges over the eyes. Some scientists thought it looked deformed or diseased. Others argued that the skull resembled that of a gorilla. In later years, other Neanderthal fossils were found in other parts of

Europe, including Belgium, France and Croatia. At the time of their discovery, the fossils were considered to be ancient. We now know that Neanderthals existed quite late in the time-frame of human development. Scientists now believe that they were a sideline branch in human evolution and that many *Homo sapiens* hold a small percentage of Neanderthal DNA, but back in Darwin's time there was discussion about whether they were the "missing link" between apes and human beings.

One of the first scientists to set out deliberately to find the early ancestors of humans, in the late 1800s, was Eugene Dubois, a lecturer in anatomy at the University of Amsterdam. His theory, similar to Darwin's, was that humans were likely to have developed either in Africa – where modern gorillas and chimpanzees inhabited the forests there – or in Asia, where orangutans were found. In November 1887, he sailed for what was then the Dutch East Indies, now Indonesia.[13] The following year in Java he found an ancient skull. He would not share it or describe it, however, for another 30 years, a pattern that was becoming more and more common in the scientific community. Many fossil finds were held back to give the person who found them more time to analyse them. Dubois spent three more years unearthing fossils throughout the East Indies. He returned to Java where the deep deposits of volcanic ash were said to hold many mammalian fossil bones and in 1891 he found a cranium and a thigh bone, which he believed were linked to one another. He named them as a new genus and species, *Pithecanthropus erectus,* subsequently referred to as Java Man. Dubois believed that Java Man was a human ancestor and a missing link between apes and humans. It took decades for scientists to fully accept the fossil as a human ancestor and only much later, in 1950, was it renamed *Homo erectus.*

When Dubois returned to Europe in 1895 and exhibited his findings, few people agreed with him on their significance. Scientists asked: Was the skull simian or human? Was it normal or pathological? Did the teeth belong to the same individual as the skull? Some scientists rejected *Pithecanthropus* and saw it of little consequence.[14]

In the second half of the nineteenth century, just as the search for fossil evidence was on, so, too, was the development of a new field of science. Anthropology, the study of humans, was distinct from the study of human history. In Britain, as the new field grew some anthropologists assumed that societies passed through a single evolutionary process from most primitive to most advanced.[15] Again the concepts of hierarchy and superiority were at play. Many European scientists saw Europeans as civilised and perceived societies outside of Europe as less evolved. They thought that studying living indigenous societies, in Australia, the Americas and Africa, might provide a link to prehistoric human life.

Of course, there were opposing points of view and counter trends at the time. For 50 years, activists in England led a campaign to end slavery, succeeding in 1833.[16] In addition to the abolitionists, there were other writers and thinkers who spoke out against the concept that the people of Europe, and its missionaries around the world, were the only ones to contribute to human civilisation. Among those offering another perspective, in the 1860s, was William Ngidi, an isiZulu-speaking translator for Bishop John Colenso of Natal in South Africa. He questioned the literal truth of the Bible, and observed that there was no greater truth in the Bible than there was from his own observations of nature or in stories from his African oral tradition and religion. Ngidi's questions turned the orthodox conventions of coloniser and colonised, civilised and heathen, superior and inferior, on their heads.[17] But these viewpoints did little to shape the overall scientific thinking about race in Europe and the colonies.

2

"The Most Interesting Specimens Were the Natives"

The predominant European scientific worldview about humans around the globe in the mid-1800s shaped the thinking of Robert Broom,[18] a young doctor from Scotland who was interested in animals and plants and fossils. He travelled to Australia to take up a medical practice and to pursue his interest in marsupials. When he encountered Australian aboriginals for the first time, Broom said he was "surprised to find that they were by no means stupid although they were supposed to be the lowest form of human being".[19]

At the same time that Dubois was excavating in Asia, Broom was exploring Australia. One day he and his friends took a picnic to the limestone Wombeyan Caves in New South Wales. After they'd finished their food and conversation, Broom went for a walk above the caves. He hiked beyond a crop of trees and came upon a wall of rock. After close inspection, Broom saw that there were fossils embedded there. He took some of the fossil-filled hard rock (known as breccia) home to examine and decided that the fossils were from small animals, some of which might have been extinct. The caretaker at the Wombeyan Caves advised Broom to write to the minister of mines for permission to remove the breccia because the area was a government reserve. Broom did write to the ministry and they refused him permission to continue his studies of the area. The ministry sent someone to the site to bring breccia back to Sydney for further study but they never followed up with any research. Still in his 20s, Broom learned a lesson that shaped his approach to government authority. After making a discovery, he told officials about it and they pushed him aside and ignored it.[20]

At the age of 30, Broom left Australia and sailed with his wife for South Africa. He planned, once again, to combine his medical practice and his interest in palaeontology, and he took up a medical post in Port Nolloth, a small dry town 400 kilometres north of Cape Town. "The most interesting specimens were the natives," said Broom. "There were large numbers of purebred Hottentots, a few black Damaras, some St. Helena natives, a few kaffirs and some possible Bushmen."[21]

This was a time of drought and large numbers of people who were indigenous to the area trekked down from the north of the Orange River to Port Nolloth. As many people died, Broom, being the physician in town, gathered their bodies in order to preserve and study their skeletons. There was a growing interest among European scientists in the anatomy of the "Bushmen" and whether they might be the "missing link" between apes and humans.[22] They thought their skeletons could provide a clue to understanding human evolution. One such scientist was William Turner at the University of Edinburgh. Broom sent several human skulls to Turner for further study; the mere description of how he prepared them makes disturbing reading.[23]

On one occasion, Broom brought a dead body into his kitchen. Taking a large kitchen knife in his hand, he cut off the man's head. Then he filled a paraffin tin with water and put it on the kitchen stove. After waiting for the water to boil, Broom placed the head into the boiling water. The skin and flesh fell away from the bones and with time the bones cooked clean. He used a large serving spoon to lift the skull out of the tin and then set it out to dry, before packing it up with others and sending them off to Europe.

Broom's work with human skeletons did not take place in isolation. In the late nineteenth century, many universities in Europe had begun collections of human skeletons and the international skeleton trade was brisk. Museums and universities in the United States were also building skeleton collections, although their emphasis was on the skeletons of Native Americans, who were referred to with contempt at the time as "Red Indians".[24]

The scientific interest was to fit these remains into the taxonomy of race. White supremacy had taken hold in science across the world.

In South Africa, in 1903 Broom was appointed professor of zoology and geology at Victoria College (which later became the University of Stellenbosch) in the Cape Colony. For seven years he gave four lectures a day, often straying from zoology to speak about religion and politics and evolution. When a group of students asked him to give a series of open lectures about evolution, there was standing room only. Broom told the audience that religion would benefit by accepting evolution. They were spellbound. Broom wasn't afraid of being expelled or attacked, but he did worry that "some narrow-minded religious Dutch" leaders might protest and hurt the reputation of the college.[25]

It was in that same year that the South African Association for the

Advancement of Science (SA AAS) was established, despite the fact that South Africa did not exist yet as a unified country. The British ruled the Cape and Natal, and the Afrikaner Boers ruled the Orange Free State and the Transvaal. Two years later, in 1905, the British Association for the Advancement of Science held their meeting jointly with the fledgling South African association. Broom attended the gathering along with close to 400 British scientists. One of the first anthropologists in England, AC Haddon, gave a lecture calling for "an accurate account of the natives of South Africa ... for scientific use, and as a historical record".[26] He also called for the need for measurements and accurate anthropometric data. Following on the early efforts of Broom and the University of Edinburgh, and other museums around the world, the South African Museum in Cape Town, led by its director, Louis Peringuey, and the McGregor Museum in Kimberley began building collections of human skeletons.

"The Bushmen are disappearing from the face of the land that once was theirs, and in a few years will be as extinct as the dodo," read an editorial in the Kimberley *Diamond Fields Advertiser*. The paper complained that European museums were collecting "specimens" and it encouraged the McGregor Museum to collect Bushmen plaster casts, skulls and skeletons. "If therefore, any reader should know of a docile Bushman who has no particular use for himself, the scientific world would be truly grateful if that same Bushman could be induced to pack himself in formalin, or something of the sort, and ship himself to Europe for the purpose of ornamenting a dust-proof show case, side by side with the mummies of Egypt."[27]

The impression that the "Bushmen" were South Africa's first inhabitants, followed by "Hottentot" and "Bantu" migrations from the north, followed by the arrival and migration of white settlers moving in from the south, was nurtured and promoted by the influential South African historian George McCall Theal in his prolific writing throughout the late nineteenth century. This racial narrative of waves of migration, which was later proved simplistic and inaccurate, became embedded in the South African psyche. Theal also provided physical descriptions of the three groups. He described Bushmen as "dwarfish and yellowish brown" with "fox-like" faces and "small deeply sunken" eyes. Hottentots were "better formed" than Bushmen, with "hollow cheeks and a flat nose". Bantu-speaking people were "better formed than the other two". Theal liked to confirm his racial hierarchy with cranial data that suggested that the average brain size of the "Bushman" was 1,228 cubic centimetres, the "Hottentot" was 1,407cc and the "Bantu" was 1,485cc.[28]

Louis Peringuey became a particularly vigorous collector of skeletons. He wrote to a local magistrate that skeletons could be "simply dug out" and shipped to the museum by rail, marked "specimen of natural history".[29] Like Broom, Peringuey also sent numerous skulls to Britain for analysis. Dr Rudolph Pöch from Austria travelled to southern Africa "to study the last remaining Bushmen of pure race",[30] while one dubious supplier known as Scotty Smith from Upington in the northern Cape "simply shot the required number of Bushmen whenever he needed their skeletons".[31] The absence of any restrictive legislation was as a result of the attitude that this practice was in support of scientific knowledge and that Bushmen skeletons were a relic of natural history. It was not until 1911 that an Anatomy Act, covering all races, was passed by the parliament of the new Union of South Africa. The act was intended to regulate the use of the bodies of paupers and prisoners that were not claimed by relatives and were then used by anatomy departments at medical schools for dissection.[32]

For the seven years that Robert Broom was at Stellenbosch he had a working relationship with Peringuey and the South African Museum. He also had a passion for the Karoo. With a travel allowance from the museum, he spent his academic holidays in the Karoo, where he developed a fascination for the mammal-like reptiles there. He became a familiar figure in the area, in his stiff collar and suit and carrying a butterfly net. Broom would spend every possible moment collecting everything. The geology of the Karoo was made of shale and sandstone and each layer offered fossil deposits from a particular geological age, beginning about 250 million years ago. In exploring the upper layers of rock, Broom estimated that there were likely billions of fossils in the area, with millions near the surface. During his time in Stellenbosch, Broom published 96 scientific papers on many of his fossil finds.

After one visit to the South African Museum, just before Broom was scheduled to leave on the train back to Stellenbosch, he heard that the museum had acquired the remains of an aardvark that had been skinned and was ready to be stuffed for display. Broom asked where the carcass was. When he was told that it had been buried, he went to the spot where the skinned corpse lay and dug it up. He wrapped the stinking carcass in newspaper and ran for the train. When he got back to Stellenbosch, he examined the aardvark's teeth and wrote a paper that was accepted for publication in *Nature*.[33] Peringuey saw the article and was shocked. He declared that museum specimens, even if dead and buried, could not be studied and written about in journals without his approval. Broom was not fond of Peringuey, who appeared to be jealous of Broom's independence and the respect Broom commanded from other fossil collectors in the field.

The minister of railways had granted Broom a free pass to carry fossils on the train provided that the South African Museum was the recipient of all his finds. However, in 1909, the new minister of railways, JW Sauer, refused to renew Broom's pass, saying that fossil collecting wasn't of any value to the country.[34] Reminiscent of his treatment by the Australian government 15 years earlier, this incident infuriated Broom.

Struggling with his inadequate income in Stellenbosch, Broom was in debt, and his relationship with the South African Museum was at an all-time low. He decided to do what he had done before – work as a doctor in a mining town and continue his fossil collecting when he could. He chose Springs in the Transvaal, one of the new gold mining towns east of Johannesburg. When he moved north the flow of fossils from Broom to Peringuey ceased and so the museum director started working with and paying other collectors in the Karoo. But by then Broom had established a strong relationship with fossil collectors and they were loyal to him, and continued to provide him with fossils. Broom always had great respect for amateurs in the field. As Peringuey started paying other collectors, he and Broom became competitors. Overseas museums looked to both scientists as a source of fossils. One amateur collector in the Karoo, Reverend John Whaits, found himself caught in the middle. He had been selling to Broom, but Peringuey wanted Whaits to sell directly to him instead. Whaits wrote to Broom, saying, "You can have the first refusal of any specimens I succeed in obtaining … I should be glad if you will send them [Peringuey and the South African Museum] specimens from my finds whenever possible."[35] Peringuey didn't like this arrangement. To make matters worse, the famous British scientist Sir Arthur Smith Woodward sent a representative to the Karoo to purchase specimens for the British Museum directly from collectors there. As the British Museum was able to pay a much higher price, they bought some specimens from Whaits before Broom could take a look.

Angry at these developments, Broom turned to American palaeontologists as a new market. He developed a warm relationship with Henry Fairfield Osborn, the president of the American Museum of Natural History and professor of zoology at Columbia University in New York. Osborn felt that there was a lack of South African fossils in their collection, so Broom suggested Osborn hire a local collector, which he did by hiring Whaits. The American began buying numerous fossils from Whaits. As a result, the British Museum and the South African Museum were upset that the fossils didn't end up with them and they both blamed Broom. Peringuey also put the word out that he had already paid Whaits for specimens that had ended up in New York, maintaining that they should have stayed in South Africa. Smith Woodward of the British Museum responded by withdrawing his support for Broom becoming a member of the Royal Society. Broom wrote to Whaits. "Here I sit with a pocketful of dollars, and not a friend in the world."[36]

Broom and his wife Mary moved to Douglas in the northern Cape, a small town about 120 kilometres from Kimberley. During his time in Douglas, Broom participated in séances at the home of the local violin teacher who was also a medium. One night, Broom found himself sitting at her round wooden table holding hands with the medium on one side and a grieving widow on the other. The lights flickered as the medium called on the deceased and brought the spirit into the room. Another night she prayed to clear a woman's cancerous cyst.[37] Many scientists admired Broom for his work in the Karoo but were uncomfortable with his belief in the spiritual realm. Broom himself believed his success with palaeontology was as a result of a larger divine plan. He also believed that evolution did not happen by chance but that "some intelligent Spiritual power" had guided the evolution of man.[38]

The lecture hall at the Geological Society in England in December 1912 was packed with people. Charles Dawson, a lawyer who was interested in archaeology and palaeontology, had found a skull in a gravel pit in Piltdown, Sussex, England. Dawson had shared his find with Sir Arthur Smith Woodward of the British Museum and the two of them presented the lecture together. "I was walking along a farm road close to Piltdown Common," said Dawson, "when I noticed that the road had been mended with some peculiar brown flints not usual in the district. On inquiry I was astonished to learn that they were dug from a gravel bed on the farm. And shortly afterwards, I visited the place." After several years of returning to the same gravel bed, Dawson didn't find much. Then on yet another return visit, he "picked up among the rain-washed spoil-heaps of the gravel pit, another and larger piece ..."[39]

Sir Arthur Smith Woodward described the skull, which had been reconstructed from nine pieces of cranium and a mandible. Two molar teeth were still in place. Smith Woodward concluded that "While the skull is essentially human ... the mandible appears to be that of an ape with nothing human except the molar teeth."[40] Twenty years after Dubois's Java Man, there was finally tangible proof that humans had an ape-like ancestry. Scientists expected that the human brain had evolved first, and that it had grown larger with greater capacity, while the jaw initially remained unchanged. These Piltdown fossils appeared to prove that theory correct. And the announcement brought enormous pride to England that it was the locale for one of the forebearers of human beings.

Once again, a fossil find brought on a great debate. Was the proposed age of Piltdown Man accurate? Was it likely that all the many pieces of the skull belonged to the same creature? Didn't the cranium represent one individual and the jaw represent another? Sir Arthur Keith, a Scottish anatomist who had developed an interest in human evolution, pointed out that there was no canine tooth in the jaw. Smith Woodward responded that if the canine tooth were ever found, it would likely resemble that of a chimpanzee, and that it would also likely show signs of wear that were quite different because it would be working in association with a more human cranium. Keith was sceptical that all of the pieces belonged to one individual. Elliot Smith, who had previously reflected on Dubois's finds in Java, also wondered, but concluded that it would be a miracle if a primitive man left behind his cranium in a certain spot without his lower jaw and that an ape left his lower jaw behind at the same place without his cranium.

For months and months, Dawson and Smith Woodward continued to search in the gravel pit for further evidence. A young French priest, Teilhard de Chardin, began digging and sifting with them. One day, as he was down on his knees working closely to where the lower jaw had been found nine months earlier, he came upon a canine tooth. After showing it to Dawson and Smith Woodward, they saw that the eye tooth showed the very signs of wear that fit perfectly with what Smith Woodward had predicted.[41] They were jubilant. The tooth helped explain the skull as a whole.

But that wasn't the end of the Piltdown story. Almost two years later, Dawson was working in another area two miles away from the first gravel pit. He came upon a cranium and a molar tooth, similar in description to the first skull. This new discovery was a blow to those who had continued to be sceptical about Piltdown Man. With the same combination of large brain and small jaw found again, it became harder to argue that the first skull was a fluke. By then World War I was under way and the news of the second Piltdown find spread slowly, but after the war more and more scientists came to appreciate the importance of the second find. Over time, Piltdown Man was broadly accepted as holding an important place in the human family tree. The fossils were often referred to as the "Earliest Englishman", an image that was enhanced when a stone tool reminiscent of a cricket bat was found at the same site.[42] English scientists believed they were onto something. They finally had a hominid fossil that could compare with those that had been found in Germany and France. Scientists would continue to look for human origins in

Europe, and possibly follow up on Dubois's research in Asia, but they were not interested in Africa. Darwin must have been wrong.

At about the same time as the announcement of Piltdown Man, in 1913 another hominid fossil was found in South Africa. It was dug up at Boskop near the town of Potchefstroom in the Transvaal and ended up in the hands of FW Fitzsimons, the director of the Port Elizabeth Museum. While little information had been recorded about the circumstances of the original discovery, Fitzsimons wanted the fossil to be thoroughly examined and analysed. He was about to send it to the British Museum when Peringuey pleaded that it be sent to the South African Museum instead. After Broom's falling out with Peringuey, the former had developed a working relationship with Fitzsimons, so Fitzsimons decided to send the skull to Broom for a closer look first. Not since his time in Port Nolloth 15 years earlier, boiling human skulls, had Broom been involved with human skeletons and physical anthropology. He examined the skull. Then he examined all the "Hottentot" and "Bushman" skulls at the Port Elizabeth Museum.

In the same year that the Boskop Man fossil was found, "the South African Native found himself, not actually a slave, but a pariah in the land of his birth".[43] These were the words of Sol Plaatje in his book *Native Life in South Africa* about the impact of the Native Land Act of 1913. Plaatje lamented the fact that "4.5 million Natives may 'buy' land in only one-eighteenth part of the Union [in the Native Locations], leaving the remaining seventeen parts for the one million whites." Not only did many Europeans see them as inferior, but also the indigenous people of southern Africa were being pushed off their land. Plaatje was a founder member of the South African Native National Congress which would become the African National Congress. He travelled to England to protest the Native Land Act, but to no avail.

Now nearing 50 years old, Broom wrote a paper about the Boskop skull. It was his first publication contributing to the discussion of human evolution. He called the fossil Boskop Man or *Homo capensis* and described it as having a heavy, thick-boned cranium that he compared to Neanderthal skulls found in Europe. Broom believed that the species could be "the direct ancestor of the more or less degenerate Bushman of recent times".[44] This derogatory language was common at the time and reflected the view of many South African scientists that the "Bushmen" anatomy could offer a clue to the relationship between apes and humans.

"It is a mistake to think that the Bushmen are dead – I have already seen half a dozen," Broom had written to his friend Osborn in New York when he moved to Douglas. "I am making a collection of skulls and skeletons from very old graves, mostly hundreds of years old. Hitherto most collecting has been very haphazard and no one seems to know what is a Bushman and what a Hottentot, and even the authorities are a hopeless mess. I hope to get one of the best collections in existence."[45] Broom wrote that he continued to dig up graves in Douglas; he had also dug up several graves in Upington. He was on his way to having a "fair collection". From 1920 to 1927, Broom served on the Douglas municipal council; for four years during that time he was also the town's mayor. There is no documentation suggesting how Broom was able to dig up graves so freely, but perhaps his position of authority helped him to do so without concern.

While in Douglas, Broom noticed that there was a man being held in the local jail that he thought appeared to be "pure Korana". He met with the man, who told Broom that both of his parents were Korana. The term "Korana" described a group of Khoi people who were distinctive linguistically from other Khoisan during the eighteen and nineteenth centuries.[46] Explorers and travellers were interested in the Korana's distinct features and Broom was interested in their anatomy. While the man was still in jail, he became ill and died. Broom wrote: "… it is unnecessary to say how I got the skeleton, but it is now in Kimberley at the Museum."[47]

This wasn't the only time that Broom asked for cadavers from the local jail. "Studying anthropology is not always a pleasant task," he wrote. "One day a very interesting native died and I wanted the skeleton rather badly so I had the body sent up to my garage for me to do a post mortem. It was in January, and the temperature was much above 100 degrees in the shade. I was called out [on] a long country journey and only got back at 10 o'clock that night … I fear that the European armchair anthropologists have little idea of the troubles we workers in the field have … If a prisoner dies and you want his skeleton, probably two or three regulations stand in the way, but the enthusiast does not worry about the regulations. I used to get the body sent up … then the remains would be buried in my garden, and in a few months the bones would be collected."[48]

On another occasion, Broom heard about a "Bushman" who had been found dead on a farm near Campbell, which was 32 kilometres from Douglas. After a post-mortem, Broom told his assistant John to bury the body at a depth of two feet.[49] A few years later, after returning from a trip to the United States, Broom asked John if he remembered where he had buried the man.

Giving him a spade and a sack, Broom told him that if he went to Campbell, dug up the grave and brought him the bones, he would pay him a pound. The next morning, John returned with the bones. It's not clear whether Broom sold the skeleton or not, but he sent it to Paris with Abbé Henri Breuil, a French anthropologist and archaeologist. Broom described the bones as "the most perfect pure Bushman skeleton in the world".[50]

Visitors to the Brooms' house in Douglas might have been startled by what they saw inside. Their book-shelves displayed rows of fossils, while the books were piled on the floors. The linen closets contained no linen; they were filled with skulls instead. At one time Mary Broom complained that she couldn't sleep on her own bed because it was the site of the reconstruction of a gigantic fossilised reptile. She was equally upset that in the kitchen she had to keep their children away from the pots on the stove – beside a pot of soup would be a pot or two of boiling skulls. And she dared not go into the garage because she knew that human corpses were being stored there waiting for post-mortem examination. She couldn't even plant vegetables in the garden because skeletons had been secretly buried there so that worms could clean the flesh from the bones. One day she welcomed Jan Smuts, the prominent South African military leader and philosopher, to their home for a scientific discussion with her husband.[51] Broom and Smuts had embarked on a correspondence about science and human history, a habit that would continue for another 30 years. Mrs Broom could barely breathe as the man who would be South Africa's Prime Minister, stood above a shallow grave, and she was relieved when he followed her through to the sitting room for tea.

While Broom was gathering skeletons and describing Boskop Man, and the British scientists were studying the Piltdown skulls, another anatomist was being groomed in Australia. This was a scientist who would play an important role in the understanding of human evolution. Born near Brisbane in February 1893, the fifth of nine children, Raymond Dart was raised in a Baptist home so it was only when he went to university in 1911 that he was first introduced to ideas about evolution. In a packed Sydney City Hall, Dart attended a lecture entitled "The Evolution of the Brain", given by the distinguished anatomist and physical anthropologist Grafton Elliot Smith. Elliot Smith's lecture marked the beginning of Dart's lifelong interest in the brain. "I fell under his spell that night and prayed that at some time I would be allowed to work under him."[52]

The White Australia Policy began when Dart was still a child. The

Immigration Restriction Act of 1901 limited people who were not defined as white from entering the country. The assumption that white people of European descent were superior to the Aborigines of Australia, to Native Americans and to the peoples of Africa and Asia was a widely held view at the time in Australia, Britain and the United States. Dart's early training, prior to World War I, had been heavily influenced by British thinking on race and evolution and the role of societies around the world. Later, studying with Elliot Smith and Arthur Keith, Dart was trained in the tradition of comparative anatomy which based racial differences on physical traits and comparisons.

During World War I Dart became a medical officer in the Australian army and served in London. The experience of working to save men with terribly injured bodies made an indelible impression on him – he described it himself as a nightmare.[53] It would also shape some of his later thinking about human behaviour. After the war, Dart followed his dream and he did get to work in London with Elliot Smith, who was spending a great deal of time building reconstructions of the Piltdown skull. This sparked Dart's interest in fossils. When Elliot Smith recommended Dart for a Rockefeller Fellowship for a year in the United States, he took advantage of the opportunity. He went to Washington University in St Louis, Missouri, where he spent most of his time in the anatomy department. The head of the department, Robert Terry, had begun to build one of the first collections of human skeletons for the purposes of comparative anatomy.

After Dart's return to London, it was Elliot Smith's influence again that resulted in his next move. There was an opening for the chair of anatomy at the University of the Witwatersrand in South Africa. Although hesitant, Dart applied. Sir Arthur Keith reviewed the application and expressed concern that Dart had written "Freethinker" when asked his religion. "Do you think that wise?" he said. "I believe the atmosphere in South Africa is strongly Calvinist." He suggested that Dart write "Protestant" instead, saying, "They're not likely to inquire what sort of Protestant you are, nor what you feel like protesting against."[54] Years later, in his autobiography, Keith said that he had had some trepidation about Dart's application. He had no doubt about his intellect, but was concerned about "his flightiness, his scorn for accepted opinion, [and] the unorthodoxy of his outlook".[55] Nevertheless Dart and his wife Dora Tyree, whom he'd met in Cincinnati in the US, sailed for South Africa shortly before Christmas in 1922. Dart was 29 years old.

The Darts' first impressions of the young mining town of Johannesburg were less than enthusiastic. They described "endless rows of red-painted, corrugated iron roofed buildings".[56] The fledgling city looked as if it could be deserted in a matter of days, they thought, especially if the search for gold dried up. The Wits Medical School building was two stories tall and stood behind a ten-foot brick wall surrounded by high grass and weeds. The place looked derelict. Inside the building, corpses that were not properly covered lay on many of the dissecting tables. The laboratories had no supplies, or even taps or plugs, and there was no library. Despite the conditions and unpromising beginning, Raymond Dart would build his department to gain a worldwide reputation for comparing the anatomy of different racial groups, and for bringing the search for human origins to Africa.

3

A Preposterous Claim

In the heat of Johannesburg in late 1924, 31-year-old Professor Raymond Dart reluctantly put on a pair of black tuxedo trousers and a white shirt. Looking in the mirror and struggling with his stiff-winged collar, his mind was elsewhere. He could hear Dora, already dressed in her most formal shoes and gown, running from room to room. The Darts had offered their home as the venue for a friend's wedding and soon the guests would begin to arrive. Dart was to be the best man. Cursing the collar, he moved to the window and glanced outside. He saw two men coming up the driveway, staggering under the weight of two large wooden boxes. Immediately his mood improved. He had been waiting for this delivery and it had nothing to do with the wedding.

Just weeks earlier, Dart had been pleasantly surprised when one of his students in the department of anatomy, Josephine Salmons, arrived at class with a fossil. He always encouraged his students to collect fossils during the holidays and Salmons had obliged. She had been visiting a family friend, EG Izod, in Taung, about 400 kilometres south-west of Johannesburg. Izod was the director of the Northern Lime Company, which was blasting in the area for lime, which was needed in vast quantities for gold processing. Salmons noticed a fossil skull, still embedded in limestone, on Izod's mantelpiece above his fireplace. She thought it might be a baboon skull and she asked Izod about it. He told her he had been given the fossil quite recently when he had visited Taung – the skull had been blasted out of a quarry there. Salmons asked him if she could borrow the skull to show it to Dart.[57]

At the time, because Dart did not know of any other ancient primate fossils that had been found south of Egypt, this one intrigued him. Intrigue grew to excitement, and caused him to to speed down the hill in his Model-T Ford to share it with his geologist friend and Wits colleague, Professor Robert Young. Dart and Young both noted that there was a hole in the roof of the skull, as if it had been struck by a sharp instrument. But this feature was less the focus of their attention than the idea that there might be more ancient skulls to be found at Taung. Young knew the limestone works there – Buxton Limeworks – and he told Dart that in fact he happened to be heading there shortly himself. He had found out that a local quarryman, a Mr De Bruyn, had been blasting at Taung for some time, and had been gathering fossil baboon skulls from the rock.

In late October or early November 1924, De Bruyn blasted out a brain cast – fossilised remains that show the shape of the brain. This brain cast was larger and stood out from any of the others that had been found so far. He thought it might not be from a baboon and that it was "possibly the skull of a young Bushman".[58] De Bruyn showed the two blocks of stone to his manager at the limestone works, who in turn showed them to Professor Young. Young decided to carry these two pieces of rock personally on the train back to Johannesburg. He stepped onto the train and held the rocks beside him on the seat in the compartment. Before he left Taung, he arranged for many other pieces of promising breccia to be boxed and sent on the train directly to Professor Dart.

When Dora saw the boxes being delivered on the afternoon of the wedding, she wasn't impressed. "I suppose those are the fossils you've been expecting," she said to her husband. "Why on earth did they have to arrive today of all days?" Knowing how dramatic and mischievous her husband could be, she looked him straight in the eye and said, "Now, Raymond, you can't go delving in all that rubble until the wedding's over and everybody has left. I know how important the fossils are to you, but please leave them until tomorrow."[59]

As soon as Dora had left the room, Dart tore off his collar and ran out of the front door to greet the men who had by then set the two boxes down on the ground. He asked them to carry the crates over to an outdoor gazebo. With the aid of a hammer, and struggling a bit, Dart prised the lid off the first box. He lifted out some of the grey, dusty rocks and set them on the ground. As he examined them, his shoulders dropped and his expression turned to disappointment. He could make out some fossilised turtle shells and a few fragments of bone embedded in the stone.

This is where Raymond Dart's telling of the story differs from the coverage in the newspapers at the time. According to Dart's memoir, he continued to ignore his wife Dora's instructions to wait until the wedding was over. After he opened the first box, he then wrestled with the lid of the second box. As soon as the lid came off, he saw, on top of a pile of stone, the mould of a skull. He knew immediately that it wasn't another baboon skull like the one Josephine Salmons had brought him. At a glance he saw that the rock in his hands held a brain cast bigger than that of any adult chimpanzee. Dart ransacked the rest of the box, looking for other pieces of rock that might fit with the first one. He found another block of stone that fit exactly with the brain cast and showed the outline of a broken part of the skull as well as the adjacent lower jaw.[60]

Both the local newspaper coverage at the time and a letter sent to Raymond Dart by Robert Young in February 1925 give conflicting stories.[61] Dart had not found the brain cast and skull in either of the two wooden boxes of rubble. These were in the pieces of breccia that Robert Young, the Wits geology professor, had carried back from Taung personally and which he personally hand-delivered to Dart on that same day. For decades, this piece of the story was lost. Dart's telling of the story in his 1959 memoir minimised the role Young played at Taung and did not mention at all that it was Young who had delivered the fossils. Dart's version of the story was retold by Dart's successor, Phillip Tobias, in the 1960s and '70s. It was only in 1984 that Tobias reassessed Young's role. "It is clear from these re-interpretations," Tobias wrote, "that history should assign a greater role to R. B. Young in the chain of discovery."[62]

The distinction between receiving a couple of rocks from a railway delivery service or from the hands of Robert Young might not be critical if it were not for the fact that what had arrived at Raymond Dart's house on that day would make him an internationally renowned scientist. That fossil, and what Tobias described as the "chain of discovery" – from the miner De Bruyn to the mine manager to Professor Young to Raymond Dart, as well as the important role played by Josephine Salmons – would turn out to be the most important fossil hominid find of the twentieth century. It would provide some of the most significant clues to our understanding of human evolution.

According to Dart, on that day of the wedding in 1924, he held "the brain as greedily as any miser hugs his gold, my mind racing ahead. Here, I was certain, was one of the most significant finds ever made in the history of anthropology. Darwin's largely discredited theory that man's early progenitors probably lived in Africa came back to me. Was I to be the instrument by which his 'missing link' was found?"[63]

Careful study of the brain cast confirmed Dart's initial thoughts, but still left him with many questions. What could an ape with a brain larger than that of a chimpanzee have been doing down here in South Africa in the midst of the grass-covered plains of the Transvaal and away from the plentiful tropical forests? There was no food for apes here. How could such a big-brained creature have survived? Dart could see the brain cast but the front of the face was buried in the rock. He wanted to reveal the face of the creature that was embedded in the stone. At first he began with a hammer and chisel but soon realised they were too big and heavy for the delicate task. In order to work on the stone, he employed one of Dora's knitting-needles, which he sharpened to a fine point.[64]

Working in every free moment, Dart scraped and scratched away at the limestone and earth, prying it away from the front of the skull and the eye sockets. He didn't find the expected eyebrow ridges of an ape. Slowly, the upper and lower jaws appeared. They did not jut forward. Instead, they were quite short and pulled back under the skull.

"No diamond cutter ever worked more lovingly or with such care on a priceless jewel – nor, I am sure, with such inadequate tools," Dart recalled.[65] On his twenty-third day of labour, December 23, the rock parted. Dart was amazed by what emerged. It was an infant's face, a small child with milk teeth and a set of molars just beginning to erupt. "I doubt if there was any parent prouder of his offspring than I was of my Taung baby on that Christmas of 1924."[66]

Dart didn't waste any time. He wrote up his findings and analysis and sent a paper off to the prestigious journal *Nature*, by boat. The process of review happened quickly. Dart's article was published on 7 February 1925, which was little more than ten weeks after he'd received the delivery of the crates of stone and Professor Young's visit on the day of the wedding, an extraordinarily short period of time for a scientific paper, then or now.

"The specimen is of importance," he wrote, "because it exhibits an extinct race of apes *intermediate between living anthropoids and man*."[67] He made this case despite the fact that the Taung child had a relatively small brain case, which ran counter to the prevailing view at the time that the transition from ape to human meant a larger brain. Dart observed that the cranium looked more like that of a human than an ape, as did the dentition and the jaw.

The placement of the opening for the spinal cord was further forward in the skull than it was in other primates, so Dart concluded that the creature had a more erect posture than apes and that it had been using its feet for locomotion more than its hands. Dart concluded that "their hands handled objects with greater meaning and to fuller purpose"[68] than did apes. Since Dart concluded that the Taung child was bipedal, he regarded it as a "man-like ape". Dart boldly proclaimed that the fossil represented a new genus and species, *Australopithecus africanus*, meaning southern ape from Africa.

In his paper Dart said that many people would see his findings as remarkable. No one, he said, would have expected such a discovery in the southern point of Africa; and no one would have thought that the environmental conditions were right in southern Africa to support pre-human stock. He made the argument that by looking for the links between apes and man in tropical countries, scientists had been overlooking the fact that luxurious forests would not provide any serious challenge for creatures looking for food. He suggested that southern Africa, with its open veld, had provided an environment for greater competition for food, meaning that a creature with greater swiftness and intelligence would have had the edge over others.

On the same date that Dart's article was published in *Nature*, Robert Young wrote a letter to Dart on a small folded card in which he congratulated Dart on the discovery and the glory it would bring to him and the University of the Witwatersrand. Three days earlier, on 4 February, an article had appeared in *The Star* with the headline "BLASTED OUT: HOW PROFESSOR YOUNG FOUND THE SKULL". In an effort to set the record straight, Young wrote in his letter to Dart: "… the part I played at Buxton in the actual finding of the skull was to select, amongst the specimens, the piece of rock containing it from some fragments of rocks and minerals laid aside in the quarry by the quarryman … I do not think it of any particular importance who 'found' the skull, and I mention the matter here merely because of the heading to the report … I had no intention of claiming anything, however small, that was not my due."[69]

Unlike Young, Josephine Salmons was consistently credited by Raymond Dart. He claimed that she was the person who inspired him to search for fossils in Taung. In any biography of Dart, no matter how brief, her name is mentioned. In his own memoir, Dart published a photo of Salmons in a graduation gown from 1925.[70] Without her input, Dart might not have become a world-famous palaeontologist. Little is recorded about Salmons' later life. Salmons completed her BSc and honours, and all but her final year of a medical degree at Wits before she married Cecil Jackson and had two children. She did not continue with a scientific career. She divorced Jackson and died of cancer in April 1950 in Scottburgh in Natal at the relatively young age of 48.[71]

It would not be the last time that someone like Josephine Salmons would contribute to the success of a prominent scientist. It would not be the last time that someone who played a crucial role in a fossil find, like Robert Young, would defer to the lead scientist. This was a pattern that would repeat itself again and again for the next century. Workers, labourers, researchers and assistants who helped build the careers of others for the most part received little attention or applause. Dart and the Taung child skull helped initiate the myth of one man, one fossil.

The reaction in South Africa to Dart's announcement was generally one of excitement. The University of the Witwatersrand was barely five years old

THINKS NEW SKULL LINKS MAN AND APE

Professor Dart Names His Discovery the Australopithecus Africanus.

OLDER THAN THE 'JAVA MAN'

London Expert Hopes It May Carry Back Human History to Period Hitherto Unknown.

Copyright, 1925, by The New York Times Company.
By Wireless to THE NEW YORK TIMES.

JOHANNESBURG, Feb. 4.—Professor Raymond Dart, discoverer of the fossil skull at Taungs, 1,000 miles south of Broken Hill, where three years ago a skull said to be of great antiquity was found, has made the following statement concerning the importance of the present find, to which he has given the name of Australopithecus Africanus:

"The geological records of the different species of man have been rendered fairly perfect. Where geological evidence has been lacking is in the specimens of that phase of pre-human existence between the most primitive and ancient of men and the most advanced of apes. This gap is now filled by the Australopithecus Africanus of Taungs.

"The individual, in brief, was not a human being and yet was a much more intelligent being than a gorilla or a chimpanzee, which is the highest of living apes. He was unable to talk, but the brain was advanced in the direction required in ancestors whose offspring were required to attain ultimately the power of communicating with their fellows by the symbolism of speech.

"He is therefore regarded not as an ape-like man but rather as a man-like ape and he reveals to us that period of human evolution more remote than the Pithecanthropus [the Java man] early in the Pliocene geological epoch, or even in the Miocene, when human stock had only very slightly advanced beyond that which led to the modern apes."

The Johannesburg Star makes the following addition to the above statement:

"The great importance of the Taungs skull lies in three things; it brings the record of the rocks down a thousand miles south of Broken Hill; it relates to a form of life really intermediate between the most advanced apes and primitive human beings, and it plainly reveals South Africa more clearly than ever as a mine of information regarding the dim past of the human race."

Headline announcing the Taung child skull in the New York Times *on 5 February 1925. Copyright:* New York Times.

and the university council congratulated Dart for his contribution to science and the distinction he brought to the university. Dart received a warm letter of congratulations from Jan Smuts, who was then the president of the South African Association for the Advancement of Science. Smuts called Dart's discovery "epoch making"; he suggested that it was "calculated to concentrate attention on South Africa as the great field for scientific discovery, which it undoubtedly is".[72] Smuts had long been interested in human pre-history and he thought Dart's discovery might be the breakthrough that would place South Africa alongside the locations of other fossil finds such as the Neanderthal skulls, Java Man and Piltdown Man.

45

Despite the scientific excitement, there were local naysayers. "The first suggestion that the Taungs (sic) skull of Professor Dart is not the missing link in man's ancestry was made last night by the Reverend William Meara of the Central Methodist Church in Johannesburg," reported the *Rand Daily Mail* of 9 February 1925. Meara claimed that numerous missing links had been proven wrong and that the talk about the new skull was misleading. He did not believe, he said, in the "monkey theory of man's ancestry".

Dart's scientific colleagues in England were generally not impressed. A week after his paper was published in *Nature*, four leading British anthropologists wrote letters in response. Sir Arthur Keith, who had been engaged in research on Piltdown Man, doubted that Dart had found anything new. He argued that the Taung child was probably in the same genus as the chimpanzee and the gorilla. When questioning where man's direct ancestors were to be found, Smith Woodward responded: "The new fossil from Africa certainly has little bearing on the question."[73]

Even Dart's own mentor, Grafton Elliot Smith, was not convinced. "We want Professor Dart to tell us the geological evidence of age, the exact condition under which the fossil is found, and the exact form of the teeth."[74]

Dart might have done well to work with Robert Young to explore the context of the fossil find. Although he never stated as much, there is always the possibility that he hesitated to share the presentation of the findings and its context with another scientist.

The British Empire Exhibition was due to take place at Wembley in London. Dart was not willing to send the original fossils for display and so instead he sent several plaster casts of the Taung skull. He prepared a tree-like chart that showed the Taung skull as the ancestor to Java Man and Piltdown Man. The chart and the casts were displayed under a banner that read: "Africa: the Cradle of Humanity". Arthur Keith and other scientists looked on the display and the banner with disdain. Keith was upset that he had to stand alongside the general public at the exhibition to examine the casts behind glass. "An examination of the casts exhibited at Wembley," he said, "will satisfy zoologists that [Dart's] claim is preposterous."[75] The popular British press was unkind. They made fun of the finding, calling the Taung skull ugly. Jokes such as "Who was that girl I saw you with last night? Is she from Taung?"[76] did the rounds. "Criticism rather than adoration of their potential ancestry [in Africa] seemed to be the overseas reaction," said Dart.[77]

Messages from New York stated that Professor Dart's theory had not convinced the legislature of Tennessee. The governor of that state had signed an "Anti-Evolution" bill in March 1925 which banned the teaching of any theory contrary to the biblical story of creation. Dart received angry letters from religious people all over the world, some warning him that he would "roast in the general fires of Hell". A letter to the London *Sunday Times* described him as a traitor to his creator and as making himself "the active agent of Satan." Another letter said that for his heresy, he should be punished by "being unblessed with a family which looks like this hideous monster."[78]

Dart's most ardent supporter was Dr Robert Broom. In a 1925 letter to Dart, Broom wrote: "The missing link is really glorious." and "Perhaps an adult skull or perhaps a whole skeleton will yet turn up."

Two weeks after writing the letter, Broom burst into Dart's laboratory unannounced. Ignoring Dart and his staff, he strode over to the bench on which the Taung skull had been placed and dropped on his knees in front of it. "What are you doing down there?" asked Dart. "I'm bowing to our ancestor," Broom replied. He stayed with the Darts over the weekend and spent almost the entire time studying the skull. "Having satisfied himself that my claims were correct, he never wavered," said Dart.[79]

Broom felt that Dart had been unfairly attacked by his British peers. He pointed out that many in England took little interest in the discovery but were more interested in whether, by using the name *Australopithecus*, Dart's Latin was correct. *Australis* means "south" in Greek and *pithecus* means "ape" in Latin, so there were complaints about the mixing of the two languages. Broom thought that perhaps Dart's most serious offence was that he had not immediately sent the Taung skull to the British Museum for examination, which would have resulted in its being held in secret for probably a decade for extensive examination and review. That Dart had been so bold as to publish an account within weeks of the discovery was possibly his cardinal sin. "English culture," said Broom, "treats him as if he had been a naughty schoolboy."[80]

Another scientist who withheld his support for Dart was the Czech-born anthropologist Aleš Hrdlička. Hrdlička was an important figure in the development of physical anthropology in the United States and the founder of the *American Journal of Physical Anthropology*. He was the first curator of physical anthropology at the Smithsonian in Washington DC, a position he held for most of the first half of the twentieth century. Hrdlička reprinted Dart's *Nature* article in the *Journal* and before the year was out, he travelled to South Africa to further investigate Taung. He went to the site where the skull had been found, even though Dart did not, and searched for more fossils. He found fossil remains of other extinct primates, but nothing like the Taung skull. His view was that in order to come to a clear conclusion, he would need to examine additional (and adult) specimens.[81]

There were two large mounds called assemblages near where the Taung skull had been blasted out of the quarry. Over time, one was named the Dart pinnacle. Given his status in science and his visit to the site, the other mound was named the Hrdlička pinnacle.

Despite Hrdlička's interest in the Taung child skull, his greatest common interest with Raymond Dart was not the study of ancient hominid fossils. It was the study of race. For years, Hrdlička had been gathering human skeletons at the Smithsonian for that purpose. He had published a widely cited textbook on anthropometry, the measurement of bones, with a focus on racial classification. Dart had already picked up on the worldwide interest in measuring human skeletons and he had started to create a collection of his own.

4

Collecting and Classifying

When Raymond Dart arrived in South Africa, he had a passion for building a collection of human skeletons. In this he followed in the footsteps of Peringuey at the South African Museum, as well as European scientists who had been gathering skeletons since the turn of the century. But it was even before he had arrived in South Africa, when he was on his Rockefeller Scholarship in the United States, that Dart first decided to collect human remains. He worked with Robert Terry there who, as head of the anatomy department at Washington University in St Louis, Missouri, was building his collection of human skeletons. Many museums and universities in the US, including the Smithsonian, built skeleton collections in the late nineteenth century. While these collections included skeletons from European or "white" people, the emphasis there was on collecting "Native American" and "Negro" skeletons in order to compare them. The scientific interest at the time was to fit these human remains into the taxonomy of race.

Terry recorded race, gender and age in order to compare the skeletons, most of which had come from destitute people who had died and whose corpses were being held at local morgues and hospitals. When Dart started his own collection at the Wits Medical School's department of anatomy, he too wanted to investigate and compare what he called different "racial types". He continued to correspond with Terry from South Africa. Taking on the American interest in race, as well as a colonial attitude towards people who were not European, Dart sent Terry six South African skeletons marked "Bantu".[82]

Bantu, or abantu, is an isiZulu word from "ntu", meaning person, and "aba", which makes it plural, hence ba-ntu means people. The term Bantu-speaking people is used to describe the linguistic group of more than 100 million people speaking Nguni languages throughout sub-Saharan Africa. In South Africa, in Dart's time, the word took on a negative, pejorative meaning with racist connotations.

It was not unusual for scientists, Dart amongst them, to believe that humans could be categorised as distinct racial types and that each type could be classified by its physical characteristics. Scientists looked for individuals who would represent a most pure racial type, and saw any mixing of characteristics as deviating from this type. That is why anatomists relied on measurements to find who represented a "pure racial type" and who did not. Dart believed it was important to measure brain size, skull shape, facial features, skin colour, hair texture, and bone length in detail. This typological method was central to physical anthropology and supported the search for fundamental racial differences in the human population. Dart emphasised difference over similarity. His search was for the distinctive "pure racial types" representing the groups that Theal had described as Bushmen, Hottentot and Bantu.

In joining the existing skeleton trade in southern Africa, Dart wrote a series of letters to Dr Louis Fourie, the medical officer in South-West Africa (Namibia), asking for help in gathering material for his collection. In one letter, in June 1923, he thanked Fourie for his "enthusiasm in the matter of securing the Bushman materials for our department" and enclosed detailed instructions for embalming the cadavers before delivery. In the same letter, Dart referred to Professor Jan Hofmeyr, the principal at Wits University and a prominent political colleague of Jan Smuts. Hofmeyr was in communication with the administrator of South-West Africa, he said, about "securing the material" and suggested that any legal difficulties could be addressed by a proclamation.[83]

In another letter to Fourie the following year, Dart acknowledged receipt of a "Bushman skeleton", reporting that it was "in splendid condition" and thanking Fourie "most sincerely for (his) kindness in remembering our needs here".[84] It is not clear how Dr Fourie acquired the skeleton. However, Dart asked about the sex and age of the individual, thinking that the information might be available if the exhumation had happened recently. This would suggest that the skeleton was taken from a grave.

In addition to the "Bushman" skeletons delivered by Fourie, Dart also acquired a loan of "Bush-Hottentot" skeletons from graves in the Orange Free State and the Cape from Maria Wilman at the Kimberley Museum. By the end of the 1920s, he had gathered 92 complete skeletons.

While some of the skeletons were taken from graves, many cadavers used in the department at Wits for dissection classes were unclaimed bodies from South African hospitals. Dart encouraged dissecting techniques that minimised the damage to the bones so that he could clean them and retain the entire skeleton for the collection.

In 1929 Dart published a paper praising Dr Peringuey's pioneering work, especially in relation to his work excavating burial sites. He gave thanks for the dry South African climate that preserved skeletal remains. He declared that this aspect of physical anthropology was in its infancy in South Africa and he called for a more systematic cataloguing of the remains that had already been collected. He wanted more research to describe the skeletal features of "the Bantu" and to find out to what extent "hybridization with Bush and other previous aboriginal inhabitants of the country" had taken place and how it affected their anatomy.[85]

One of Dart's senior lecturers, LR Shore, worked with a group of the skeletons and published an article describing "abnormalities" in the spinal column of what he described as "the Bantu". Another senior lecturer who joined Dart, JC Middleton Shaw, focused his work on the skeletons and published *The Teeth, the Bony Palate and the Mandible in the Bantu Races of South Africa*. The term "Bantu" was used freely at the time, whereas it later became associated with apartheid policies. Soon after Wits officially had been established in 1922, the university added a department of Bantu studies, in 1923, which published *Bantu Studies: A Journal Devoted to the Scientific Study of Bantu, Hottentot and Bushmen*.[86] Reflecting concerns about the meaning and intention of the term, in 1942 the journal *Bantu Studies* was changed to *African Studies*.

This kind of research was not unique to South Africa. The central focus of the American Association of Physical Anthropology at the time was also on questions of racial classification and anatomy. A 1932 meeting at the Smithsonian, led byAleš Hrdlička, included papers such as "The Nose of the American Negro" and "The Plasticity of the Japanese Physical Type".[87]

But in the 1920s in South Africa, the term "race" was changing. Previously, at the turn of the twentieth century, the term was often used to describe "nation"; people referred to the British race and the Afrikaner race rather than using the term to describe a difference between black and white people.[88] Prime Minister Jan Smuts began to embrace the reconfiguration of the term "race" to refer to colour because he wanted to see the British and the Afrikaners work together to develop a united white race in South Africa.[89] Smuts had been a revered Boer general during the Boer War and had been a driving force behind setting up the League of Nations after World War I.

Alongside his own increasing interest in scientific matters, Smuts saw scientific discovery in South Africa as an important strand in his strategy for putting white South Africa on the international map. When he lost political office to Barry Hertzog in mid-1924, he devoted considerable time to scientific matters, and in 1925 he served as president of the South African Association for the Advancement of Science.

Smuts proclaimed that the SA AAS should "bring together and unite all South Africans irrespective of race and language". Again, "race" here referred to Anglo-Afrikaner relations, at at time when the treatment of black South Africans was becoming increasingly restrictive and discriminating.

In a public speech that year entitled "South Africa in Science", Smuts drew on the geological theory that Africa had been at the centre of continental drift, and that South America, India and Australia had all broken away from Africa. He suggested that increased knowledge of local geography, flora, zoology and astronomy could shed light on an understanding of these sciences throughout the southern hemisphere. He spoke excitedly about the new field of human palaeontology in which he claimed South Africa had a "central position".[90]

Smuts applauded Dart's discovery of *Australopithecus africanus* and the theory that the Taung skull provided a "transitional form between the ape and the human", showing that humans could trace their distant ancestral line to Africa. "South Africa," he said, "may yet figure as the cradle of mankind, or shall I rather, say, one of the cradles?"

Yet in the same speech, Smuts said: "Our Bushmen are nothing but living fossils whose 'contemporaries' disappeared from Europe many thousands of years ago." In anthropological terms, he said, South Africa was "possibly ten thousand years behind the time as measured by the standards of European cultures".[91]

Smuts was interested in two objectives at the same time. He wanted to cast off colonial domination in science and place South Africa at the centre of global knowledge creation. At the same time, he was concerned that South Africa could be marginalised, and he looked for approval from his European colleagues. Despite his disdain for British colonial control in South Africa, he had no interest in ending the colonial domination of the country's black people. It was white South Africa he wanted to unify. And he wanted to solidify the country's role in the field of science. Smuts was well positioned to influence the fields of archaeology and palaeontology and he used his influence in an effort to place white South African science in the forefront. His interest in these new fields of science was responsible for much of the institutional support for the search for human origins for another 25 years.

The newly developing field of social anthropology, as distinct from physical anthropology, also had an impact on how the term "race" was understood. Social anthropology provided a context in which social science could pronounce on cultural differences and the "native question" in an allegedly detached, scientific manner. The anthropological thinking of the time helped pave the way for the arguments of white supremacy in support of segregation. Arguing that African people were inferior, science generally and anthropology specifically began to make the case that Africans could not live on equal terms with white people. But it was physical anthropology more than any other field of study that helped to promote the racial paradigm in South Africa and the social engineering that flowed from it.

Dart thought the South African skeletons in his collection might hold a clue to understanding human evolution. He thought it was possible to identify the prehistoric racial types from which modern African populations originated. The first article he published after arriving in Johannesburg was in *Nature* in 1923 and it included a review of the existing evidence in support of the "Boskop race", about which Robert Broom had written years earlier. What Dart described as Boskop-like features in several living individuals led him to believe that Boskop Man was an ancient precursor to the Bushmen. He argued that South Africa "harboured the pre-human stock from which the human race itself was derived".[92]

This paralleled similar thinking in the United States' scientific community. In addition to their interest in comparative anatomy, and the hierarchy of racial classification, many US scientists in the 1920s and '30s, including Hrdlička, began to explore how evolutionary change occurred within defined racial groups. The interest in skeletons, in terms of their racial differences, began to expand to an interest in how race related to human history. Therefore, some aspects of the early interest in human evolution in the US grew from the human skeleton collections and the idea of distinct racial types, and were themselves built on racist foundations.[93]

One of the most striking illustrations of the convoluted interest in both racial differences and human evolution within Dart's department came from the work of one of his students, Lawrence Wells (who later went on to become the head of the department of anatomy at the University of Cape Town). Wells studied the bones of Europeans, "Bantu" and "Bushmen" skeletons in the late 1920s and authored a scientific article based on his master's thesis entitled "The Foot of the South African Native". Wells wrote: "In both the Bushman and the Bantu the anterior articular facet of the first cuneiform faces not directly forward as in the European, but obliquely forward and medially so that the first metatarsal in these types is directed more medially, and separated by a greater space from the second metatarsal, than in the case in the Europeans."[94] In other words, Wells argued that African feet were anatomically closer to the feet of apes than European feet, showing that they had more recently lived in trees and less recently begun to walk on the ground. He concluded that "The foot of the Bushman, both in structure and function, forms a most remarkable connecting link between the feet of the apes and those of the higher races of man" and "The Bantu foot forms an almost ideal connecting link between that of the Bushman and that of the European."[95]

This is an example of science being shaped by the scientist's racist expectations. Bernhard Zipfel, a podiatrist, fossil curator and researcher of hominid feet at Wits today, says that Wells' conclusions were outrageous and wrong.[96]

These ideas about feet coming out of Dart's teaching were as wrong as the idea of a Boskop race as a precursor to the "Bushman". Nevertheless the concept of a Boskop race would shape the thinking of many students and would take many decades to be proven wrong.

Dart's theory that the Taung child skull proved that human evolution began in Africa would take just as long to be proven correct. The prevailing view at the time, argued by Sir Arthur Keith and Sir Grafton Elliot Smith in England, was that the races of Africa had branched out from Europe and Asia and had quite recently arrived on the African continent. The thinking was such that white supremacy would win no matter where the fossils were found. If modern humans were shown to have evolved in Europe, it would be used to prove the theory that Europeans were superior. However, the thought that human evolution might have begun in Africa was also used as an illustration that Africans were not as advanced as Europeans – a scientific catch-22.

In general, for many European scientists, southern Africa was considered unlikely territory for palaeoanthropologists. Given Java Man and another *Homo erectus* fossil found in China, popularly known as Peking Man, the search for human origins had focused instead on Asia – in particular the village of Choukoutien outside of Peking. It became the focus of major scientific expeditions looking for fossils. Between 1915 and 1951 the Rockefeller Foundation donated close to $45 million, an enormous amount of money then, to establish a medical school in Peking, from which onging expeditions were supported.[97] For more than 20 years, until after World War II, the scientific community, and the media in Europe and the US, focused attention on the search for hominid fossils predominantly via explorations in Asia and the major excavations at Choukoutien. "What dissonant squawkings were these from the puny South African infant at Taung?" wrote Dart, sarcastically imitating the view from the north. "Could anything good emerge from the Kalahari Desert?"[98]

Robert Broom, however, believed the Taung child skull was the most important ancestral human fossil ever discovered. Despite his strong defence,

Broom didn't embark on any of his own research on human origins. Instead, he continued working in Douglas as a doctor, mayor and skeleton collector; it seems he didn't want to interfere with Dart's territory in physical anthropology. But Dart was not doing any further fossil research either, saying he was too occupied with his work at Wits Medical School "to sit brooding on it". Broom believed that Dart backed away from the field because of the hurtful treatment he received from fellow scientists in Europe and lamented the fact that for nearly a decade the search for hominid fossils in South Africa had ground to a halt.

Despite Dart holding firm to his argument that southern Africa held the clues to the origins of humankind, he did not believe cultural development began there. He believed culture was largely determined by race. And he took the assumption about racial types one step further: he asserted that these types were linked to behavioural characteristics and cultural practices. He assumed there was a racial hierarchy, not only in terms of physicality, but also in terms of cultural development. He had been greatly influenced by Sir Grafton Elliot Smith, who had promoted the theory of cultural diffusion, which suggested that ancient migration and travel from other parts of the globe was the only possible explanation for cultural development among the indigenous peoples of Africa. One month after Dart published his momentous article about the Taung child skull in *Nature*, he published another article declaring his diffusionist views, stating that the people of southern Africa were influenced by ancient visitors from the Near East who "not only visited these territories and carried off their denizens, particularly their women, but also intermarried with them and settled down amongst them, bringing to them novel arts and customs."[99]

Dart was one of the many scientists who did not believe that cultural achievements such as Great Zimbabwe, the large stone buildings in Rhodesia (Zimbabwe), could have been built by the local African population. Many scientists claimed that Great Zimbabwe had been built by foreigners, such as the ancient Phoenicians. Gertrude Caton Thompson, an influential archaeologist, was one person who clashed with Dart on this issue. She was commissioned by the British Association to conduct research on the origins of Great Zimbabwe and her presentation to SA AAS, in 1929, was supposed to bring an end to the long-running debate. She declared that the civilisation that produced Great Zimbabwe was African. Dart was in the audience. With great indignation, he stood up and lashed out at Caton Thompson, refuting her findings and stating that the foreign influences for the buildings had come from ancient Egypt and Phoenicia. Caton Thompson responded calmly, holding firm to her view that the construction was produced by African people, which resulted in Dart storming out of the room. He felt strongly about the argument, despite the fact that he had never been to Great Zimbabwe himself.

This mix of thinking about skeletons, race, cultural hierarchy, and human evolution did not stay in the laboratory at Wits. Dart took his thinking out into the public realm. In the same year as his fight with Caton Thompson, Dart was called to give evidence in court. On the witness stand, he gave a technical statement on "the question of 'colour' in Europeans and natives". The *Rand Daily Mail* reported that Mrs Christie Neff had been charged with possession of a jar of sherry, which was illegal because she was "coloured". The term was used to describe someone who possessed ancestry from more than one group, namely Europeans, Asians, "Bushmen" or other Africans. Dart examined Mrs Neff and declared that she had "coloured blood in her veins". He could see the "tawny hue of her skin" on her shoulders, the back of her arms and her hands. "Some Europeans are darker than Mrs Neff, but they are not yellow," he said.[100] The hearing was adjourned.

A second case involved a Mrs Batty, who had testified she was "coloured". *The Star* reported that Professor Dart "swore that she was not coloured", thereby defending the three bottle-stores that had sold her alcohol. As the first witness for the defence, Dart declared he "could find no physical feature in her constitution which could be considered diagnostic of a coloured person".[101] He produced a skin colour chart used by ethnologists and concluded that Mrs Batty's skin colour proved that she was European.

While Dart was gaining a reputation in South Africa, he continued to struggle to find the international respect he hoped for. He made one trip to London in 1931 to present the Taung skull. On his arrival, scientific colleagues, including Smith Woodward, who had presented Piltdown Man, Sir Arthur Keith, and Grafton Elliot Smith, were friendly but they seemed most interested in talking to Dart about the recent discovery of Peking Man in China rather than talking about the Taung skull. Elliot Smith gave a presentation at the Zoological Society of London about his recent trip to Asia, complete with lantern slides. He showed that Peking Man was bipedal and appeared to know how to make fire. He finished his presentation to resounding applause.

Then Dart stood up to make his presentation. "I stood in that austere and chilly room, my heart bounding with hope that the expressions of polite attention on the four score faces before me might change to vivid interest as I spoke." He didn't have any casts to pass around nor any lantern slides to show. All he had was the small skull in his hands to illustrate his points. "I realised that my offering was an anti-climax."[102]

At the time, many scientists surmised that the initial step from early primates to more human ancestors was the enlargement of the brain. The theory was that bipedalism was a later development, resulting in popular images of a stooping creature with a big head. Piltdown Man was in keeping with this view as it represented a large-brained creature with an ape's mandible and teeth. Sir Grafton Elliot Smith was particularly committed to the idea that the brain developed first, which then led to other advances later. The Taung skull, on the other hand, had a relatively small brain, and a more erect posture, which confused many scientists. Dart argued that there were features of the Taung skull that pointed to a smaller but more human-like brain. Scientists also critiqued Dart on the basis of the need for more information about the geological context of the fossil as well as an estimate of its age. They called for a search for adult specimens of *Australopithecus* that would confirm or disprove Dart's claims.

As if the general disdain wasn't discouraging enough for Dart, while he was in London, the 300-page monograph he had worked on for years describing *Australopithecus africanus* in detail was rejected by The Royal Society of London. Sir Arthur Keith had recently published a lengthy account of *Australopithecus* based on his own investigation of casts – not on the original fossil – in his book *New Discoveries Relating to the Antiquity of Man* in the same year. The society rejected Dart's monograph and only agreed to publish a small portion of it related to the dentition. Dart carried his lengthy monograph back to Johannesburg where it still sits today, unpublished, in the Wits Archives.

Dart believed he personally owned the Taung skull. Back in 1925, after he had described the fossil, the Witwatersrand Council of Education offered to pay for his travels so he could compare the Taung skull to others held in England while he was working on his monograph. They made this offer provided Dart donate the skull to the university. Dart declined, and as a result, technically, the skull remained his personal property.

Dora Dart had a disconcerting experience with the Taung skull on the same trip to London. Dart had gone back to South Africa ahead of his wife, who was continuing her medical studies in London for a few months. He left the Taung skull with Elliot Smith to have model casts made for further distribution, with the plan that Dora would bring the skull back with her to South Africa. The night before she left, Dora visited Elliot Smith to pick up the skull. They spent the evening together and then Elliot Smith escorted her back to her hotel in a taxi. When they arrived at the hotel, they had coffee together in the lounge. It was only when Dora returned to her room that she realised she had left the skull in the taxi. She just about fainted with stress and anxiety. She called Elliot Smith and together they went to the police station where the officer on duty agreed to put out an alert to all the police stations in the area.

It was after 4am that the taxi driver received the message to look for a little box in a brown paper bag in the back of the taxi. He brought it into the police station, where the local officer had a fright when he opened it up to find a skull inside.

After his return from London, Dart largely set the Taung child skull aside and pursued other interests. It wasn't long thereafter that Dora and Raymond Dart's marriage started to deteriorate and they were divorced in 1933. Dart married Marjorie Frew, the head librarian at the University of the Witwatersrand Medical School Library, and they remained together for the rest of their lives. Dart continued to lead the department of anatomy and became the dean of the medical school. "The man who put the medical school and indeed the University, truly on the map, was Raymond Dart," wrote the Wits historian, Bruce Murray.[103]

In the meantime, Robert Broom had been struggling to find his place in the scientific community. His sale of the Karoo fossils to the American Museum of Natural History, fossils that were meant for the South Africa Museum, continued to haunt him. In 1924 Dr Peringuey died and Broom was considered to replace him as director, but the job went instead to another candidate, Edwin Leonard Gill from the Royal Scottish Museum. Nevertheless, Broom was awarded a Royal Medal by the Royal Society for his extensive work on the origin of mammals. He was better recognised in the US and the UK for his contribution to science, he felt, than he was in South Africa. Dart even had to work hard to convince Broom to attend the important 1929 joint meeting of the British and South Africa Association for the Advancement of Science (SA AAS).

In 1933, Broom was elected president of SA AAS. He was pleased with the recognition, but he could not afford his train fare to Johannesburg for the meeting. By then he had published close to 350 scientific papers but did not hold a scientific research post of any kind. He had also left his medical practice in Douglas, and was moving from place to place to find work as a doctor, struggling to make ends meet.

Jan Smuts returned to political office in 1933, as Deputy Prime Minister, and Dart wrote to tell him about Broom's situation. Smuts had long admired Broom and felt it was not right that a man of his age and stature should have so little support for his research. They had long been in correspondence, and that year, Smuts wrote to Broom about evolution: "Our knowledge is too fragmentary, our interpretation of what we do know too vague and uncertain to entitle us to definite conclusions. It appears to be a great Plan – with a Planner in the background ... And so I continue to look round for clues ... I look for something in the nature of the universe to account for what has happened. And that something must be both physical and organic and mental and also much more."[104]

Less than a year later, with Smuts's recommendation, Broom was appointed, at the age of 68, as curator of fossil vertebrates and anthropology at the Transvaal Museum in Pretoria. "I am so glad that you are at last at the Museum where I have long wished you to be," wrote Smuts. "I hope you will now have some time to devote to the work which is peculiarly your own, and which has brought you world fame. The salary is a pittance but they say that wise men don't want much to be happy. I wish you many long years of fruitful labour."[105] Broom finally had the opportunity to move away from medical practice completely and more squarely focus on his scientific interests.

Just as Broom's career was shifting, so too was Dart's. Dart had decided to embark on new areas of research in the Kalahari that focused on his desire to locate a "pure Bushman race". His interest in race typology, in pre-historic fossil discoveries such as Boskop Man that pointed to pure racial types, in racial hierarchies amongst current populations in southern Africa, and in the importance of external influences on cultural progress had shaped his thinking. Bruce Murray wrote that Dart "virtually created the University's international reputation for the study of the peoples of South Africa, with his discovery of the first australopithecine skull, the Taung skull."[106] However, this quote blurs the distinction between indigenous people living in South Africa and ancient pre-humans. As a result of the common belief in evolutionary anthropology and the hierarchy of race within the field of physical anthropology, this lack of distinction between two different disciplines in science, and two very different time-frames millions of years apart, would occur again and again, to cloud thinking. The research Dart was about to pursue in the Kalahari in the 1930s would continue to confuse the distinction between the palaeo-sciences, which explored ancient fossils, and anthropology, which explored living people. It would also support a race paradigm in South Africa that contributed to the thinking about segregation and the development of reserves for African people.

5

"Living Fossils"

On stage, wearing a black top hat and a cravat with his three-piece suit, Raymond Dart stole the show in the Wits production of *The Anatomist* by the British playwright James Bridie. First published in 1931, the play was based on the life of the infamous English anatomist and surgeon Dr Robert Knox, and the Burke and Hare murders. In 1828, Knox had bought sixteen cadavers from William Burke and William Hare, who had murdered their victims with the intention of selling their bodies to Knox and others like him who wanted them for teaching anatomy. Hare gave evidence to the state and Burke was hanged for murder. Dr Knox served no jail time but he was deemed guilty by association and his reputation was sullied for the rest of his life.

Five years after the play was first published, Dart was perfectly suited to portray Knox in the Wits production – and the play was performed in the dissection hall. The audience gathered in the authentic setting to watch the macabre production. "Send him out to rob the graveyards," proclaimed Dr Knox, throwing his hands out towards the audience and referring to Walter Anderson, the other character on stage. "It is all the ass is fit for." "Churchyard raids," said Walter Anderson. "I disapprove of them very strongly." "You do, do you?" responded Knox. "You are a prig Mr. Anderson."[107]

Later in the play, Knox appears on stage with two female students. His booming voice declares, "Ladies, I will not be treated like a naughty schoolboy. Do you know who I am? Do you know that I am the apostolic successor of Cuvier, the great naturalist? Do you know that I am a comparatively young man, much of my work is already immortal? ... I am strong in the knowledge that the name of Knox will resound throughout the ages." "Yes," said one of the students. "For bullying and blustering."[108]

The scene ends when a body is brought on stage. Accusations of murder begin to fly and Knox remains arrogant. He argues that in accepting the cadaver for his anatomy class, he has done nothing wrong.

Raymond Dart was often described as a showman. In the Wits Senate discussing university affairs, he "put on some dazzling performances ... as the occasion demanded, he could be angry and sarcastic, earnest and emotional, or gentle and persuasive."[109] We don't know what Dart believed about Knox's practices, but the play illustrates the short step from grave robbing to anatomy murder.

In the same year that Dart had performed in *The Anatomist*, he also entered into correspondence with Donald Bain, a South African former farmer and hunter, who was the great-grandson of Andrew Geddes Bain, the well-known nineteenth-century South African geologist, road engineer and explorer. In the 1920s, Donald Bain had been involved in the skeleton trade and, just as Dart had done, he too had written to Dr Louis Fourie in South-West Africa. In one letter Bain wrote that he had met with limited success because he had only secured "two fairly complete Bushman skeletons".[110]

By May 1936, Bain had shifted his attention from skeletons to living people in the Kalahari. He wrote to Dart asking him to travel to the Kalahari to examine the "Bushmen" there, some of whom he described as "very fine specimens" and others as "useless from my point of view".[111] He told Dart that he was preparing to take a group of "Bushmen" from the Kalahari to the British Empire Exhibition in Johannesburg later in the year and he hoped that in advance of the exhibition, Dart would travel to a camp he was preparing for them to conduct scientific research. Bain wanted Dart to confirm the quality of the "specimens" and Dart was keen to go. He convinced the University of the Witwatersrand to sponsor a large expedition to Bain's camp at Tweerivieren in the Kalahari, about 250 kilometres north of Kimberley. Dart, like Bain, had expanded his interest from skeletons to living people.

Donald Bain stated he had the best interest of the "Bushmen" at heart. The research, he said, would help make the case for the South African government setting aside land for a "Bushman" reserve at a time when much of the land that had been previously occupied by the Bushmen had been taken over by farmers. Bain promoted the growing myth that Bushmen were isolated and timeless, when in fact they had interacted with other communities throughout the region for millennia. His campaign had a decidedly paternalistic tone. "Much that is of such interest among the uncontaminated Bushmen of the north is missing in these people," he wrote to Dart. "Those that are farthest away wear the typical Bushman costume and those that are living in the neighbourhood of the outlying farms wear any old thing they can pick up ... Their music, with its endless monotony, will drive me mad before I am through."[112]

The trip to the Kalahari was not Dart's first expedition. During his sabbatical in 1930, he had spent eight months as part of the Italian Scientific Expedition from Cape Town to Cairo under Commander Attilio Gatti. Just as in the case of England, Belgium, France and Germany, Italy had colonised parts of Africa, including parts of Eritrea, Somalia and Ethiopia for a time. Gatti's expedition had introduced Dart to the entire continent for the first time. After many years of speculative commentary, it was on his trip with Gatti that Dart first saw the Great Zimbabwe ruins, which only confirmed his belief that they had been constructed by visitors from afar and not by Africans. It was also on this trip that he furthered his interest in primates. Dart's "greatest thrill" from the trip was tracking a gorilla in the Congo. Attilio Gatti had been granted a permit by the Belgian authorities to shoot a gorilla for an Italian museum. After trekking for three days, Dart said "our Pygmy guides" led the group to where five gorillas were resting.[113] Gatti shot a huge male gorilla that measured almost 2 metres in height and weighed over 180 kilograms. Dart proudly had his photograph taken standing next to the dead animal.

Raymond Dart with the gorilla shot by Attilio Gatti on the Italian Scientific Expedition in the Congo in 1930. Courtesy of Goran Strkalj. Copyright: School of Anatomical Sciences, University of the Witwatersrand.

It was on this same expedition that Dart was introduced to the process of making face masks. Professor Lido Cipriani, an Italian physical anthropologist, had carried out an earlier expedition in Africa during which he had begun to gather face masks from across the continent by molding plaster of paris onto the faces of living people. Cipriani believed in the superiority of the Italian race and the inferiority of Africans and supported the colonial project of the Fascist regime. He later worked for the Italian Race Office and signed the Racial Manifesto in 1938. Cipriani's face mask technique was aggressive, and in many cases he lied to people, telling them that the procedure had healing qualities.[114] After acquiring several of Cipriani's masks, Dart saw this process as a significant new methodology in the field of physical anthropology that would allow important comparisons between different racial types. The making of face masks was a technique European anthropologists were using around the world to document and describe people different from themselves. From then on, through to the 1980s, almost every expedition from the Wits department of anatomy to study the human biology of living people across Africa included making face masks. Compared to the larger skeleton collection of 2 605 human skeletons, the lesser-known Raymond Dart Collection of African Life and Death Masks holds 1 110 face masks.

The University of the Witwatersrand Kalahari Bushmen Expedition left Johannesburg on 23 June 1936. As professor of anatomy and dean of the Faculty of Medicine, Dart led the team of eight. He was joined by Professor Maingard, who studied languages, Professor Doke, head of the department of Bantu studies, PR Kirby, a professor of music, Professor ID MacCrone, Mr Eric Williams, who was Dart's technical assistant, Mr Hall, who was a medical assistant, and Professor Maingard's son. James van Buskirk, a member of the Royal Geographic Society, was the official journalist on the trip and he sent regular articles to the *Rand Daily Mail* to describe the research activities at the camp. Van Buskirk also participated in the research himself by acting as the official photographer. Like face masks, photography was becoming another method of research for anthropologists. This trip was one of the first institutionally sponsored research expeditions based from a South African university. Just as Jan Smuts had encouraged at the 1929 SA AAS meeting, the aim of this type of expedition was to produce world-class science that could be offered up to the Empire as proof of South Africa's scientific worth. At the same time, the new science consolidated the power of white scientists in their ability to observe and measure people they thought were inferior, both culturally and physically.

In preparation, Bain sent Dart photographs of the people in the Kalahari so that Dart could assess whether they were appropriate specimens. Bain thought that from a physical point of view, this group, as opposed to "Bushmen" from other geographic areas, was perfect for the Empire Exhibition. He wrote to Dart: "They are what the public will expect to see ... I did not believe that so pure a type existed."[115]

Three cars and an Albion lorry drove across a thousand kilometres from Johannesburg to Tweerivieren. The lorry was loaded with measuring instruments, supplies to make the face masks, discs for sound recording, and photographic equipment. They drove over dirt roads, tracks through the veld and dry river beds. On several occasions the team had to dig the truck and the cars out of the sand and mud. When they finally arrived at the camp, the scientists set up their equipment so that "the whole place took on the appearance of a great outdoor scientific laboratory", as Van Buskirk described it for the *Rand Daily Mail*.[116]

The reality of contemporary "Bushman" life was migratory throughout a large geographic area, but it had increasingly become affected by encroaching farmers. The scientists conducted their research in the stylised setting of the camp, created by Bain, who had brought people together from various places around the Kalahari. Knowing that people were struggling to find food and water, Bain had offered rations of both to those who joined him at the camp. Dart and the other scientists largely ignored the social organisation of the group. His main interest was to document their physical morphology and to learn more about what he thought were the last surviving members of an ancient racial type. He was also interested in comparing these living people to the fossil record in order to determine if Bushmen and other living populations in Africa could be related to pre-human ancestors.

The patriarch and acknowledged leader of the group in the camp was a man named !Gurice, who was almost 100 years old. Because the scientists found his name too difficult to pronounce, they used his English name, Abraham. It is not clear how !Gurice came upon the name Abraham, with its obvious Biblical connotations, but the name was in use prior to the arrival of the Wits scientists. Missionaries to countries all over Africa, with the intention of converting the local people to Christianity, would often assign a "Christian" name to their converts. White people also often wielded their power by ignoring an African person's name and giving him or her a new English name, a colonial practice that continues today. As with Linnaeus and his naming of all living things, the power of naming continued into the twentieth century.

Bain had enlisted !Gurice to bring his wives and several of his children to the camp with him and to recruit other research subjects as well.

The group of scientists began their work by gathering the names of the 77 people in the camp and then issued each of them with a cardboard identification tag and a number. !Gurice/Abraham was tagged as Kal 4 (Kalahari 4). His daughter /Khanako was Kal 5. His granddaughter /Keri-/Keri was Kal 51. The scientists saw the tags as important because each professor could write information on the tag related to their subject of interest. Van Buskirk said that a need arose for a ceremony to tie the tags around each person's neck as if they were "a distinct badge of individual honour" in order to convince people not to exchange them or tamper with them.[117]

Dart and his assistant John Maingard started measuring the physical characteristics of each person, including their faces, bodies, skin and eye colour, and hair features. They took cranial measurements as well as measurements of each limb. Dart gave special attention to the external genitalia and the steatopygia, the term used to describe the accumulation of fat on many of the females' buttocks. He believed the photographs would contribute to the effort to confirm racial types. Reflecting a sexual obsession that Europeans had back to the 1600s, Dart brought this demeaning practice into the twentieth century.[118]

Dart and Maingard entered information on each tag and from the data collected, Dart put each person into different categories according to their degree of racial purity. Several families had up to four generations at the camp. Dart's main interest was examining the inheritance of physical characteristics, gazing and measuring. Despite !Gurice having been alive for close to 100 years, very little if any information was gathered about his and /Khanako's life histories.

Once each person's body measurements had been taken in one tent, Dart sent them to another tent to have a face mask made. He adapted the methodology from his Gatti expedition, replacing Cipriani's piece-moulding technique with a whole-face technique. The whole face negatives were taken in the field and then the positive masks were made; this was often done back at the university lab. Eric Williams, Dart's technical assistant, took the lead and became an expert in making face masks.

!Gurice was the first to go through the process. He lay down on a table in the tent, where Dart and Williams inserted reeds into his nostrils, which allowed him to breathe while his entire face was covered with damp plaster of paris. Condescending rather than sympathetic, Van Buskirk wrote: "Not an easy operation to explain to a primitive Bushman!" The plaster was allowed to dry and then it was removed, taking some of !Gurice's whiskers with it. Referring to him as Abraham throughout the article, Van Buskirk described !Gurice as a "marvellous subject" who seemed to take it all with an "amused tolerance ... as long as he was being fed regularly he didn't mind a bit."[119]

After the mask was taken, !Gurice left the tent and returned to the area of the camp where he was living. To test the reaction to the face mask after it was finished, the scientists showed it to several of the young people at the camp first. Van Buskirk records that a young girl took one look at the mask and was hugely frightened. She showed an expression of fear in her eyes and then ran away from the mask "shouting at the top of her voice that the old man was dead! She evidently thought that we had cut off his head."[120] The photograph Van Buskirk took of !Gurice standing next to his face mask gives a frightening impression.

!Gurice heard the report from the young girl and walked over to the camp to find out what had happened. The scientists showed him the mask. According to Van Buskirk, !Gurice was first incredulous and then amused. He took the face mask in his hands, looking at it closely, turning it over and over and talking to himself. Then he asserted his control over the situation. After thanking Professor Maingard, he walked off with the mask and could not be persuaded to give it back. Van Buskirk wrote that !Gurice felt that he "was more entitled to carry it around" than anyone else.[121] Dart and Maingard were not concerned because they could make another mask from the original "negative" mold. At this stage, there was no standardised procedure at the University of the Witwatersrand to obtain a research subject's consent.

Dart and Williams made 70 face masks of nearly all the adults and some of the children in the camp. They then compared these masks with over two hundred "European, Bantu and Bush" facial masks that they had already gathered in the department of anatomy at Wits. Making these comparisons was Dart's effort to conclude racial purity or admixture. And it was in the hope that he would be able to gather information about human evolution. Taking body measurements and making face masks became standard for physical anthropologists. It reflected the assumption that people from around the world who deviated from the European norm were objects to be studied.

Dart returned to Johannesburg, where he began to review his data and write up his analysis. He continued to believe in a racial hierarchy in which "Bushmen" were seen as the most primitive. "They are, as it were, living fossils, representative of the primitive state of all mankind, mementos of our own primaeval past."[122] In the journal articles Dart published in *Bantu Studies*, he referred to the Boskop racial type that had first been described 25 years earlier and which he had publicised ever since. Pure Bushmen were the link he was looking for between the ancient Boskop fossil and living humans. He concluded that "the living Bushmen were proven, contrary to our expectations, a mixture in virtually equivalent parts of two South African racial stocks: Bush and Boskop".[123] Dart concluded that he could isolate the fundamental anatomical features of the Bush type on the one hand and the ancient Boskop type on the other, and show how living people were a link to the ancient past. One of his journal articles included a comparative chart of Bush and Boskop features.

Dart referred to other racial types in his descriptions. He described one man as having "another Boskop type of head with broad frontal region. The face is Boskop-Armenoid with a beard of Mediterranean type". He described another man as having "a genuinely Boskop head of the type which is very broad frontally. His face is also Boskop-Armenoid and the beard follows the Armenoid type of distribution."[124] Dart was looking for, and found, certain features in his subjects. Conforming to his diffusionist expectations, he concluded "that Mediterranean, Armenoid and Mongolian people were hybridizing with the Bush-Boskop population of Southern Africa *before it was invaded by the Negro*."[125] Eric Williams later wrote about the racial types that Dart had allegedly found in the Kalahari. The language he used was extreme. As late as 1954, Williams wrote that Dart had distinguished between the two racial types in 1936, namely the "Bush, a foetal, pygmoid type of face, and Boskop, a muzzlelike, chimpanzoid type of face".[126]

In addition to his anatomical measurements, Dart also concluded that the Bushmen were childlike in body and behaviour. "This is exactly the sort of form which the European child's head has at birth and it is this retention of child-like characters in adult physique that characterizes the whole frame work of the Bushmen."[127] Dart described this feature as "pedomorphic (literally child-form). The face is flat like that of a child; the small body with its delicate limb-bones, the diminutive hands and feet, the tiny ears, their simple needs, their incessant playing and dancing in which old and young participate alike."[128]

Dart wrote specifically about his interest in /Khanako's children: "/Khanako had borne children to three different fathers (/Keri-Keri and Marta by a Bushman, Klein/Khanako and Kuskai by a Hottentot and Lena by a European). She made no effort to conceal the facts from us, her associates or the children themselves."[129] Dart showed particular interest in /Keri-Keri and Marta, who he thought to be most "pure".

Given Dart's study of human skeletal remains and living people, he was asked to contribute an important introductory chapter entitled "Racial Origins" in I Schapera's widely known 1937 ethnographic survey of African cultures of southern Africa entitled *The Bantu Speaking Tribes of South Africa*. Dart admitted that he could no longer argue that the Europeans or "Bantu" or "Bushmen" in South Africa were a pure race, an indication that race typology was fraying at the edges. However, he continued to argue for the existence of a "Boskop" race and a "Bush" race.

Beginning the practice of using the term "Negro" interchangeably with "Bantu", Dart wrote that "every intelligent person is familiar today with the obtrusive physical characteristics of the typical Negro ... the skull is infantile in form, being long and relatively narrow."[130] In describing the composition of southern Africa's peoples, Dart used the term "Negroid", saying that the population was 51.2 per cent Negroid, 25 per cent Bush, 22.3 per cent

Caucasoid and 1.5 per cent Mongoloid. How Dart established this exact set of statistics, however, is unclear.

The confusion of trying to maintain the ideal of pure racial types showed clearly in Dart's chapter. His language became more and more convoluted. "Some writers affirm that the Bushmen and Pygmies are Negroes; others separate all three from one another and proceed even further to manufacture Hottentot, Strandloper and Korana and other branches of the African human stock; alternatively suggestions have been put forward that Hottentots are hybrids between the Bush and the Bantu."[131] Despite the contradictions, Dart was not ready to give up his search for pure racial types. In fact, his chapter in Schapera's reference book continued to be reprinted through to 1959 and Dart's race typology was disseminated well into the 1960s.

In May 1938, Jan Smuts wrote to Professor Drennan, who was Dart's academic equivalent at the University of Cape Town. Smuts was preparing to give an evening address to the Science Association in Durban about "human types" and he was anxious because he did not have any slides to show. He asked Drennan if he had any photos or sketches that he could use during the talk. "They help to keep the attention of the audience during the long strain of an address."[132]

Despite Dart, Drennan and Smuts's hold on race typology, there were seeds of a critique growing from the relatively new fields of social anthropology and archaeology. Mapungubwe, long seen as a sacred site by local people on the banks of the Limpopo River, sat in the far north of the Union of South Africa, near the border with the Bechuanaland Protectorate (Botswana) and Southern Rhodesia (Zimbabwe). The University of Pretoria led an excavation of the area in 1933. Jan Smuts encouraged the development of archaeology as an important field of study in his continued support of South African science, and he personally visited the site. The differing interpretations of the Mapungubwe site illustrated growing strain between the concepts of race and culture. Gold, ivory, and glass artefacts were found in the site, which led archaeologists to believe that the site was populated between the tenth and twelfth centuries. When the first results of the university research were published in 1937, the consensus was that Mapungubwe was inhabited and created by Bantu-speaking inhabitants.

One exception to this view came from Dart's department at Wits. One of his senior lecturers, Alexander Galloway, studied skeletons and skulls that had been excavated from the Mapungubwe site and concluded that the bones displayed no "Negro" features and that instead they showed evidence of a "Bush-Boskop" population. Galloway was insistent that the physical analysis took precedence and that there was no validity to the conclusion that the site showed cultural and linguistic links to a Bantu-speaking civilisation. Dart fully supported Galloway's views.

Dart had been asked to analyse the findings at the site and write a report but he had refused. He was concerned that his involvement with Mapungubwe research would be seen as biased because of his earlier public altercation in 1929 with Gertrude Caton Thompson over Great Zimbabwe. Dart suspected "quite rightly as it turned out, that I would be accused of trying to prove my own theories".[133] He agreed with Galloway that the skeletons were of the "Bush-Boskop type". He concluded that this earlier race had cultural characteristics that were later adopted by Bantu-speakers in the area. In addition, building on his diffusionist thinking, he believed that the accomplishments of the inhabitants of Mapungubwe were accounted for by the cultural influence of "superior races" from Asia.

While this debate was under way, !Gurice, /Khanako and /Keri-Keri and many other people who had stayed at the Bain camp at Tweerivieren, travelled with Donald Bain to Johannesburg in late 1936 to be part of the Empire Exhibition, which celebrated Johannesburg's fiftieth anniversary. It took place near the Wits campus and close to Empire Road (what is now part of the Wits campus), but the group resided at the Wits farm called Frankenwald, where they were subjects of further research. James Laing, a young dentist, examined members of the group and he remembers being quite chummy with the teenage /Keri-/Keri and trying to communicate with her. /Keri-/Keri, with her laughing eyes and warm smile, sat down on the makeshift dental chair at Frankenwald. Laing looked in her mouth and saw that her wisdom teeth were about to erupt, which suggested that she was younger than nineteen. Several of /Keri-/Keri's sisters gathered around and Laing sat down to chat with them as best he could. /Keri-/Keri had learned quite a bit of Afrikaans, but Laing was slow to learn her language except for phrases related to his dental research such as "open your mouth".[134]

/Keri-/Keri wouldn't go to the display at the Empire Exhibition every day because there were only about 30 Bushmen "on display" there at any one time. Speakers at the open-air pavilion invited visitors to view the Bushmen's physical distinctiveness and predicted the extinction of their race. They declared their distinct role in the evolutionary chain from ape to man. More than half a million visitors, including politicians and high-profile visitors, filed past the display, taking photos. In contrast, another pavilion next door depicted mining and industrialisation to illustrate a modernising South Africa. Science and spectacle worked together at the exhibition to place /Keri-/Keri and her family on display as part of "the traditional, even savage past" as compared to the "civilized present".[135]

Part of the motivation for the "Bushman" display at the Empire Exhibition was that Donald Bain wanted to make the case that "Bushmen" needed their own reserve. Internationally, this was a practice that had existed for many years in the United States and Australia. In the US, as early as 1830, government policy had supported forcibly removing Native Americans from areas that had come to be populated by Europeans. The concept of reservations for Aboriginal people in Australia began in the mid-1800s. Cattle farming by Europeans who had arrived in Australia forced many people off their land. By the 1870s the US had established reservations. The Indian Reorganisation Act of 1934 encouraged land management by tribes, which might have sparked some interest in South Africa. The concept of a Bushman reserve in southern Africa in the 1930s may have followed from the US example where colonisers took land from the indigenous people and then wanted to give small portions of it back.

In support of this idea for a Bushman reserve, Bain took 55 people from Tweerivieren and Frankenwald, including !Gurice and /Khanako, to Cape Town to march to parliament. It's not clear if /Keri-/Keri made that trip or not. Her mother /Khanako, however, went to the University of Cape Town where Matthew Drennan made casts of /Khanako's head, hands, feet, genitalia and the bottom half of her body. As with the process of making face masks, taking these casts of /Khanako was a form of violence against her that was never questioned by the scientists.[136]

Bain and the march to parliament caught the attention of the South African senator Thomas Boydell, who took up the cause for a Bushman reserve. He formed the Committee to Promote the Preservation of the Union's Bushmen. Jan Smuts was supportive of a Bushman reserve, and in May 1937 he spoke to the senate and referred to the Gemsbok Game Reserve, which had been established in 1931, as a possible location. The National Parks Board had the authority to allow Bushmen to live on that land and Boydell asked Smuts to introduce a bill to parliament to put the reserve into law.

However, members of the senate were concerned that "bastard Bushmen" rather than "pure Bushmen" might take advantage of a reserve. Estimating there to be about 200 genuine "Bushmen" in the country, Senator Boydell put forward that they should be carefully examined by a professor of anthropology, suggesting that Dart might assist. He recommended that the "Bushmen" who passed the purity test be given an identification disc, which would allow them to claim government protection and the authority to live in the reserve. "I know there are some in authority who would rather preserve the gemsbok than the Bushmen," said Boydell, "but the gemsbok are amply protected in various parts of South Africa and there is no reason why they cannot both be preserved."[137]

Raymond Dart wrote extensively for the press in support of Bain and Boydell's proposal. He professed that he failed to understand why South Africans were so interested in preserving wild game but not as interested in "preserving a group of human beings who by virtue of their physical structure, primitive culture and curious customs represent one of the greatest national monuments which any country could possibly possess".[138]

There was a significant contradiction between Dart's public statements offering respect and dignity to the "Bushmen" as human beings and his scientific research of "Bushmen" as specimens.

In keeping with his thinking about providing reserves for "Bushmen", Dart supported racial segregation more broadly. He wrote: "The policy of segregation being adopted in both the Union and Rhodesia as well as further north, is of great value to human conservation; it represents a break with the traditional practice of Europeanizing the native and of destroying by neglect most of what was valuable and otherwise in the native's own cultural heritage."[139] Ultimately, Bain, Boydell and Smuts failed in their campaign to establish a "Bushman" reserve. However, broader segregation legislation did proceed. While more recent scientists such as Phillip Tobias have argued that physical anthropologists in South Africa did not contribute to the racist policies of apartheid, it is not unreasonable to conclude that Raymond Dart's prominence as a scientist, and the respect offered him by Jan Smuts, meant that their thinking about segregation and their ideas about race were in keeping with the prevailing attitudes in the 1930s, and the implementation of segregation legislation in 1936.[140]

When !Gurice, /Khanako and their family returned to Tweerivieren in the northern Cape, they found their homes had been destroyed and their relatives chased away from the area. They went searching for them and found them scattered and living on the edges of various farms. One of /Khanako's first cousin's grandchildren, /Una Rooi, said that she and her father arrived home to find that their shelter had been burned down. Before they left, they had packed together clothes, tsamma pips and crockery in a group of grass huts. They had covered them with umbrella thorn, acacia branches, and fastened them together with wire so that their belongings would be protected while they were gone. When they returned, the entire structure had been burned to the ground. They, too, had to disperse and join the other members of the community who worked on farms. In effect, they had been evicted off the land. They ended up living on the fringes of Swartkop, a township outside Upington.[141]

At the same time as /Keri-/Keri's family was being dispersed and destroyed, dark clouds were gathering over Europe. Hitler was on the rise in Germany.

In Johannesburg a Society of Jews and Christians was formed and they sponsored a series of public symposiums. The second in the series, "The Problem of Race", was held in October 1937 and Raymond Dart and several other Wits professors were asked to participate. Hundreds of people attended in the newly opened library building near City Hall in the centre of downtown Johannesburg. The Library Theatre had a curved ceiling, flip-up auditorium-style seats and a large stage, where the professors sat facing the audience. Dart told the audience there was a worldwide interest in the problems of race and that "the once cloistered solitude of science has been disrupted by political violence." He said that "the living races of mankind today [are] so mingled that the puzzle is to separate them" – which was one of the goals of the physical anthropologist.[142] Dart went on to describe the hierarchy of racial types, including the European, the brown-skinned Hamitic, the yellow-skinned Mongol, the Negro and the Bush, a classification system largely echoing what Linnaeus had set out over 200 years earlier.

Dart's voice projected out into the theatre as he proclaimed that Germany was "attempting to legislate upon a biological matter" and he pointed out that both the Jews and the Germans were extremely mixed racially. He suggested that there were Jews with light and dark eyes, straight and curly hair, fair and dark skin. He proceeded to describe many different groups of Jews and their physical features. Then he asked, "Why is it that there exists amongst the white-skinned peoples and especially amongst the Nordic groups such an intolerance of colour?" He pointed out that there was both "racial hatred" and "racial fusion" taking place across the world. Then he proceeded to explain that "the white-skinned stocks may be dominant intellectually but when they cross with the juvenile black, brown or yellow races, their offspring are coloured." He concluded that the white race was being diluted; he called this "racial submergence". The Nordic race was trying to "preserve their purity and to resist by every means at their disposal their inevitable racial destiny of submergence".[143]

Two years after Dart's lecture, and a week after World War II began, on 9 September 1939 Donald Bain sent Dart a telegram. "Kanacos (sic) eldest daughter perfect specimen bushwoman dying Oudtshoorn Hospital. If interested communicate Dr. Nel local."[144] Dr Nel was the superintendent of the Oudtshoorn Hospital. Two days later Dart sent him a telegram: "Re: Bushwoman. Please send the information for government authorities concerning name, nature of illness, expectation of life, and possibility of being claimed by relatives."[145] Dr Nel responded the next day. The information he gave Dart was that the young woman's name was /Keri-/Keri and that she was also known as Katrina. She was suffering from septic pneumonia and her prognosis was uncertain.

/Keri-/Keri was barely 20 years old, could hardly breathe, and was lying in a hospital far from home.

"I wish to thank you for your wire concerning the bushwoman, Keri Keri (alias Katrina) and her complaint," wrote Dart.[146] He informed Nel that he had secured approval from the acting secretary for Education, Mr Van Zyl, for the University of the Witwatersrand "to obtain from Oudtshoorn the body of a female bushman when it becomes available".[147] Dart immediately contacted Dr BM Clarke, the inspector of anatomy in Johannesburg, to authorise the move. Clarke was responsible for the treatment of cadavers in Johannesburg, Krugersdorp, Germiston, Boksburg, Benoni and Pretoria. Dart arranged for Clarke's jurisdiction to be extended to include Oudtshoorn, which is in the Cape Province and over 1 100 kilometres from Johannesburg. A letter signed by Van Zyl, which stated: "The following Government Notice will be gazetted on the 22nd instant in accordance with section four of Act No 32 of 1911", gave Dr Clarke the broadened authority. In the course of two days, Dart had secured the extension of the geographic area for the inspector of anatomy, gained approval from the acting secretary for Education, and received the co-operation from the Oudtshoorn Hospital. One wonders how he was able to clear these many hurdles in such a short period of time.

"The only point about which we are uncertain here," wrote Dart, "is whether or not there is any likelihood of the body being claimed by a relative. We have presumed that this was unlikely but calculations in that respect can be easily upset."[148] It was Donald Bain who had informed Dart that /Keri-/Keri was ill. There is no documentation to confirm whether Bain informed /Keri-/Keri's mother /Khanako, with whom he had previously been in communication. He made it a priority to inform Dart.

The documentation of the events leading up to /Keri-/Keri's death is sparse. It is possible that before she got sick, she was travelling with the controversial Mr CF MacDonald, who took a group of fifteen Bushmen on a tour throughout the western Cape, hoping to make money from the display. It is not clear if Bain was travelling with MacDonald at the time, nor how Bain found out that /Keri-/Keri was ill.[149]

Anat.
17   14th September, 1939.

            The Superintendent,
                Oudtshoorn Hospital,
                    Oudtshoorn,
                    Cape Province.

            Dear Dr. Nel,

                    Enclosed please find copy of a letter I
            have received from the Acting Secretary for Educa-
            tion.
                    I wish to thank you for your wire concern-
            ing the bushwoman, Keri Keri (alias Katrina) and
            her complaint.

                    In themeantime I have instructed my
            Assistant to forward you information relative to the
            method of embalming, in the hope you will be able
            to make arrangements with some member of your staff,
            or a competent local undertaker to do the work prior
            to despatching the body by train.

                    The only point about which we are uncertain
            here is whether or not there is any likelihood of
            the body being claimed by a relative. We have pre-
            sumed that this was unlikely but calculations in that
            respect can easilt be upset. Will you please let
            me know what the prospects are likely to be in this
            respect and also communicate with me about any detail
            of procedure concerning which you may be in doubt.

                                    Yours sincerely,

                                    RAYMOND A. DART,
                                    PROFESSOR OF ANATOMY.

            Enclosure.

Letter from Raymond Dart to Dr Nel, 14 September 1939. Courtesy of the Raymond Dart papers, Wits University Archive.

After Dart exchanged telegrams with Dr Nel in Oudtshoorn, while /Keri-/Keri was still in the hospital with pneumonia, Dart asked his chief technician Eric Williams to write to Dr Nel about how to embalm her body. Williams complied. "I have been instructed by Professor R. A. Dart in the event of the death of the Bush woman, now an inmate of the Oudtshoorn Hospital, to furnish the following details on the technique of embalming to be followed if possible ... Raise the axillary artery and vein. Insert a canul both ways into the axillary artery and also insert a canul or drainage-tube into the vein. Let out as much blood as will drain from the body. Now close the drainage-tube and inject the embalming fluid into the artery, allowing blood to escape from the body through the vein every now and again until clear fluid escapes. Now finally close drainage-tube and inject until body obtains a good degree of firmness in all extremities." In one and a half pages of detailed instructions, Williams offered an alternative embalming process if the first method was not applicable. He also provided a recipe for the embalming fluid and suggested that "the body will probably take from 2 to 4 gallons."[150]

Two days later, on 16 September, Dart received the news he had been waiting for. "Bushwoman died last night. Nearest embalmer available Port Elizabeth. Estimated cost embalming 25 pounds. Wire university authority for expenditure and government authority for removal of body," said the telegram from Dr Nel.[151] There is a note in the Wits Archives on a University of the Witwatersrand notepad that says: "Keep body in cool place. Leaving Johannesburg today by road and will bring it back." It's not clear whose handwriting it is because there is no signature, but it is likely to have been Eric Williams. It was Williams who drove to Oudtshoorn to retrieve /Keri-/Keri's body and sent a signed telegram to Dart on 18 September saying: "Arrived Safely Everything OK." He drove back to Johannesburg with /Keri-/Keri in the back of his bakkie. According to lore within the department of anatomy, her body was damaged during the long journey. But Raymond Dart had achieved his goal. Within ten days of receiving Bain's telegram alerting Dart to /Keri-/Keri's illness, her body was in the possession of the department of anatomy at Wits University. In his effort to understand pure racial types, Dart saw /Keri-/Keri's body and her skeleton as a "specimen" to be studied. In his desire to piece together the story of ancient human ancestors, and "the missing link", in human evolution, his priority was that he had attained a valuable addition to his human skeleton collection.

There would be no burial for /Keri-/Keri. For her family, there would be no closure. For 50 years, her skeleton would remain on a shelf in the Raymond Dart Collection of Human Skeletons. At some point in the 1980s or early 1990s, her skeleton went missing. It is not clear if it was stolen, or misplaced. One theory is that it was taken out for teaching purposes, and never returned. For over six decades, her body cast would stand on display at the department of anatomy before it was put away in storage. For 75 years, the story of /Keri-/Keri's life and death would remain hidden in the archives, until 2014 when I went to the department of anatomy at the Wits Medical School to try to find out more.

6

The Search for /Keri-/Keri[152]

The large room that houses the Raymond Dart skeleton collection in 2014 is about a quarter of the size of a FIFA football field, and is lined with floor-to-ceiling shelves, with not a square inch of empty space. Thousands of wooden boxes stand on the shelves, looking like small caskets in a mortuary. On closer inspection each box is marked with a white sticker, labelled with the essential information about its contents.

Date of death. Sex. Race.

"Zulu". "Ovambo". "Ndebele". "European".

Each box also has an accession number – A1, A2, A3 etc.

The boxes contain skeletons of human beings. There are no names. No stories.

The medical school moved in 1982 from its location in Hillbrow to Parktown, where the school of anatomical sciences is a maze of artificially lit passages in the basement. This is where the collection now resides, and where Brendon Billings, its curator, has his office. Billings was born in 1980 and grew up in Coronationville, a township on the western side of Johannesburg that was designated under the Group Areas Act for people who were classified "Coloured". He grew up under apartheid until he started high school in 1994. It was that year that he first went to a previously whites-only government technical school. "That was the first time I was in a newspaper," recalls Billings. "There were six white guys that brutally assaulted one of the black guys at my school and I ended up in the photo too." He also remembers taking the bus to school and having the Afrikaner boys spit at him when he got off at his stop. "There was a lot of conflict. We were still familiarising ourselves with each other."

In 2000, Billings started a science degree at Wits and subsequently enrolled in an honours degree in paleoanthropology and forensic anthropology. In 2005, he began a master's degree in the History of Science and wrote his dissertation about the South African scientist and author Eugene Marais. In 2009 he became curator of the Raymond Dart collection. "They knew I loved bones," he says, his voice filled with zeal. But at that point, Billings was still fairly junior in the department. "I was always seen as a technician. There's this huge dichotomy between support staff and academic staff. I wasn't respected within academic circles." Soon his position expanded so that he became curator of all the collections within the school. This included cadavers, the paediatric collection, the histology collection, X-rays, bottled specimens and plastic models. In addition to the Raymond Dart skeleton collection, he also had oversight of the over a thousand face masks in the Raymond Dart Gallery of Face Masks.

The masks, which were made by Dart and his successor Phillip Tobias, are no longer hung in public but they are displayed on the walls of several long hallways in the department. "Each collection has its own set of hidden issues and problems," explains Billings. "I sat down with each one of them and developed a plan for each collection. We've made a lot of progress. If it's human tissue, it's sensitive in terms of display and in terms of research."

At the beginning of 2014, Billings was promoted from associate lecturer to a tenureship lecturer position. He began pursuing a PhD in neurology and the brain.

"Initially I was in awe of Dart's work," he says. "The Taung child. He set up all these collections. He was ahead of his time in his thinking. But when I started understanding the fundamentals of why he wanted these collections and what he was trying to prove, I realised he was a typologist. He was a racist. Then I went through an angry phase. Now I'm taking all the different elements of him. I no longer put him on a pedestal."

Billings knew that Dart had led an expedition to the Kalahari in 1936 and was aware of his research methods, but he did not know much about the details of the trip to Tweerivieren and the group Bain had gathered together. It leads him to speculate. "We have a full body cast of a Bushman lady," he says. "I removed it from display. It's a naked cast. For the life of me, I couldn't figure out where this came from, so I asked around. Someone mentioned to me that this woman had died and that scientists had made a cast to preserve her. I thought it was derogatory but then Professor Tobias explained to me that it was about preserving this individual." After a pause, Billings offers: "I can show you the body cast."

When I show him a photograph of Donald Bain and /Khanako and tell him that Dart was particularly interested in /Khanako's daughters with a Bushman father because he wanted to find a "pure Bushman", his interest is piqued. And when I ask about /Keri-/Keri's face mask (Kal 51), which should be somewhere in the department, he is determined to locate it. The information I have about her skeleton is that it had been in storage in the Dart collection as item A43, but that when researchers went looking for it in 1996 it had "gone missing".

"Let me check what I've got," says Billings, turning to his computer with a sense of excitement and beginning to scroll down his screen. "All I've got on the system for A43 is 'Bushman female, specimen absent, de-accessioned'."

"Do you have a box A43 in the collection?" I ask.

Billings shakes his head and stands up. "No, if it's not in the system, it won't be out there," he says. "But I want to check that body cast I mentioned in relation to A43."

Billings strides out of his office, across the passage and into a large dissection hall, where a group of medical students are talking loudly and hovering around a cadaver. There is a strong smell of formaldehyde. Down the length of a long hallway hundreds of masks hang high on the wall. "This is the remainder of my face mask collection," says Billings. They all have a CF number – cranio facial – so that Kal number you have doesn't make sense to me." Each mask is eerily lifelike. Each of the people, whose features were documented in this visual way, has a story to tell. One woman's mask appears so real it looks as if her eyes might blink open at any moment and she might begin to talk.

In another large room there is an embalming session taking place. A cadaver is being prepared for dissection. The pale bloated body lies on a steel gurney with a stream of water flowing above it. Billings explains that they pump all of the blood out and then pump formaldehyde in.

In an adjacent room filled with large boxes and storage cupboards Billings stops. The body cast should be here, he says. "If we find it, I'm going to see if there is a number on it." Across another hallway he knocks on a door and without waiting for an answer, opens it and puts his head inside. The squeak of the hinges echo loudly. "Do you remember we had a cast of a naked Bushman lady?" he says to someone inside. "I had it in the plastination room. Do you guys know where it is? I can't find it."

After a minute Billings backs out of the room, looking defeated. This is probably the end of the search. Not only is /Keri-/Keri's actual skeleton missing, but it seems her body cast – if indeed it is hers – is missing too.

On the way back to his office, Billings runs into a colleague and he asks him about the cast. The colleague says it's still there – "behind the cupboards". Back in the storage area, after moving a pile of boxes, and poking between the cupboards and the wall, Billings lets out a sigh. "Ahhhhh, there she is," he says.

The woman is standing with her face to the wall, right into the corner. She is short in stature, which makes the cast look to be the size of a teenager. All of a sudden, I am devastated that we continued the search and that we might have found her. Without proof, but assuming that this is the body cast of /Keri-/Keri, the weight of history sits with her in that corner. Billings turns her around and the cast looks ghoulish; it captured her face at death. Horrified, I think back to all the male scientists who thought themselves entitled to probe /Keri-/Keri's anatomy. The culmination of her oppression, and that of her people, was that they were prevented from burying /Keri-/Keri with respect.

A student of Philip Tobias in the 1970s, Alan Morris, remembers that /Keri-/Keri's skeleton, A43, was still in the collection when he left Wits in 1979. This means that her skeleton disappeared sometime between 1979 and 1996. What had happened in those 17 years? Anatomists regarded the skeleton as particularly valuable. Did someone steal it deliberately?

A previous curator of the fossil collections at Wits (from 1997 to 2004), Mike Raath, also had oversight of the Raymond Dart skeleton collection. Raath said that skeleton A43 was not present when he took over the collection in 1997 and that there was no explanation for its absence. His theory was that the skeleton was taken out for teaching purposes and never put back. His records showed that there were two other numbers associated with that skeleton, which suggests that it had arrived in the department as a cadaver. If this was the case, it would have been recorded in handwriting in the cadaver catalogue.

The Raymond Dart papers are housed in the archive at Wits University. It was a box marked "Dart Correspondence – department of anatomy" that yielded critical documents. Not far from the top of the pile of folders and papers was the telegram from Bain to Dart about /Keri-/Keri's illness and, deeper down, the telegrams, letters and notes that told the story of /Keri-/Keri's death of septic pneumonia in Oudtshoorn. These documents confirmed the date of her death and how she had died, but beyond these stark facts there was nothing further – nothing about her life while she was living it. Did she have a favourite food? What made her laugh?

Phillip Legodi currently manages the cadaver process at Wits Medical School and he has access to the historic, handwritten cadaver book. He confirmed the date of /Keri-/Keri's death as 15 September 1939. Legodi agreed with Mike Raath's theory that skeleton A43 was taken out of the Raymond Dart collection for teaching purposes and never returned. He also thought the skeleton might be fully articulated and in someone's office somewhere.

Billings was shocked when he learned more about /Keri-/Keri's death, even more so that Dart had arranged for her body to be sent to Wits even before she died, and that Eric Williams had driven down to Oudtshoorn to fetch her body and loaded it into a bakkie. He questioned the legality of what had happened. Then he wondered whether she might have donated her body. From the little we know and are able to glean from the handful of documents in Dart's archive, the paperwork that passed between the university and the hospital might have given the process a semblance of legality, but the likelihood of /Keri-/Keri herself or a member of her family giving consent seems extremely doubtful.

Over the years, Billings said, he had got the impression from stories he had heard in the department that whoever it was who went to get /Keri-/Keri's body had known her personally. "As soon as this guy heard that she died," said Billings, "he jumped into his bakkie and drove down to bring her back." In talking through the possibilities, Billings got visibly distressed. No doubt Billings had preferred to imagine that this person had cared for /Keri-/Keri in some way, but the grim reality was underscored by the actual telegrams and letters in the Dart archive.

There was also the question of /Keri-/Keri's face mask. Again, Billings turned to his computer. He scrolled down his list of face masks. Each listing had an A number and a CF number. He scrolled down and down. And there! There was A43, the accession number for the skeleton, and next to it was the face mask number: CF401. This led him to an old card file for face masks, where he found the card that was marked CF401. There, on the same card, were: "/Keri-/Keri" and "Kal 51", the number that the Wits scientists had put on the identification tag during the Kalahari expedition. !Gurice/Abraham was Kal 5. /Khanako was Kal 4 and /Keri-/Keri was Kal 51, and her face mask number was CF401.

Further careful study of the face masks on the walls of the long hallway revealed that many were from the Kalahari expedition but CF401 was not among them. /Keri-/Keri's face mask was missing.

On the same card as /Keri-/Keri's face mask number were two other numbers: C12 and CS131. Billings knew right away that they were numbers for other casts. He went to the back of his office and pulled out an old volume marked "Accession Book", with the words "Casts of hands, feet, etc" on the spine. He opened it to the first page that began with C1. The second page showed accession number C12: "Cast of Bushwoman". This seemed to be confirmation that it indeed was /Keri-/Keri's body cast standing with her face to the wall behind the cupboard.

Billings looked for the final cast recorded on the card, CS131. Going to a cabinet in the room next to his office, he said casually: "This is where we store casts of hands and feet and other things." On a shelf beside the cabinet was a yellowish plaster cast in the shape of a pelvis with large folds of labia in the centre with a small penis protruding forward. When he saw me looking at it, he said, "Probably a hermaphrodite."

The thought that there might be a cast of /Keri-/Keri's genitals somewhere was horrific to me. In many ways her story was reminiscent of Sarah Baartman's. As a Khoisan woman held in slavery, Sarah Baartman was taken from South Africa to England at the age of 21 by a Dutchman, Hendrik Cezar, in 1810 and displayed at Piccadilly Circus in London as "The Hottentot Venus" and then later at exhibitions in Paris. The Empire Exhibition in Johannesburg in 1937 was an anachronistic continuation of colonial exhibitions of indigenous peoples that had occurred in Europe over a century earlier.

In 1815, the French anatomist Georges Cuvier examined Baartman physically and described her features as "monkey-like" and "orangutan like". He was particularly interested in her genitals and her steatopygia, just as Dart had been when he travelled to the Kalahari and examined the group at Tweeriveiren. Both /Keri-/Keri and Sarah had been pulled away from their home and family in South Africa to be placed on show at an exhibition and to be studied by scientists. Both women died young, Baartman on 1 January 1816 in Paris of unknown causes, /Keri-/Keri on 15 September 1939. Cuvier proceeded to study and dissect Baartman's body after her death. He made a total body cast in wax which, along with her skeleton, was on display at the Musée de L'Homme in Paris until 1974. /Keri-/Keri's cast was also on display for over 50 years at Wits University; we know that her skeleton disappeared from the collection in the 1980s or early '90s. To many South Africans who have honoured Sarah Baartman, the mention of Cuvier's name brings a wave of anger, and yet he is celebrated in France. The grand avenue Rue Cuvier is named in his honour. Similarly, Raymond Dart is considered a hero of science, but the story of his macabre call for /Keri-/Keri's body before she died is hardly known. It remained buried in the archives among his papers for 75 years.

After a long period of negotiation with the French, Sarah Baartman's remains were returned to South Africa and she was at last given a dignified burial in 2002. Books have been written about her difficult life and documentary films made. The details of /Keri-/Keri's life, however, remain largely unknown. While her body cast is at Wits, the location of her skeleton and another cast remain a mystery. It would take another 18 months, and the efforts of a post-doctoral research fellow, Dr Tobias Houlton, to track down the missing face mask, in 2016. Once he found the mask, he was able to confirm that it matched the body cast, but questions remained.[153] In addition, the only photograph with a caption referring to /Keri-/Keri that was published in Dart's *Bantu Studies* journal article in 1937 does not match with the physical build of the body cast. Did the scientists get confused about her name? Was the photograph marked incorrectly? Despite an extensive search, no other photos have been found that can be confirmed as /Keri-/Keri.

Plate 81 in Dart's Bantu Studies *journal article 1937. Dart's caption to the photo reads, left to right: No 28 //Khaku, No 42 /Kaap, No 60 Klein Taki, No 7 Marta, No 33 /Ues, No 6 Klein /Khanako, No 51 /Keri-/Keri, No 40 Tamtam, and No 56, !Nansi. Courtesy of Wits Historical Papers.*

Several large, framed black-and-white photographs hang on the walls in Billings' office. "Those were taken in the Kalahari," says Billings. In one of them there is an older man whose features look familiar. Billings reaches up and takes it down. On the back is a handwritten note: "No 4 Abraham (identified by Ouma /Una)." I had read that Ouma/Una was a descendant of !Gurice (Abraham) and /Khanako, who passed away in 2012. The fact that she had identified Abraham in the photograph means that at one time she must have made a visit to Wits.

Another large photograph has a similar handwritten note on the back: "No 28 /Khaku (identified by Ouma /Una on 14/10/2004)" – so Ouma /Una must have been in Johannesburg, visiting the university, in 2004, which meant that she likely met with Dart's successor, Phillip Tobias. Perhaps they talked about Raymond Dart's research in 1936 and 1937. Perhaps they discussed /Keri-/Keri's death in 1939 and looked for her skeleton, only to find it was not in the collection. Did Ouma /Una tell Tobias about the impact of all this on her family? And what did Tobias think? Unfortunately, this is one of the episodes from his life he did not write about before he died.

7

Sterkfontein, Kromdraai and Makapansgat

In 1936, the same year in which Raymond Dart led the Wits expedition to the Kalahari, in his capacity as Prime Minister in the coalition government, Jan Smuts implemented new legislation that formalised racial segregation. This was the Native Trust and Land Act and it went further than the Native Land Act of 1913. It confined black people to a smaller portion of land in the country, and prevented them from buying land outside of these reserves.

For much of his life, Smuts had been a vocal supporter of segregation. In a major speech in London in May 1917, he had argued for separate "parallel institutions" for white and black people. "Instead of mixing up black and white in the old haphazard way, which instead of uplifting the black degraded the white, we are now trying to lay down a policy of keeping them apart in our institutions. In land ownership, settlement and forms of government, we are trying to keep them apart."[154] In 1929, at a Rhodes Memorial Lecture at Oxford, Smuts had encouraged European settlement across Africa, as had Rhodes. He referred to the "Negroid Bantu" as a "distinct human type". He spoke of African people as having "a happy-go-lucky disposition, but with no incentive to improvement" and described them as "a child type, with a child psychology and outlook".[155] Smuts put forward that the solution in South Africa was to implement a policy of "segregation". In addition to this new law, his support for a "Bushman" reserve also illustrated his belief in this policy.

Segregationist policies sat comfortably with Smuts's thinking about prehistory and human origins. To him, these themes related to the difference between Africans and Europeans. In 1932 he had published a paper called "Climate and Man in Africa" in the *Journal of South African Science*. It was a long, technical explanation of the climate changes in Europe and Africa and the archaeological stages of human development in both regions. "As they were racially and physically not very different 15,000 years ago," he wrote, "what has caused the immense difference between the Europeans and the Bushmen of today? We see in the one the leading race of the world, while the other though still living, has become a mere human fossil, verging to extinction."[156]

Not only did Smuts view these two groups of people as being on different scales of evolution, but he also made the case that Europeans had every right to return to their ancestral home in Africa. He argued that Europeans "now return to find a very different situation from that which their African ancestors left some 15,000 years ago". His conclusion was that Europeans, having evolved further, could rightfully return to their place of origin. He had the timing wrong, and he was creating a past that would reflect his world-view. And his worldview was shaped by his interpretation of human origins. Palaeoanthropology provided the rationale he needed for Europeans to be present on the continent and to take control.

In securing Robert Broom's position at the Transvaal Museum, Smuts had another avenue by which he could shape South African science on the world map. He was hopeful that Broom's work would lead to new insights into the understanding of human evolution. After 18 months at the Transvaal Museum focusing on mammal-like reptiles, Broom decided it was time to shift his attention to hunt for an adult specimen of the same species as the Taung child. He hoped that such a find would support Dart's claims for the australopithecines. Two of Dart's students encouraged Broom to investigate the caves at Sterkfontein, outside Johannesburg, where they had recently found several baboon skulls.

The Sterkfontein Caves were originally uncovered in 1897 by Guglielmo Martinaglia, an Italian who had been looking for a source of limestone that could support the fledgling gold mining industry in the area. He set explosives down on the ground, pressed the plunger, and caused an explosion that set rocks and dust flying into the air. The hole he had blown out on the side of a hill turned out to be the entrance to a large underground cave.[157] Since the first public announcement of the existence of the Sterkfontein Caves, for close to 40 years no geologist or palaeontologist had spent much time there. It was the lime quarrying that had drawn people to the area and much of the breccia that was blown out of the cave was discarded.

Sterkfontein, like Taung, was a site that came to the notice of scientists only as a result of the mining industry in South Africa. Without the quarrying first, it would have taken a lot longer for scientists to investigate. The author of a little guidebook published about Sterkfontein in 1935, Mr R Cooper, might have had an inkling that important fossils were being lost as the breccia was blasted out and burned in the lime kilns when he wrote: "Come to Sterkfontein and find the missing link."[158]

Broom went to see the supervisor of the Sterkfontein quarry, George Barlow, who had also worked at Taung in the 1920s. They met at the rondavel that served as a tea room near the mining site and Broom asked Barlow if he had seen anything at Sterkfontein like the Taung child skull. Barlow said he hadn't, but he pointed to a set of fossils he had gathered, and which had been arranged on a large table, and he promised Broom that he would keep a look-out for other interesting specimens.

In August 1936, about a week after his first visit to Barlow, Broom had reason to go back to Sterkfontein. This time Barlow had something for him. He handed him a clean brain cast in perfect condition that had been blasted out of the quarry that very morning. Broom examined the fossil, and couldn't believe that within nine days after his first visit to Sterkfontein he had found what he was looking for – an adult skull of *Australopithecus*. He and his team hunted around the blasting area for several hours to see if they could find additional pieces of the skull, but to no avail. The next month, Broom published his analysis of the skull in *Nature* and hoped that the world would be convinced. The headline in *The Illustrated London News* read "A NEW ANCESTRAL LINK BETWEEN APE AND MAN".[159] Broom was certain the owner of the new skull was similar to the Taung child. He was a "splitter" (someone who would split fossils into different categories, as opposed to a "lumper" – someone who lumps different fossils into the same category) and initially he placed the skull in a distinct but related genus called *Plesianthropus transvaalensis*. Years later, the skull would be reassigned as an *Australopithecus africanus* specimen together with the Taung skull.

The day after the find, Broom returned with Dr Herbert Lang to have photos taken of himself and two technicians from the Transvaal Museum. One of these men was named Saul Sithole, and he is in one of the photographs confirming the location where they found the adult skull. In the picture Broom is seated, dressed in a suit and tie with his white shirt-sleeves rolled up to his elbows. His note on the back of the photograph reads: "Taken by W. Herbert Lang at Sterkfontein the day after the Sterkfontein skull was discovered. I am indicating with my left hand the spot where the skull lay. With me are Mr. Barlow and behind two museum boys, Saul and Jacobus. The other boy is one of the quarry boys."[160] Barlow is dressed more informally than Broom in work boots and an open-necked shirt and jersey. Saul Sithole and Jacobus are wearing shirt, tie, vest and hat. The un-named man is holding a spade.

Many people in the field of anthropology and archaeology over the past century have been invisible. We don't know anything about the nameless "quarry boy" with the spade and not much about Jacobus the "museum boy" in the photograph with Broom on that day in 1936. The social and political context of racism and segregation at the time led scientists to welcome the labour of black workers, but did not offer them opportunities for education, or recognition.[161] Robert Broom did not mention Saul Sithole in any written form, but there is some documentation at the Transvaal Museum that gives us a sense of what Sithole's life was like, working as a black man in the field of science.[162]

Saul Sithole had his first day of work at the Transvaal Museum in Pretoria on 11 November 1928. He started out as a cleaner but within his first year he began to work on setting up exhibits, helping to mount the enormous elephant that is still displayed in the museum's entrance hall today. In less than two years, he began to specialise as a technician and was chosen to participate in the Vernay-Lang Kalahari expedition in 1930. Arthur Vernay, an antique dealer from New York City, funded the trip. Herbert Lang (who took the August 1936 photograph of Broom and Sithole at Sterkfontein) was a German naturalist and photographer from the American Museum of Natural History in New York City. He had led previous expeditions, including one to the Congo and another to Angola.

The focus of the 1930 expedition had been birdlife. Herbert Lang taught Saul Sithole to skin birds, a skill he would put to use at the Transvaal Museum for decades to come. In his expedition report, Lang mentioned Sithole: "This native, by his ready willingness to render himself useful and by his good example, assisted in maintaining an excellent discipline and great industry among the skinners." Another member of the expedition, Austin Roberts, who went on to become a well-known expert on the birds of southern Africa, confirmed Sithole's role in his report of the expedition: "Lang is training the boys to make good skins ... Saul is far better than the other skinners and is doing very good skins." The expedition yielded 30 000 bird skins that supplied four museums, one in New York, one in Chicago, the British Museum in London, and the Transvaal Museum.

Sithole had grown up in the city, so this trip into the bush was his first. He was born in 1908 in Standerton, a town in the Transvaal that was under British rule at the time. When he was young, Sithole's family moved to Lady Selborne, a freehold black township near Pretoria. The family spoke isiZulu and Sithole's father was a Methodist lay preacher.

The second of three sons, Sithole was the middle child between Solomon and Amos and his education was interrupted when his father died in 1923.

Saul and Solomon left school and went to work to help their younger brother Amos to continue his studies. By that time, Saul had completed Standard 6 (Grade 8). He found work as a bus driver while Solomon found work at the Transvaal Museum. Five years later, in November 1928, Saul joined his brother at the museum and found the job there as a cleaner.

On his return from the expedition in 1930, Sithole continued to work with the bird skins at the museum. As a technical assistant, he supported important scientific research with expertise that he had refined over many years of detailed work. Observing living birds on those expeditions helped him to reconstruct the bird skins back at the museum, but it took considerable skill to arrange the feathers, the skin, the muscles and the structure of the body so that each species would come to life.

Saul Sithole travelled with Austin Roberts to Vryburg, Grahamstown and Zululand where he served as a translator for Roberts, gathering invaluable knowledge from the local people about the birds and mammals in the area. Roberts' reports rarely make mention of his African assistants, but in his February/March 1933 report of the Transvaal Museum bird department, he wrote: "Native Assistant Saul has continued his useful assistance in cleaning skulls and re-making up badly prepared skins for birds. I may record his marriage in March."

Saul's wife Sophia Nomvula was also from Lady Selborne, and their only child, Zondi, was born in 1931 when Sithole was 23. Zondi Zitha recalled how meticulous her father was about his work. As a child she was interested in his bags that were filled with tools, knives and other supplies. Once while he wasn't looking, she opened one of the little bags to take a closer look at the cotton-wool and the needles inside. What she saw instead – which gave her such a shock that she jumped back and let out a small yelp, causing her father to turn around to see what she was doing – were eyes! Lots of them, staring up at her. These were small round balls of glass – artificial eyes for the birds. Sithole scolded his daughter and told her not to interfere and Zondi Zitha learned that the bag was the "'Don't Touch' bag. It belonged to Dad."

When Roberts travelled to the United States in 1934, Sithole's career as his assistant was interrupted. He continued to work at the Transvaal Museum, but there was a period of time in which he worked in the palaeontology department with Robert Broom, who had recently been given his post at the museum. That was how Sithole happened to be with Broom on 18 August 1936 when they found the first adult *Australopithecus africanus* skull.

As Sithole began working with fossils, he learned new skills. At Sterkfontein, he began cleaning the breccia from the fossils, chipping the rock carefully away from the bones. The conditions at Sterkfontein were very different from the work he had done with skins. With the birds, there were plenty of specimens and opportunity for practice. Each one of the *Australopithecus* fossils, however, was a unique specimen and invaluable. There was no room for mistakes. The work with the fossils, just as with the bird skins, required great attention, patience and care and Sithole became skilled in this new field. Robert Broom never recorded working with Sithole, other than in that one photograph, and Sithole never had the opportunity for further study. This attitude was reflected in the language used to describe black assistants at the time such as "quarry boy" and "museum boy" and the fact that workers in photographs were identified by their first name only, or not identified at all. The politics of colonial racism was reflected in the world of science. In the era of segregation, before apartheid, Sithole's opportunities were severely constrained.[163]

Broom's work at Sterkfontein had stirred up hard feelings in the Wits department of anatomy. Dart's students who had suggested to Broom that he head out to Sterkfontein felt that he did not acknowledge their contribution. Dart was upset because he thought his students should have come to him instead, so that he could have been the one to explore Sterkfontein first. But he was also aware that Broom's find contributed to the vindication of his own claims about *Australopithecus*. This wouldn't be the last time that Dart and Broom would spar over work in the field.

For months, Broom travelled from Pretoria to the quarry at Sterkfontein every week and sometimes twice a week, following Barlow's quarrying. "We had all the quarry boys engaged in the hunt," wrote Broom, "and every visit cost us many shillings in tips to the natives. But it paid us well."[164]

Almost two years later, on 8 June 1938, Barlow showed Broom a mandible with a molar tooth attached to it. When he saw the fossil Broom was excited. He thought it might be a hominid, but he was quite sure that it was not from Sterkfontein because the surrounding rock was different from what he expected in the area. A few days later, Broom insisted that Barlow tell him where he had found the mandible. He could see that several teeth had recently broken off and he wanted to search to see if they might be lying on the ground somewhere.

Barlow apologised for initially not sharing the information and he told Broom that a schoolboy, Gert Terblanche, had found the fossil in an area nearby called Kromdraai. Broom didn't waste a moment. He drove straight to the boy's house where he found Terblanche's mother and sister. They told Broom where Gert had found the fossil so he drove up the hill about a half a mile. Despite being 72 years old (and in a suit), Broom crawled around on the ground in the area and found several small skull fragments and a couple of teeth. Although Sterkfontein and Kromdraai are only five kilometres apart, the two areas have different geological formations and fossils of different ages. The animal remains in the two areas are different as well.[165]

Gert's sister told Broom that Gert had also found four beautiful teeth. Broom could not wait until the end of the school day, so he decided to go to the school to find him. At first they drove, but the road was so bad that they had to abandon the car and walk another mile over rough ground to the school. As Broom wrote in his memoir, "Gert was soon found, and drew from the pocket of his trousers four of the most wonderful teeth ever seen in the world's history. These I promptly purchased from Gert, and transferred to my pocket." Broom found that two of the teeth fit onto the original palate piece.[166]

Gert told Broom about another fossil, which he had hidden away, and Broom decided to wait with Gert until the school day finished. After discussing it with the headmaster, Broom gave an impromptu lecture to 125 people – all the teachers and students at the school that day – about bones and fossils, and about caves and how they are formed. He illustrated his talk with chalk drawings on the blackboard and spoke for over an hour. At the close of school, back in Kromdraai, Gert led Broom up the hill behind his house to his hiding place. There he showed Broom a complete lower jaw with two teeth in position.

In the 1990s, palaeontologist Francis Thackeray, who became the director of the Transvaal Museum, tried to locate Gert Terblanche but he was unsuccessful, so it is unknown whether Gert followed his interest in science as an adult, but as a "schoolboy", he was memorialised in Robert Broom's memoir.[167]

Over the coming days, Broom arranged for his assistants to carefully excavate the entire area at Kromdraai with a sieve and gather every piece of bone or tooth. When all the fragments were cleaned and joined together, Broom had pieced together nearly an entire skull, which was larger and more human-like than the Sterkfontein skull he had found in 1936. The face was flatter, the teeth larger and the jaw more pronounced. Broom saw this new Kromdraai skull as part of a different genus and species so he called it *Paranthropus robustus*. Broom was fairly certain there were two species of hominids in the area. An account and description of the discovery appeared on 20 August 1938 in *The Illustrated London News*. Reinforcing the incorrect idea that there was one linear line from apes to humans, the article's headline read: "THE MISSING LINK NO LONGER MISSING".[168]

With the start of World War II, the Sterkfontein quarry closed in 1939. Broom's discoveries at Sterkfontein in 1936 and Kromdraai in 1938 had helped to bring some public attention back to South Africa, but recognition from other scientists was still slow in coming. Broom used his time during the war to review the many specimens of fossils he had gathered over the previous three years. He worked on a book, which he co-authored with Dr Gerrit Schepers (and which would be published in January 1946), *The South African Fossil Ape-Men – the Australopithecinae*. Just as with the erroneous concept of a "missing link", the term "ape-men" was used widely, fuelling the stereotype that there were ape-like men in Africa.

Broom and Schepers' book firmly set out the relationship between the Taung child, the Sterkfontein find and the Kromdraai skull, suggesting that different species of "ape-men" might have been living millions of years ago in South Africa all at the same time. Broom was pleased Jan Smuts wrote the preface for his book: "I think Smuts is delighted to see that South African science has again, Phoenix-like, risen from its ashes."[169] Smuts applauded Broom's earlier work that had aimed to bridge the gap in evolution between reptiles and mammals but he confessed he found "Broom's apes, from a human point of view, much more intriguing, than even his mammalian reptiles" and concluded that many readers would share his interest in "the discovery of those long-lost relatives of the human race divine".[170]

Building on the racial paradigm constructed by George Theal's books in the late nineteenth century that described waves of "Bushmen", "Hottentots" and "Bantu" and "Europeans" into southern Africa, Smuts was intrigued by the possibility Broom put forward that multiple species of ancient hominids might have been living together in the same geographic area at the same time. In 1945 he wrote to a colleague that Broom's work showed that "manlike apes, very close to man, were living ... cheek by jowl". Smuts had used the same phrase in 1942 when he had addressed the Institute of Race Relations on the modern "native question". Again, Smuts revealed how his thinking about palaeoanthropology could provide a rationale for his modern vision for South Africa.[171]

By the time Broom's book was published in 1946, it had been 21 years since Raymond Dart's article about the Taung child had appeared in *Nature*. With Broom's discoveries at Sterkfontein and Kromdraai, more scientists were beginning to consider that perhaps Dart and Broom were onto something after all. Dart's mentor Elliot Smith had passed away in 1937, so he could not provide further comment. However, Sir Arthur Keith, the scientist who had called Dart's claim "preposterous" back in 1925, wrote a letter to Broom in 1944 saying, "No doubt the South African anthropoids are much more human than I had originally supposed."[172] Broom and Dart and Smuts had hope that South Africa was back on the scientific map. Darwin's original hunch might finally gain some traction.

While Broom's star was rising in the early 1940s, Dart's life hit a difficult chapter. In 1941 his wife Marjorie gave birth to their second child, a son, Galen, after their first child Diana had been born in 1940. Galen was born prematurely at six and a half months, weighed just over one kilogram, and spent his first six weeks in an incubator. The Darts weren't sure that their son would survive. He did, but as the boy grew, he suffered from extremely rigid muscles and poor co-ordination. The term cerebral palsy was not yet in use at the time, and the condition was poorly understood.

Dart was shocked at the predicament facing his son. For much of his career, he had been searching for clues regarding how it was that humans came to walk on two feet instead of four. He had argued for years that he could tell from the fossil that the Taung child had walked upright. Dart believed that becoming bipedal was part of the process of being human and he wondered how Galen's brain development might have affected his physical challenges.

Dart suffered a nervous breakdown in 1943. The various biographies of Dart, and his own memoir, mention it in passing, with some saying it was the impact of World War II. Dart himself wrote in 1956: "The outbreak of the war in 1939 finally led to such an impact of other duties, both academic and military, that by 1944, I was sent on long leave to recover from the strain."[173] Others, however, suggest it was Dart's ongoing disappointment and pressure regarding the acceptance of the Taung child that resulted in his emotional and physical collapse.

Dart was forced to leave all of his duties at the university. After a year of rest at home, focusing on his garden and paving his driveway with slate, he had sufficiently recovered to enable him to return to work as the head of the department of anatomy. However, his very personal disappointment at Galen's disability would continue to be the source of great sorrow for Dart for years to come. It would also be the source of his longstanding preoccupation with human posture and poise.

Dart was intrigued by the German anatomist Ernst Haeckel, who had promoted the idea that "ontogeny recapitulates phylogeny". In other words, he believed that the development of an organism from embryo to adult followed the evolution of all organisms from the earliest life forms to the most developed. Haeckel believed that the human embryo passed through different developmental stages in the womb that mirrored the development of life on earth. A human embryo would initially resemble the embryo of a fish, then develop into a reptile and then into an early mammal and finally a human. This concept had an impact on Dart's understanding of the stages of a human embryo's development and he believed his son had failed to pass through some of those final stages during those months he missed in the womb. Dart developed a series of exercises and movements for Galen that mirrored what he may have missed before birth. Once Galen was a toddler, he helped him with an exercise called the anthropoidal pose where Galen put all of his weight on his knuckles on a chair seat. Another exercise, called spiral rolling, had Galen looking to the right, as if he was trying to see something behind him, while Dart helped him to spiral his body to the left.[174]

Here was a case in which Dart's understanding of science and human evolution came too close to home. For decades, Dart had believed different societies were at different levels on an evolutionary scale. It had not been that long ago when he had expressed the idea that Europeans were superior and that "Bushmen" were evolutionary fossils. With these ideas in mind, it must have been very difficult for him to question how it was that his son might have not evolved sufficiently in the womb. However, on both counts, his understanding of human evolution was mistaken.

In 1944, when Galen was three, the entire student body of the medical school gathered in the Great Hall at Wits to pay tribute to Raymond Dart. He would continue as head of the department of anatomy, but after 18 years, he was stepping down as the dean of the medical school. The vast auditorium, which had a stage at the front and could hold over a thousand people, had officially opened four years earlier. Heavy-set and balding by now, but still with his bushy eyebrows, Dart stood up on the stage. As he stepped forward to accept a gift from the Students' Medical Council, he began to cry. On another occasion, still quite fragile emotionally, he hosted a party in the department of anatomy to celebrate Robert Broom's 80th birthday. As he raised his glass and offered some remarks, Dart began weeping. He took a moment to compose himself and then explained to the party-goers that his tears were as a result of how much he appreciated Broom's support in the tough years when the world had rejected his claims for the Taung skull.[175]

It was one of Dart's students, the young Phillip Tobias, who led the way for Dart to return to palaeoanthropology and find new meaning in his work. In 1945, Tobias led a trip to the Makapansgat Valley, roughly 300 kilometres north of Johannesburg. A network of eight caves threads across the narrow valley, which is about two kilometres long. In two of the caves that had been explored in the 1930s – the Cave of Hearths and the Rainbow Cave – many stone tools had been found. However, in another section of the valley there was a limeworks, just as there had been in Taung and Sterkfontein. Once again, a lime quarry for the mining industry served as a clue for the scientists. The team found grey, fossil-bearing breccia in the valley, and a fossil skull of a baboon. At both Taung and Sterkfontein, the presence of an ancient baboon skull had been the first step that led to the discovery of *Australopithecus* specimens. Tobias's team wondered if the same could be true at Makapansgat.

Dart had encouraged the student expedition. When Tobias came back with exciting findings, his interest was piqued and he started working with his students and assistants to gather tons of breccia from the limework dumps at Makapansgat. They brought grey and pink breccia back to the department, where they got to work with hammers and chisels to pry loose the fossils of many extinct types of baboons, antelopes, pigs and giraffe.

It was Dart's dream that the area might be an australopithecine site. One afternoon in 1946, he attended a meeting of the Palaeontological Research Committee, where Robert Broom was speaking. In attendance was Dr Bernard Price, a Scottish-born electrical engineer. Dart succeeded in convincing Price to fund his work at Makapansgat to better understand "the man-apes".[176]

But before Dart's initiative at Makapansgat gained momentum, there was another period of tension between Broom and Dart. During the initial student expedition, it was Joseph Jensen, Tobias's deputy leader of the team, and a student of Broom's, who had found an ancient baboon skull. Broom and Jensen co-authored a paper about the fossil in the *Annals of the Transvaal Museum* in 1946. Dart was upset. He thought Jensen should have shared the find and written the paper with him. Broom bristled, saying that Dart had known about Makapansgat for 20 years but had never been there, while Broom had worked on the site briefly in 1936 so he felt he could claim credit for having been there first.[177]

A half a century later, in 1997, Tobias would reflect on the tensions that flared occasionally between Dart and Broom, in 1936 over Sterkfontein and in 1946 about Makapansgat. "Happily these problems seem to be well behind us," Tobias wrote, "and inter-institutional and interpersonal relationships have been kept on a somewhat more even keel by the later generations of palaeontologists."[178] But the interpersonal tensions in palaeoanthropology at Wits University in the 1990s would look like a raging blaze compared to the mild flare of competitiveness between Broom and Dart.

Despite the tension between Broom and Dart, they were moving closer towards their mutual goal of gaining international recognition. They continued to look for approval from their European colleagues and they especially hoped to gain it from Sir Wilfred Le Gros Clark, a well-known English anatomist from Oxford. Like many of his colleagues, he had not been convinced the australopithecines were hominids and assumed they were more closely related to apes. However, he had followed Dart and Broom's work and decided to travel to South Africa to take a direct look at the fossils himself on his way to the first Pan African Congress of Pre-History, due to be held in Nairobi, Kenya in January 1947. The congress was an indication that scientists were open to hearing more about the recent finds in Africa. Before he left England, he examined the skulls and teeth of over a hundred apes from museum collections so that he would be well prepared with observations about ape variation in order to compare these with the fossils in South Africa. He was quite certain that what he would find would not be hominid, but he was in for a surprise.

After spending days with the original fossils in December 1946, Le Gros Clark came to a new conclusion. The human-like characteristics showed themselves much more clearly on the originals than they did on the casts, he realised. The result was that he completely reversed his position, concluding that the australopithecines were hominid. And he decided that he would say so publicly at the upcoming congress.[179]

Dart and Broom travelled to Nairobi together, with funding from Prime Minister Jan Smuts. They wondered whether other scientists would be convinced by Le Gros Clark's interpretations or if they would continue to reject the conclusion that australopithecines were hominid. One plenary session was entitled "Symposium on Fossil Apes in Africa". After Dart and Broom's presentations, Le Gros Clark stood up to speak. He explained that he had just come from South Africa where he had spent several weeks with the australopithecine fossils. Then, with a confident voice and great conviction, he told the gathering of scientists from 26 countries that he had no doubt the fossils were hominid. He described eight characteristics that showed clear human resemblance. There was murmuring from the audience which rose to something of a commotion. Most of the other scientists present continued to question Dart and Broom's conclusions, but Le Gros Clark was highly respected and so his support for their analysis had an impact on many people's willingness to take the idea seriously. After over 20 years, Dart and Broom must have thought that perhaps their beliefs were finally coming closer to acceptance by their peers.[180]

Prime Minister Smuts was scheduled to travel to England and the United States for several months. Before he left, he telephoned Broom and requested a meeting with him. "Please continue your search for 'missing links'," he requested.[181] His concern was not only for the broadening of knowledge about human origins; he also wanted to cement worldwide respect for South African science. Smuts told Broom that whatever funds he needed could be provided by the South African government and Broom promised to continue the search. In support of Broom, the Transvaal Museum hired a full-time assistant, a young man from the eastern Cape named John Robinson. The 23-year-old Robinson had studied zoology at the University of Cape Town and brought new energy to the task. Broom by then was close to 80.

After the request from Smuts, Broom decided to return to Sterkfontein. Almost immediately, he received correspondence from the Historical Monuments Commission, which was responsible for giving out permits for excavation. They refused to give Broom a permit to work at Sterkfontein because they believed he would not take care of the stratigraphy (layering) of the caves. Perhaps they were concerned about the blasting he had done at Sterkfontein in the 1930s. They also required that he work in association with a "competent field geologist". Broom was highly offended. He had been a professor of geology at Stellenbosch for seven years, had studied the caves of Australia, and knew the quarries and caves at Sterkfontein better than most. He refused to agree to the proposal and carried on with his work, ready to pay a fine if necessary. "I have no compunction whatever about breaking the law," wrote Broom. "I considered that a bad law ought to be deliberately broken."[182]

Broom wanted to appeal the commission's decision to Smuts, but it was difficult to communicate with him during his busy trip in the US. The board at the Transvaal Museum did not want Broom to continue without permission and they asked that he wait until Smuts returned to South Africa. When Prime Minister Smuts came home and learned about the stand-off, he told Broom to continue with his work. The Historical Monuments Commission then sent Broom a permit to work at Kromdraai, but they did not send him a permit to work at Sterkfontein.

At first Broom worked with John Robinson and another assistant, Daniel Mosehle, at Kromdraai and made many discoveries – including two baboon skulls and the skull of a sabre-toothed cat. However, Broom thought that his prospects would be better at Sterkfontein so he started working there again – without the permit. He decided to go back and clean up the old site where he had found the first adult *Australopithecus* skull, the skull he had named *Plesianthropus* 11 years earlier, in 1936. Within a few days, they found the crushed front of a skull with several teeth; Broom thought it to be of the same type. A day or two later, they found some upper molars. Encouraged, they kept searching.

Within two weeks of starting their search at Sterkfontein, Broom decided to blast open a large piece of breccia about a metre away from where the 1936 skull had been found. When the smoke cleared, they saw immediately that a skull had been blasted in two. "As the top of the skull had been split off, we could see into the brain-cavity, which was lined with small lime crystals," recalled Broom. After admiring the beauty of the fossil, he drove to a farm about two miles away to phone *The Star* newspaper. He told the editor about the discovery and asked him to send a photographer and a reporter out at once. In about an hour and a half, the photographer arrived and took photos with Broom, John Robinson and Daniel Mosehle. The photographs would appear not long afterwards in *The Illustrated London News* and the American Museum magazine *Natural History*.[183]

They used crow-bars to remove the base of the skull from a large block of stone. Within an hour, they had driven back to the Transvaal Museum in Pretoria with the two pieces of the skull. It was 18 April 1947. Broom named the skull *Plesianthropus*, and the specimen quickly became known as "Mrs Ples". His naming of the new fossil was later changed and most scientists agreed the fossil was in the same genus and species as the Taung child skull, *Australopithecus africanus*. But the nickname Mrs Ples stuck.

In Broom's opinion, Mrs Ples was "the most important fossil skull ever found in the world's history" and it was true that this find would have long-lasting implications for the understanding of human origins. The Historical Monuments Commission, however, was furious. As Broom described it, "the fat was in the fire."[184] The commission held a meeting at which Broom was unanimously condemned. They wrote to Smuts stating that Broom had blasted in the area illegally, that he had destroyed valuable evidence and that he must be stopped immediately. Smuts was more interested to hear whether this was the find that would put South Africa on the map.

The discovery did put South Africa in world news again. Despite the slap on the wrist, Broom had the impression that the local press and much of the South African public were on his side. The Historical Monuments Commission sent a professor from Pretoria University to investigate the cave at Sterkfontein where Broom had been working, who reported that there was no stratigraphy in that area, and therefore Broom's work hadn't done any harm or destroyed any important aspects of the cave. As a result, the commission relented and sent Broom a permit to continue working.

One of the photographs taken at the time of the find shows Broom pointing to the skull with his right hand. As Broom would write about the photograph later, "Mr. J. T. Robinson, my excellent assistant, is beside me. Behind me is my quarry-man, Mr. Van der Nest and my chief native assistant, Daniel."[185]

Daniel Mosehle, who was on site when Broom found Mrs Ples, had worked at the Transvaal Museum for about 20 years by then. There is even less information about Mosehle on record than there is about Saul Sithole. In Broom's memoir, he referred to Mosehle by his first name only. He described him as "having natural ability and knowledge that most Europeans did not, for example being able to distinguish between a sabre-tooth tiger's skull and that of a lion." Broom wrote: "As a fossil hunter, he is worth his weight in gold."[186]

Robert Broom points toward the point where Mrs Ples was found on 18 April 1947. Standing to the right is quarryman Van der Nest. John Robinson is kneeling, and Daniel Mosehle is standing behind Broom. Courtesy Becky Sigmon in "The Making of a Palaeoanthropologist: John Robinson," originally published in The Star.

Some months after the discovery of Mrs Ples and the worldwide excitement that followed, a church pastor drove out to Sterkfontein to visit the site of the find. Mosehle was on site that day. The pastor asked, "Is it true that the skull came out of the rock there?" When Mosehle confirmed this and showed him a photograph of the skull in the breccia and his copy of *The Illustrated London News*, the pastor took the paper from him. He looked at it for a moment. Then "I don't believe it," he said, and threw the newspaper on the ground. In South Africa, and globally, there was still enormous resistance not only to the idea that humans had evolved in Africa, but to the idea of human evolution at all.

Robert Broom lived long enough to support Dart's analysis of the Taung child skull as best he could. He was so committed to Dart's point of view that shed light on the "strange course of evolution", that he had spent years looking for the proof.

Smuts, too, finally had the confirmation he had been looking for. He had spent much of his life seeing palaeoanthropology not only as a fact-finding mission, but as a mission to confirm his political point of view. With the discovery of Mrs Ples, both men had more evidence that South Africa was a cradle, or one of the cradles, of humankind. On 31 October 1949, Smuts presided over a ceremony at the Transvaal Museum to unveil a bust of Broom holding Mrs Ples.

Smuts passed away on 11 September 1950. Broom died on 6 April 1951. Neither man lived long enough to see a shocking development that would further bolster Dart's case for human origins in Africa. The unexpected breakthrough would not occur until November 1953, and it would come about because, socially and politically, the world was changing, and scientific assumptions were starting to change as well.

8

A 40-Year Setback and the Steadying of the Pendulum

Joseph Weiner, a South African and one of Dart's former students, had become a professor of physical anthropology at Oxford. At a conference in London in July 1953, Weiner met Kenneth Oakley, who was also a physical anthropologist and palaeontologist, and well known for his work on trying to date fossils. During the conference, the group of scientists toured the British Natural History Museum, where the vaulted central hall held the 32-metre long skeleton of a dinosaur, the *Diplodocus*, that still stands there today. Weiner and Oakley chatted together throughout the tour. When they entered a smaller room, which housed the fossils of Piltdown Man, the group was silent for some time as they examined the bones. Most of the scientists present had not seen the original fossils before. They began a lively debate about whether the fossils fit into the growing fossil record of human evolution. Some pointed out that the hominid fossils found in the previous 40 years did not align with Piltdown Man. Others were reluctant to reject Piltdown Man and continued to look for ways in which it could fit into the picture. According to Weiner, the conversation "provoked the usual tail-chasing discussion".[187]

That night at dinner, Weiner and Oakley had an intense discussion about the mystery. Weiner reminded Oakley that there were only two possible explanations – that Piltdown Man was either a "composite man-ape", as Smith Woodward and Dawson had declared, or that the fossils of two distinct creatures had been found side by side. Weiner wasn't happy with either of these explanations but wondered what other explanation might exist. He asked Oakley to think about the likelihood of someone, either by error or by chance, dropping the fossil of one creature into the area that already held the fossil remains of a completely different individual. And how likely was it that it could have happened twice?

Oakley told Weiner about the fluorine test he had conducted on the Piltdown fossil four years earlier, in 1949. By observing how quickly the bone absorbed fluoride, the fluorine method could help determine how long the object had been underground. Oakley had published an article in *Nature* claiming the fossils were relatively modern. However, Oakley agreed with Weiner that Piltdown Man was an oddity. Both men were suspicious of the unique wear on the canine tooth that had never been found on any other fossil or living ape. Weiner had only examined casts of the fossils but Oakley had access to the originals. When Weiner asked him to examine the canines of the original fossils, Oakley did. His conclusion was somewhat shocking. He said he believed the teeth had been deliberately ground down so as to resemble a more ancient jaw.

With recent improvements in the fluorine technique, Oakley got permission to perform the test again. It confirmed that the jaw and teeth were younger than the crania. He and Weiner were then given permission by the Natural History Museum to conduct further tests and to take a limited drill sample from the fossil. In 1953 for the first time, there was an array of new chemical tests and methods available to scientists. Heading back into the lab, Oakley tested for iron, nitrogen, and organic carbon. His new tests confirmed what many had conjectured for years, but could never prove. In every single case, there was one set of results for the mandible and another set of results for the crania. They also found a distinct contrast between the staining of the crania, which penetrated throughout the bone, and that of the mandible, which was only on the surface. They continued their examination with new X-ray technology and found that the jaw contained no sulphate, whereas the cranium did. The scientists came to the conclusion that the entire fossil was a forgery.

In order to confirm their conclusion, Oakley took the specimens out of the safe and handed them not only to Weiner, but also to Le Gros Clark. Without sharing any of their former findings, they asked Le Gros Clark what he thought of the jaw and the worn-down teeth. "The evidences of artificial abrasion immediately sprang to the eye," said Le Gros Clark. "Indeed so obvious did they [the scratches] seem it may well be asked – how was it that they had escaped notice before?" He went on to quickly answer his own question. "They had never been looked for ... nobody had previously examined the Piltdown jaw with the idea of a possible forgery in mind, a deliberate fabrication."[188] Weiner, Oakley and Le Gros Clark wrote a full report for the British Natural History Museum. The next day, it was all over the papers.

"THE BIGGEST SCIENTIFIC HOAX OF THE CENTURY!" read the *London Times*. "PILTDOWN MAN HOAX IS EXPOSED" announced the *New York Times*.

Weiner's opinion was that "The skill of the deception should not be underestimated, and it is not at all difficult to understand why forty years should have elapsed before the exposure; for it needed all the new discoveries of palaeontology to arouse suspicion and completely new chemical and X-ray techniques to prove the suspicion justified."[189]

Once the truth was out, there was enormous embarrassment and speculation about who had been responsible. Someone had taken a mandible, molars and a canine from a modern orangutan, ground down the teeth, stained them, and then stuck them together with parts of a modern human skull. The hunt was on to identify the culprit, or culprits. The reputation of several revered British scientists was up for question. Forty years of study of Piltdown Man had been wasted trying to analyse and fit these fossils into the growing fossil record of human evolution, and all the while they had been fake. Phillip Tobias recalled that Raymond Dart had edited a series of radio broadcasts for the South African Broadcasting Corporation in 1953. Tobias gave one of the talks and remembered referring to Piltdown Man, indicating that the fossil had not been rejected until the final blow.[190]

This revelation had enormous consequences for Raymond Dart and the Taung child skull. Dart reflected in his memoir that one of the reasons the Taung child skull had not been accepted was because of the emphasis on the importance of brain size having been one of the first changes in hominid ancestors. This theory had been promoted by Dart's mentor, Sir Grafton Elliot Smith, and it was confirmed by Piltdown Man. If scientists believed that a large brain came first in human evolution, then the Taung fossil made no sense. "In spite of Piltdown Man's being thoroughly discredited," wrote Dart, "the argument about brain size continued for many years."[191] Dart lamented: "Piltdown was my downfall."[192]

The history of human evolution was thrown off track for 40 years from the time Piltdown Man was announced in 1912 to the time it was declared a hoax in 1953. And it delayed by close to 30 years the acceptance of *Australopithecus africanus* as a genuine contribution to the search for the ancestry of humans in Africa. The hoax succeeded because it fit in with the theories about race and human evolution, and fulfilled the expectations at the time.

One of the reasons why the fraud had been so successful was that the credentials of the team who had found the original fossils, Charles Dawson and Arthur Smith Woodward, were untarnished. Each man was held in high esteem. Teilhard de Chardin, who later found the canine tooth, became a highly respected palaeontologist. Sir Arthur Keith and Sir Grafton Elliot Smith, both highly regarded, had spent years of their lives studying this fossil.

Writing in 1980, Stephen Jay Gould's theory was that it was Dawson who was responsible and that he had drawn in the young Teilhard de Chardin.[193] For a Frenchman, it could have been a great joke to plant a fossil in England, which had no human fossils at all, while France was already the site of both Neanderthal and Cro-Magnon fossils. More recently, Francis Thackeray also asserted that it was De Chardin, and suggested that he was involved in a prank he could never reveal. In 1992, Tobias argued that the blame sat with Sir Arthur Keith; Keith, he believed, had a hidden agenda in his "vehement and sustained opposition to the acceptance of Dart's and Broom's claims for *Australopithecus*".[194]

But certainly a major factor was that British scientists and the interested public in Britain and the US were more comfortable with the idea that humans had ancestors in Sussex, England, rather than on the continent of Africa. Smith Woodward supported Piltdown for his entire life and wrote his final book, *The Earliest Englishman*, in 1948 about the fossils before he died. Sir Arthur Keith had used Piltdown for 40 years as the foundation of his thinking about human evolution. Piltdown Man complemented these scientists' worldview.

Although one of the reasons that the hoax came to light when it did was certainly the new technology available to Weiner and Oakley, another reason was the growing amount of fossil evidence emerging from Africa. The incompatibility between the new thinking and the Piltdown skulls helped force the revelations.

The hoax illustrates how false scientific data can be accepted quite easily when it fits within existing expectations. Scientists expected that humans evolved in Europe, so they readily accepted the evidence. Mistakes are clear with hindsight, but false information presented as fact can be welcomed because of assumptions that are incorrect. Racist thinking at the start of the twentieth century was certainly a part of what created the environment in which a fallacious fossil find in England could be readily accepted.

In 1950, a few years before the revelations of the Piltdown hoax, Phillip Tobias was completing his medical degree at Wits, at around the same time as Raymond Dart began looking ahead to retirement and applying his mind to finding a successor. Tobias had spent several years as a student and a part-time teacher and he was trying to decide if he wanted to practise medicine. When Dart told Tobias that he'd like him to take up a post as a lecturer in the department Tobias said yes.

Tobias always saw Dart as a father figure. Whimsically, he liked to link his own conception to the finding of the Taung child skull. Tobias was born on 14 October 1925 and he calculated that if his mother's pregnancy was of an average length, "I would have been conceived on or about 3 February, the very evening on which *The Star* of Johannesburg carried the first exciting announcement of the discovery of the Taung skull."

Tobias's father was born in England and his mother was from the Orange Free State. He grew up in Durban.[195] It was his family experience that drew him toward the field of genetics for his PhD. When he was 16 years old, his sister Valerie died from diabetes at the age of 21. Traumatised and upset, the teenage Tobias went to speak to the family doctor. On his way there, and in the waiting room, Tobias thought about the fact that his maternal grandmother had also died of diabetes. When he was face to face with the doctor, he asked him this question: if the disease had been carried from his grandmother's genes to his mother's to his sister, why had his mother not been sick? Impressed by Tobias's question, the doctor said he didn't have an answer for him. Not to be put off, Tobias asked who in South Africa had studied medical genetics and could help him with the answer. When the doctor was unable to come up with a name, "There and then," recalled Tobias, "I secretly resolved to become the first medical geneticist in South Africa." One of the reasons that Tobias had been aware of the role of genes in the first place was that he spent many afternoons after school in the Durban Natural History Museum. One display illustrated heredity by explaining human eye colour. Another showed mealies of different colours.[196]

By studying chromosomes at the medical school, Tobias continued his interest in genetics but he also followed another interest – palaeoanthropology. He took classes with Dart and Broom where he was exposed to their work with fossils and their attempts to understand evolution. In his third-year science class in 1945, Tobias went to Sterkfontein and Kromdraai with Robert Broom. He knew that genes and mutations were the building blocks of evolution, but it just wasn't clear how it all worked.

Dart often tried to illustrate to Tobias and his classmates the process of evolution of upright posture and how humans became bipedal. Dart was in his early fifties at the time. To demonstrate how a crocodile and other reptiles walked, Dart lay on his stomach on the long, marble-topped table at the front of the hall. He moved his arms and legs forward and back and curved his spine, creating an S shape that moved from side to side. Dramatically, he lifted himself up to show how mammals evolved from reptiles, showing that they learned to lift their stomachs off the ground and stand on all fours. Dart walked up the centre of the aisle of the lecture hall, getting closer and closer to the ceiling as he went. Continuing his lecture all the while, when he reached the back of the auditorium, Dart again turned around towards the front and jumped up to grab one of the ceiling pipes. Hand over hand, he made his way back to the front, his legs dangling above the steps. Then he announced to the class how primates, who had been living in the trees, "came down to earth", which he proceeded to do himself with a flourish.[197]

Toward the end of June 1951, Dart told Tobias that the South African government had approached the university and requested that a qualified person be sent on a French expedition to the Kalahari desert. One of its aims was to study the Bushmen in the Kalahari and Dart proposed that Tobias join the expedition as an anthropologist and take measurements. Tobias didn't feel he had the experience for the trip, especially in the field of anthropometry, the measurement of the human body, and he told Dart as much. But Dart was not dissuaded and sent Tobias anyway. Just as Dart had gone to the Kalahari in 1936, nearly 15 years later Tobias followed in his footsteps. He travelled with Francois Balsan, a French explorer and geographer, on what was called the Panhard-Capricorn expedition. Over a period of 11 weeks, the group travelled 6 800 kilometres through Bechuanaland (Botswana), South-West Africa (Namibia), Mozambique and South Africa along the line of the Tropic of Capricorn. It was the first time the 25-year-old Tobias had travelled outside of South Africa. "The enchantment, the lure, of our continent was borne in upon me ... I encountered teeming humanity, from the light, yellow-brown San to the taller, dark-skinned black Africans, from hunters and gatherers to pastoralists, agriculturalists and urbanites."[198] After returning from the expedition, Tobias published a number of articles on the Bushmen, especially from the point of view of human biology.

When Tobias travelled to Paris, the curator of the Musée de L'Homme set out 14 adult Khoisan skulls for him to examine. He examined hominid fossils as well, including some classified as Neanderthal and Cro-Magnon. He then went to the Museum National d'Histoire Naturelle, where he was shown Sarah Baartman's remains on display. Although Tobias was not familiar with her story at that time, his guide wanted to show her skeleton to him because he knew that Tobias had been studying "Bushmen" on the French Panhard-Capricorn expedition. He recalled seeing Baartman's "well-preserved, articulated and mounted skeleton" as well as two parts of her soft tissue that had been preserved, "her brain and her external genitalia". It is striking that Baartman's remains were on display then, not in the Museum of Man with other displays about human culture and anthropology, but in the National

Museum of Natural History alongside exhibits of biodiversity and ecology.[199]

In the mid-1940s, when Tobias first arrived at Wits, he became a member of the National Union of South African Students (NUSAS) and was soon elected a national office bearer, becoming its national president in July 1948, two months after the National Party was elected to power in South Africa. During the election, the National Party message to voters had been that apartheid would more systematically maintain white supremacy than the policy of segregation under the Smuts government had done. The party had campaigned for the election arguing that universities needed to exclude black students and Tobias wrote a lengthy pamphlet "The African in the Universities", making the case that the government should not interfere with university management. Not surprisingly, following the election victory, the National Party ignored Tobias's argument and implemented policies to prevent the admission of black students at universities such as Wits.[200]

Tobias, who was Jewish, was greatly affected by World War II. Hitler's genocide raised scientific questions for him about race. Prompted by the Holocaust, UNESCO drafted its Statement on Race in 1950 which began: "Scientists have reached general agreement in recognising that mankind is one; that all men belong to the same species *Homo sapiens*." Race, from a biological point of view, was defined in the document as one of the groups of populations in the species. The statement made reference to the Mongoloid, Negroid and Caucasoid groups of mankind, noting, "It is now generally recognised that intelligence tests do not in themselves enable us to differentiate safely between what is due to innate capacity and what is the result of environmental influences, training and education."[201]

The UNESCO statement drew criticism because it had been drafted predominantly by social anthropologists. A second group was convened that included physical anthropologists and geneticists, to review and revise the statement. The second UNESCO Statement on Race of 1951 declared, "Scientists are generally agreed that all men living today belong to a single species, *Homo sapiens*, and are derived from a common stock, even though there is some dispute as to when and how different human groups diverged from this common stock." The statement said that race as a "classificatory device" was useful in terms of "studies of evolutionary process" and clarified that physical differences are at times the result of genetics and at other times as a result of differences in the environment. Unlike the sociologists, the physical anthropologists and the geneticists in the US and Europe had not come to any consensus on how human evolution occurred and how it related to the concept of race. The UNESCO statements were an effort to reach greater consensus, but without complete success. The second statement did declare that there were no scientific grounds whatsoever for the "racialist position regarding purity of race and hierarchy of inferior and superior race". However, it also declared that physical anthropologists and "the man in the street both know that races exist".[202]

In South Africa, along with the Immorality Act of 1950 which outlawed sex across the colour line, and the Group Areas Act of 1950 which proclaimed residential and business areas for particular race groups as defined in South African law, the Population Registration Act of 1950 was one of the cornerstones of legislation of the new apartheid government. This law required every individual in South Africa to be registered and classified by race from birth. The classification was based on subjective characteristics such as hair, skin colour, facial features, home language, place of residence and employment. The act defined a "native" as "a person who in fact is or is generally accepted as a member of any aboriginal race or tribe of Africa". A "white person" was defined as "a person who in appearance obviously is, or who is generally accepted as a white person, but does not include a person who although in appearance obviously a white person, is generally accepted as a coloured person." The Act then defined a "coloured person" simply as someone "who is not a white person or a native".[203] Soon Indians (or Asians) were added as a fourth category.

Given Tobias's background in genetics and his work in physical anthropology, he was increasingly asked about the scientific understanding of race. His first ever publication on the topic was an article in *The Journal of Forensic Medicine* in 1953 when he was 28 years old. In it he examined the question of whether the South African government was justified in assigning a racial label to every individual. "The purpose of this paper is to stress some scientific aspects of the problem created by the co-existence in one territory of different racial groups." Tobias wrote that because the idea of race had been taken up by political leaders, it was up to him as a physical anthropologist "to elucidate the basis on which the concept of race is built". He pointed out that race was a biological concept, not sociological, but that socio-cultural influences had come to play a role as well.[204]

Tobias outlined five "races of man" as Caucasoid, Mongoloid, Negroid, Bush and the Aborigines of Australia, and explained that these classifications were based on appearance. He indicated that genetics was beginning to play a role in the process of classifying people. He said that he was not sure what accounted for mental, as opposed to physical, variations. "That mental difference may parallel the physical differences is a definite possibility, but we do not yet possess adequate techniques for separating off those mental differences which are due to the impact of the environment from those which are inborn or hereditary or racial."[205]

Despite the fact that Tobias condemned the invalid interpretation of race purity by the Nazis, he said that the scientific reaction to the events of World War II might have "led the pendulum to swing almost too far in the opposite direction. We find today a tendency to deny the validity of the race concept for anthropology ... A steadying of the pendulum is required; this will centre on the idea that race in human biology is as useful a concept as it is in other biological fields."[206]

At the end of the paper, Tobias extended his argument to the specific circumstances in southern Africa. Holding onto the concept of a Boskop race, he suggested that Bushmen retained "primitive and foetal features" as a "mutant offshoot of Boskop Man" and that "Bush and Boskop elements [had] entered into the racial composition of the Negroes". He gave scientific credibility to the development of a new "race" in saying that "where miscegenation [had] created a large hybrid population as in the Coloured folk of South Africa, a relative stability does tend to emerge; for a new race is being born". His paper gave credence to the South African government's decision to develop a racial classification called "Coloured".[207]

According to Tobias, the concept of race was more valid when speaking about groups of people, rather than individuals. "A Coloured man with a light skin and a high-bridged nose may pass as a European more easily than a White who has high cheek bones and dark colouring." Therefore, he concluded that it would be exceedingly difficult to place every individual into a natural racial category. "No physical anthropologist could set himself up as a scientific arbitrator over these boundaries in a multi-racial community like that of South Africa."[208] Tobias might not have been aware that Dart had acted as a scientific arbitrator with his declarations on race on the witness stand over 20 years earlier. As would later become apparent, Tobias, too, would arbitrate on racial boundaries and assist with the process of racial classification.

At the same time as the newly elected National Party government in South Africa was imposing new race legislation, and protecting "racial purity", scientists around the world were increasingly questioning the biological existence of race at all. None the less, a theory that had been put forward years earlier about the relationship between race and human evolution continued to hold influence in the 1950s. Franz Weidenreich was a Jewish German anatomist and physical anthropologist. He spent the 1930s and 1940s in China studying Peking Man, a group of fossils found in the 1920s described as *Homo erectus*. Weidenreich saw morphological characteristics of Peking Man that he believed were common to modern Asian people, which led him to develop what he called a "polycentric" model of human evolution. This theory proposed that there was biological continuity in each of the regions of the world and that the different races of humans developed separately. He argued that evolution may have occurred at different paces in different geographic areas. "From the morphological point of view ... the Australian bushman are less advanced human forms than the white man; that is, they have preserved more of their simian stigmata. Whether they have 'entered' into evolution at a later time than the whites, or their evolution 'rested' or was 'retarded' while that of the whites went on, we do not know."[209]

While interested in regional variation, Weidenreich, however, asserted that there had been gene flow between the different regions of the world since the early stages of human evolution. He used a trellis-like drawing to describe the distinct evolution in each region as well as the crossing of genes between them. Weidenreich died in 1947, but aspects of his theory continued to be considered by scientists for decades, especially in relation to human evolution and race.

Back in South Africa, as Tobias grappled with the implications of race and embarked on his career in physical anthropology and palaeontology, John Robinson, who had been Robert Broom's assistant, would also play an important role in the 1950s. Unlike Tobias (and Dart and Broom), who came to the field of human evolution through the study of medicine, anatomy and physical anthropology, Robinson's interest began as a naturalist. Growing up on a farm in the eastern Cape, he explored what he found in the outdoors and went on to study zoology at the University of Cape Town. When Robinson was hired in April 1946 by the Transvaal Museum to work with Broom, he began to focus on human evolution and the role of the australopithecines. In the late 1940s, australopithecines had only been found in South Africa and it wasn't until 1959, with Mary Leakey's discovery at Olduvai Gorge in Tanzania, that an *Australopithecus* fossil was found anywhere else in Africa. For more than a decade, Robinson was one of the only scientists in the world who focused on the australopithecines and tried to draw attention to their importance.[210]

Robinson had been on a research trip to the UK and the US from March through September 1951, so he was out of the country when Broom passed away. The director of the Transvaal Museum wrote to Robinson confirming his formal appointment in Broom's place, as head of vertebrate palaeontology and physical anthropology. While in London, Robinson spent time in museums with fossil collections doing his own research and he was in great demand for lectures in which he discussed the 26 fossils he had brought with him to illustrate the richness of the South African fossil sample. Robinson believed there were three species represented at Sterkfontein, Kromdraai and Swartkrans: *Australopithecus*, *Paranthropus* and *Telanthropus*, another fossil specimen that Robinson then described as "intermediate between ape-man and true man". Newspaper accounts in London pointed out that this was the first time since Dart's 1931 visit 20 years earlier with the original Taung skull that London had seen original australopithecines. Robinson was struck by the academic snobbery of some of the English scientists, who hadn't offered much respect to South African scientists over time; he felt they were poised to offer final judgement on his work.

By the time Robinson returned to South Africa in September 1951, Raymond Dart was no longer centre stage in palaeoanthropology and Phillip Tobias had not yet taken on the spotlight. The Transvaal Museum, not Wits University, held the bulk of the hominid fossils from the Sterkfontein Valley. With the dawning recognition that South Africa had a significant role to play in the field of human origins, a growing number of scientists wanted to visit the museum. And after Le Gros Clark's presentation at the Pan African Congress in 1947 supporting the important role of australopithecines, more and more scientists became interested in examining the fossils first-hand. In 1954 Teilhard de Chardin wrote to him, saying: "You happen to be at the most sensitive place spotted so far for the unveiling of human origins ..."[211]

At the age of 28, without his mentor Broom, Robinson stood alone to represent those fossils in South Africa. He found himself hosting scientists from abroad, all of whom had many and varying demands, and he was also required to give lectures and speak to the media. "I was pitchforked by circumstances into a prominent position scientifically and that was perhaps a heavier burden than could easily be borne by a kid who grew up on a farm in the sticks in South Africa," Robinson recalled.

Robinson was thrown into the limelight in another way as well. Since the National Party came to power in 1948, the term "evolution" was being questioned more regularly than it had been previously. Unlike Jan Smuts' United Party government, the National Party leadership, including Prime Minister DF Malan, had no interest in palaeontology and human origins and was not supportive of the theory of human evolution. Christian National Education was imposed and evolution was not taught in schools. Growing Christian nationalism and the Calvinist principles within the Dutch Reformed Church were both influential in shaping Afrikaner identity, policy making and political thought.[212]

In August 1951, months after Broom's death, the Transvaal Museum opened a memorial exhibition entitled "Man's Family Tree". It described "Man's Place Among the Primates" and offered a visual interpretation of Robert Broom's discoveries and their place within what was known about human evolution at the time. In September the following year, three religious leaders from three Dutch Reformed churches wrote a letter of protest to the museum about the display, asking it to remove the display because it implied that humans had evolved from animals and in so doing offended the religious convictions of the majority of the country's Christian population.

"The Family Tree of Man" display at the Transvaal Museum, prepared by Robert Broom, and protested by three ministers from the Dutch Reformed Church in September 1952 asking the museum to remove the display. Copyright: Museum Africa.

The theory of evolution is the theory of change over time within species on earth across generations. Darwin suggested that these changes occur by "natural selection", which means that a small change turns out to be useful to the organism in some way and then is passed on to its offspring. This was a theory to explain how very simple forms of life began to change into more complex organisms, and then how one species could change over time to become a new species. Broom believed that small changes in reptiles over time resulted in their becoming mammals. Creationists and other people around the world who were opposed to the theory of human evolution were against the idea that primates evolved and changed over millions of years to become human.

Dr Antonie Janse, the acting director of the Transvaal Museum, responded publicly. "I have not set out to attack the churches," he said, "but I am ready to defend science and its search for truth to the last ditch." In a formal response to the protest letter, the museum's board of trustees said the exhibit had not been presented as fact but as an interpretation of a scientific theory. A newspaper report described the exhibit by saying that "the skulls of man-

110

like apes, ape-men or missing links, and of primitive man still leer from their cases".[213]

The Sunday Times stoked the coals of controversy by referring to an upcoming talk by John Robinson. The headline read "SCIENTIST TO REPLY TO CHURCH OVER APEMAN THEORY", suggesting that the lecture had been designed as a response to the clergymen's protest and would answer questions about the depiction of evolution in the museum. In fact, the lecture had been scheduled long before the protest letter was written. It had been organised as a memorial lecture honouring Dr Broom and it had been intended to focus on the evolutionary history of vertebrates.[214]

Robinson wanted to continue with his planned lecture on vertebrates but he also knew that there would likely be high attendance because of all the publicity over the exhibit and that it would not make sense for him to ignore the questions about human evolution. The lecture became an opportunity for Robinson to present his own beliefs about science and religion publicly. It took place on 1 October 1952 at the Pretoria Technical College and was sponsored by the South African Biological Society. Robinson gave his prepared lecture and then he spoke to the brewing controversy. As his mentor Broom had said 50 years earlier, Robinson believed there was no conflict between religion and the science of evolution. Building on his own beliefs, he stated that science was concerned only with the physical universe and was not designed to explain non-material matters. "The raw material of evolution consisting of the fossils studied by palaeontologists," said Robinson, "gave no clue as to whether the animals being studied had a soul, or whether there was such a thing as a divine spirit."[215]

Robinson's lecture was well received. The board of the museum met to consider how to respond to the protest letter. They decided that the evolutionary display of the "Tree of Man" would remain but that the labelling would change. The new labelling stated that the exhibit was based on an interpretation of the theory of evolution, instead of presenting it as fact. The clergy accepted the labelling change and the matter was resolved. One of the three clergymen, Dr De Beer, who was also the commissioner of Public Morals of the Dutch Reformed Church, stated in the *Rand Daily Mail* that he had been successful in pruning man's family tree, by forcing the museum to label the exhibit as a "theory rather than a fact". It was "a retreat by the Museum authorities that represents a notable victory for Mr. De Beer over Dr. Robert Broom". "...the path of science is strewn with discarded theories," he declared. "Evolution is probably one of them."[216]

In March 1952, *The Star* published an article entitled "Misgivings About

Nat Outlook on Science", suggesting that the new National Party government had a "narrow bigoted attitude" that would have a negative impact on the country's scientific reputation. The article stated that a government official had requested that the word "evolution" should not be mentioned in an exhibit about the origins of mankind at the South African Museum in Cape Town. The article also stated that government was not going to allow any official delegates to travel to the international geological conference that year, nor to the pre-history conference in Algiers in September. In fact, the pre-history conference, the follow up to the congress in Nairobi that Dart and Broom had attended in January 1947, was originally scheduled to take place in South Africa. However, the National Party government refused to provide funding for the conference, so South Africa could no longer host, and the location had to be changed to Algeria. The South African government led by Jan Smuts in 1947 had wanted to spread the reputation of South African scientists by sending them off to the pre-history conference. But five years later, the apartheid government prevented South African scientists from attending. Just as South Africa was beginning to be recognised in the scientific world for its role in palaeoanthropology, in the wake of Le Gros Clark's support of Dart and Broom and the discovery of Mrs Ples, and on the eve of the exposure of the Piltdown hoax, government policy had radically shifted. *The Star*'s article stated that South Africa had built an international reputation in science, especially in anthropology and archaeology, and expressed concern that the current South African government did not want to participate in the conference because it might deal with evolution. The article's conclusion was dispiriting. "If that impression is correct, South Africa will further recede into the darkness in the eyes of the world."

TOP: *Raymond Dart with the Taung child skull in February 1925. Copyright: Museum Africa.*

BOTTOM: *The Raymond Dart Human Skeleton Collection in the early 1960s. Courtesy of Goran Strkalj. Copyright: School of Anatomical Sciences, University of the Witwatersrand.*

Photo taken by Herbert Lang of Robert Broom indicating the location of the 1936 fossil find at Sterkfontein. Broom's caption at the time read: "With me are Mr. Barlow and behind two museum boys, Saul [Sithole] and Jacobus. The other boy is one of the quarry boys." Copyright: Illustrated London News Ltd/ Mary Evans.

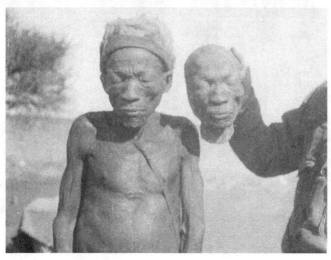

!Gurice, also known as Abraham, alongside his face mask at the Bain camp in Tweeriverien in 1936. The photo was taken by James van Buskirk. Courtesy of the Raymond Dart papers, Wits University Archive.

This photo is similar to one hanging in the Origins Centre. Likely taken in 1937 at Johannesburg Empire Exhibition. No names are provided, but because of their likeness, it is assumed that from right to left are /Khanako and her daughter Klein /Khanako. Photographer unknown. Copyright: Rock Art Research Institute, University of the Witwatersrand.

RAND DAILY MAIL, TUESDAY, JULY 21, 1936.

BUSHMEN SHOULD BE SAVED, SAYS PROFESSOR DART

Useful Scientific Work in Kalahari

IMPORTANT scientific results have been achieved by the expedition that penetrated the Kalahari to study the Bushmen.

"If the net result of this expedition, quite apart from its scientific value, should amount to nothing more than the preservation of this small group of people, every effort put into it will be well worth while," Professor Raymond A Dart, the leader, told a "Rand Daily Mail" reporter yesterday on his return to Johannesburg.

Professor Dart thinks that this expedition is merely a beginning in the study of the Bushmen. Others should go to remote parts of Bechuanaland, South-West Africa, and the south-western part of Rhodesia.

"I fail to understand," he said, "why South Africa as a whole should be vitally interested in the preservation of wild game and not as yet interested in the preservation of a group of human beings who represent one of the greatest monuments any country can possess."

FURTHER INVESTIGATION NEEDED

News clipping from **Rand Daily Mail**, *21 July 1936. Courtesy of the Raymond Dart papers, Wits University Archive.*

TOP: *Telegram from Donald Bain to Raymond Dart, 9 September 1939. Courtesy of the Raymond Dart papers, Wits University Archive.*

BOTTOM: *Telegram from Oudtshoorn Hospital to Raymond Dart, 16 September 1939. Courtesy of the Raymond Dart papers, Wits University Archive.*

TOP: *Raymond, Galen, Marjorie and Diana Dart in the garden of their home, circa late 1940s, at 26 Park Street, Oaklands, Johannesburg. Courtesy of the Raymond Dart papers, Wits University Archive.*

BOTTOM: *Raymond Dart and a portion of the Gallery of African Faces, circa early 1950s. Copyright: Africa Media Online.*

TOP: *Robert Broom and his colleagues on the occasion of finding Mrs Ples at Sterkfontein on 18 April 1947. Behind Broom is Daniel Mosehle, next to him is quarryman Van der Nest, and kneeling is John Robinson. The photo originally appeared in* The Star. *Courtesy of* Physics, Evolution, God: Mass and Nomass *by Becky Sigmon and Richard Dowden.*

RIGHT: *Robert Broom at the unveiling of a bust of him at the Transvaal Museum on 31 October 1949. Seated on the left are Professor DE Malan and General Jan Smuts. Copyright: Ditsong National Museum of Natural History.*

LEFT: *Phillip Tobias measuring an unnamed person during an expedition to the Kalahari in the early 1950s. Copyright: School of Anatomical Sciences, University of the Witwatersrand.*

BOTTOM: *Phillip Tobias with schoolchildren at the Man in Africa exhibit at the Johannesburg City Hall in 1963. Copyright: School of Anatomical Sciences, University of the Witwatersrand.*

"Gallery of faces" at Wits. is world's finest collection

Mrs. H. Eriksen, a technician at the Medical School, looks at a life mask of a Bushman, one of more than 500 masks representing diverse African peoples in a collection said to be the best of its kind in the world. Mrs. Eriksen helped to make many of the masks.

By a Staff Reporter

THE "GALLERY OF FACES," it might be called. Lining both sides of a corridor at the University of the Witwatersrand Medical School, are more than 500 life-size models of faces—men and women of many different colours and racial features.

They are life-masks of specimens of nearly every major ethnic group in Africa from the Cape to Cairo and represent, according to Prof. R. A. Dart, head of the department of anatomy at the university, the finest anthropological collection of its kind in the world.

About 40 or 50 tribes are represented, with skin colours ranging from the coal-black of West African Negroes to the dusty yellow of South African Bushmen. There are faces of Berbers and Fezzanese from North Africa, Kikuyu from East Africa, the pygmies of Central Africa and the Hottentots and Bantu of the south.

The masks are made from plaster-of-paris casts, taken from living subjects on field expeditions by Professor Dart, his colleagues and staff over the last 25 years. The collection is continually being added to and the 500 masks are not all—elsewhere the anatomy department has another 300 or so death-masks.

BECOMING EXTINCT

Pride of the collection are more than 100 faces of Bushmen from Angola, South West Africa and the Kalahari. The masks form a permanent record of tribes which are rapidly becoming extinct.

The technique is being refined and skin details, such as facial hair and wrinkles, are reproduced faithfully.

Technicians at the Medical School are experimenting with a method of taking the impressions with the subject's eyes open.

To complete the collection with European types, masks have been taken of a few lecturers. Some Indian and Native students have also volunteered as subjects.

"The collection is an invaluable record which permits detailed study in the laboratory," said Dr. Tobias, senior lecturer in anatomy. "It has opened a vast new field in anthropological research."

Hertha Erikson (later De Villiers) alongside the Gallery of African face masks, **The Star,** *28 February 1952. Courtesy Wits Historical Papers.*

Part Two
Searching for Humanity

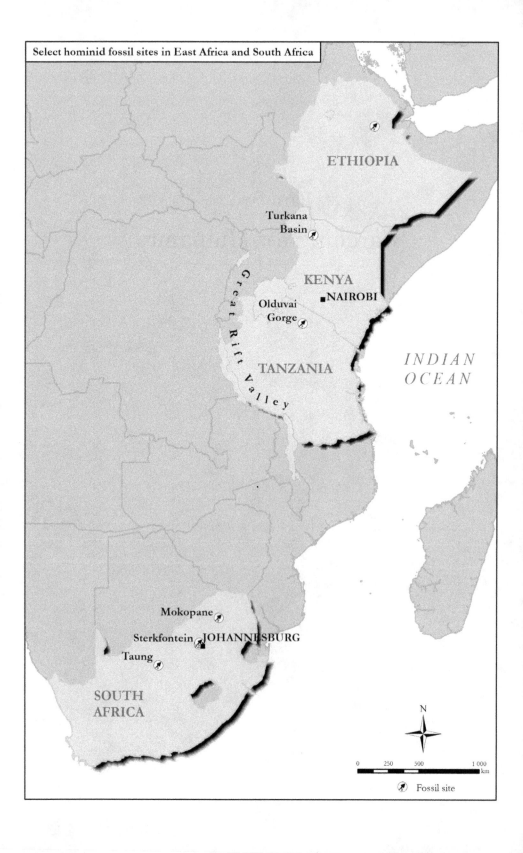

Select hominid fossil sites in East Africa and South Africa

9

A New Generation:
Moving Away from Type

In January 1959, in the very week that Tobias took over from Raymond Dart, all the pipes running down Hospital Hill towards the Wits Medical School burst. His predecessor had kept his collection of human skeletons – over a thousand of them – in boxes on the lowest floor in the department of anatomy. With the water waist deep down there, all the boxes were now floating on the surface. To make matters worse, not all of them were tightly sealed and many of the lids had come off. Bones and sodden boxes floated around in the foul-smelling water. It took the fire department the entire day and into the night – 14 hours – to pump the water out of the building.

Two technicians in the department, Hertha de Villiers and Alun Hughes, painstakingly laid the bones on the roof of the dissection hall to dry. It was an enormous job to sort the bones and to make sure they went back into the correctly labelled boxes. Tobias was almost in tears. "In my heart of hearts and in the depths of my mind," he said, "I wasn't sure how much mixing of bones occurred at that second stage." For years, there were problems. Researchers would report that a femur or a humerus didn't seem to match. Close to 50 years later, Tobias reflected that "it took years to try to straighten that out and I doubt whether it was ever done completely."[217]

Tobias took heart from the fact that during his time as head of the department, he added another 2 000 skeletons to the collection and that these later additions were all correctly catalogued and labelled. As a result of the disaster, individual bones, as opposed to the skeleton as a whole, began to be marked with their correct accession number.

The way the skeletons were catalogued were often ambiguous. For example, some skeletons were marked "Zulu" and others were catalogued as "South African Negro". Hertha de Villiers, who studied the skeletons, wrote

115

that many individuals represented in the collections were migrant workers who had come to Johannesburg and whose origins were unknown or had not been accurately recorded. In some cases, a person's ethnic category or race had been solely decided by scientists in the department based on his or her surname.

In addition to the Raymond Dart skeleton collection, Tobias inherited the Raymond Dart Gallery of African Faces, which included over 600 life masks and 800 death masks. Just as with the skulls and skeletons, Dart had developed a system of categorising the masks, describing them according to racial type based on anatomical features. Tobias inherited this way of describing the masks, which he did annually with his students. Tobias also inherited the Taung skull. Dart had considered the skull his own private property since 1925 but when he retired, he told Tobias that he was to be its new guardian.[218]

For much of the 1950s, Tobias's research focused on the study of living human beings and human measurement was a major aspect of his career, which included decades of research in the Kalahari and with other peoples of southern Africa. In the late 1950s, he participated in a survey of the people living in the Tonga Valley around the Zambezi River, where the imminent construction of the Kariba Dam was going to affect thousands of people. The governments involved were planning to resettle more than 34 000 people on the Zambian side and about 20 000 on what is now the Zimbabwean side. Tobias gathered data from the population that was going to move and compared it to a control group of people who were living beyond the future water line. He studied 400 adult males and over 600 children. It is unclear why Tobias did not measure any women at the time. He only began gathering data from women in the area in the early 1960s. Tobias took fingerprints, height and weight and measured close to 200 different physical characteristics on each adult.[219]

Five years after Tobias's participation in the French-led Panhard-Capricorn expedition, Wits established the Kalahari Research Committee (KRC), with Tobias as chairman. The intent was "to study the ecological and ethological aspects of Bushmen, with a view of throwing some light on the possible behaviour of early man". Early funding came from the English Nuffield Foundation as well as smaller amounts from the Council for Scientific and Industrial Research (CSIR), the Medical Research Institute, and the Chamber of Mines. The first expedition was led by Dr John Weiner from Oxford, who had worked on uncovering the Piltdown fraud. Over the next two years, seven expeditions produced 25 publications.[220]

After Tobias's expeditions to the Kalahari and to the Tonga Valley, he began to compare these two groups of people. He remarked that the "Bushmen" were "short, light-skinned and relatively hairless, small-boned and delicately built", as compared to "the somewhat bigger and definitely darker-skinned, rather more robustly constructed Tonga". He wondered whether these distinctions could be accounted for by environmental differences or whether they were due to genetic variation. This was a question he carried with him from the 1950s throughout the rest of his life.[221]

At the time Tobias became Dart's successor, he was steeped in Dart's typological paradigm. He had absorbed it fully during his studies in the 1940s and '50s. In a 1958 article, Tobias continued to write about "Europoid features at Mapungubwe". He wrote about the "Bush-Boskop race" and the history of the "Boskop concept".

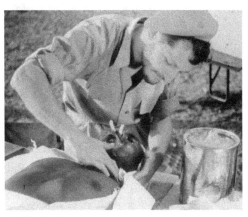

Phillip Tobias preparing an unnamed person for a face mask in the early 1950s. Courtesy of Tobias in Conversation, *published by* Wits University Press. Copyright: School *of Anatomical Sciences, University of the Witwatersrand.*

Hertha de Villiers had begun working in the Wits anatomy department as a technician with Raymond Dart in 1951. She went on to become a professor. When she arrived at Wits, she had already studied anatomy and received her BSc in zoology and physiology from UCT. Originally, she was working with the cadavers in the department, but one day Dart noticed that she looked quite ill so he offered that she start working with his bone tools instead. She began to organise many of Dart's fossils from Makapansgat. About a year into this work, she was thinking of leaving her post in order to become a doctor but when she discussed her plans with Dart, he responded, "Do you want to spend the rest of your life looking at rotten tonsils and ingrown toenails?"[222]

When De Villiers wondered what she might do instead, Dart again had the answer. "Stay here and work with my skeletons" – which was what she did. By the end of 1955, she had measured 315 skulls. She wrote to Tobias, who was in England at the time, saying that she had begun work on the collection of Bantu skulls since "it might be possible to establish differences in the main tribal groups – Zulu, Xosa (sic) and Sotho". She noted features such as the "Glabellar prominence, superciliary eminence, shape of forehead, shape of nasal bones, etc" and she asked Tobias to advise her on how to limit the scope of her study. "I know that I am asking much and perhaps my little panic is premature." By May 1956, she had measured 763 skulls and was keen to begin her statistical analysis for her MSc. "But when I showed the Prof [Dart] what I had completed," wrote De Villiers to Tobias, who was in the US by that point, "he suggested that I not do any statistical analysis for the MSc. but leave that part of the work for a PhD." De Villiers' PhD thesis, which she completed in 1963, was entitled "A Biometrical and Morphological Study of the Skull of the South African Bantu-Speaking Negro". In 1968, she published her thesis as a book: *The Skull of the South African Negro*.[223]

Despite having been greatly influenced by Dart, De Villiers' research was groundbreaking in the sense that she concluded there were no distinguishing features between different cultural or tribal groups in the African, Bantu-speaking populations in southern Africa. She did not try to create an imaginary type and she did not try to explain any variation as differing from that type. Instead, she concluded that there was a "range of variation in a cluster of closely related populations".[224] Slowly, she was starting to turn the department of anatomy's Titanic away from the iceberg of race typology.

De Villiers' methodology also differed greatly from Dart's. Although she followed him in her reliance on measurement of physical characteristics, she relied much more heavily on statistical analysis, which was groundbreaking. She was not as groundbreaking in her use of terminology. While she and Tobias began to become uncomfortable with the National Party government's use of the word "Bantu" instead of the term "African" or "Black", both of them, throughout the 1960s and 1970s referred to the South African "Negro" in their writing. Some 46 years after her book was published, when Hertha de Villiers was asked in an interview I had with her why she had referred to the "South African Negro" in her title, she said: "Bantu is a term describing a linguistic group and I was not working in linguistics."

One of the most striking features about Hertha de Villiers was that she brought rigour and statistical analysis to her science. In addition, despite her lab coat, she carried a sense of glamour and sophistication. One of her Wits students from the early 1980s said: "We all respected her. We were in awe of her."

De Villiers' parents were born in Germany but they met and married in Pretoria, South Africa. When De Villiers went to high school in Cape Town in the 1940s during the war, there were very strong anti-German feelings in the country. She remembers the police arriving at her house one day and accusing her of being a spy and of sending messages. When they saw she was a schoolgirl, they said there must be some mistake and left her alone.[225]

De Villiers' first husband was from the Netherlands and they had a son, Adrian, in 1948. Adrian's father was an alcoholic and, according to De Villiers, had an eye for women, and so the marriage eventually led to divorce. Her second husband, Otto Ericson, was Norwegian (much of Hertha's work at the department of anatomy in the 1950s is credited to Hertha Ericson). When Hertha got the job as a technician, Otto bought her a little biscuit coloured Fiat so she could get to work at Wits from their flat in Bellevue, Johannesburg. One night, Hertha woke up to the smell of gas. Whether it was jealousy or rage or both, Otto had intended for them not to wake up. Fortunately, Hertha got Adrian and herself out of the apartment in time. She alerted the department of anatomy that she wouldn't be coming in that day and found another apartment. Otto visited the next day as if nothing had happened. At one stage, he said to De Villiers, "Don't you think you should pay your own income tax?" "No," she said. "I think I should get a divorce."[226]

Hertha married Bungy De Villiers in 1955. They were married for 40 years until his death in 1995. "He was the only one who lasted," she said, adding that she had learned a lot about human behaviour by then and that she never argued and never shouted. Perhaps it wasn't De Villiers who had changed, but that her new husband accepted her as she was – strong-willed and with a career of her own. Bungy de Villiers was a skilled carpenter and furniture maker so perhaps he didn't feel that he had to compete with his wife the scientist.

In the 1950s and early '60s, Raymond Dart frequently brought his son Galen into the department and De Villiers spent time with him. He was quite mobile and went to the King Edward School for boys. Sometimes the boy would bring his pet budgerigar into the department with him and Dart would give a lecture with the small bird sitting on his head. "The students thought he was mad," said De Villiers, "but there was nothing mad about Dart." Once when Galen was visiting De Villiers' office, he took her cigarettes and threw them out of the window. He became quite aggressive and De Villiers was frightened. One of the technicians came and helped her to take Galen to his father. By the time he reached puberty Galen had become very difficult and the Darts were advised to investigate. They found that he was cystic – that there were gaps in his brain. It was about that time that they decided they had to place him in a hospital.

In addition to Hertha de Villiers, the other person who made the most significant contribution to critiquing typology in South African anthropology was Ronald Singer, who had studied at the University of Cape Town, and who joined the department of anatomy at UCT in 1949 as a lecturer. Given his experience in genetics, he began to question race typology as completely off

119

the mark. In 1958, he published an article in *Man* entitled "The Boskop 'Race' Problem". He complained that on the basis of one isolated skull fragment found over 45 years before, in 1913, physical anthropologists had created a "race". The exact geological location of the fossil had not been clarified and the surrounding fauna had not been documented, but scientists had continued to speculate and give enormous importance to this one piece of bone. Not only had scientists constructed a "race" around it, he said, but they also continued to "detect occasional features of this 'race' in the faces of living South African *individuals*".[227]

Quoting a paper that Tobias had written in 1955, Singer asked: "What does it mean when we read: 'When the Kakamas folk arrived in Southern Africa, they found a numerous pygmaeopedomorphic people, the Bushmen, compounded mainly of the Bush and Boskop physical types, with a slight admixture of gerontomorphic strain and possibly an earlier Europoid strain; they also found some relatively unhybridized Bush, Boskop and Gerontomorphic types.'" What did a "'relatively unhybridized' type" look like, he asked. What was a "'slight admixture of a gerontomorphic strain'"?[228]

Singer also argued that physical anthropologists had reached general conclusions about an imaginary race and type based on a very small collection of fossil material that had been gathered across thousands of miles of southern Africa. "So called Boskopoid features", such as excessive thickness of the skull or a wide palate, he maintained, could not be ascribed directly to the one Boskop fossil find. Singer's paper cited Tobias's writing, as well as other South African scientists such as Broom and Dart, and he suggested that physical anthropologists based in England such as Arthur Keith had also based conclusions about the Boskop race on faint evidence. He correctly placed the responsibility for the original concept of a Boskop physical type with Dart, and rightly suggested that it had been Dart, and then his students, who had promoted the concept for more than 30 years.

Singer concluded that any further reference to a Boskop race as the ancient ancestor of the Bushmen was untenable. "It is now obvious that what was justifiable speculation (because of paucity of data) in 1923, and was apparent as speculation in 1947, is inexcusable to maintain in 1958."[229]

Tobias was reluctant to give up the concept of typology and its implications; he recalled that Singer's remarks "stung at the time".[230]

Despite Tobias's hesitance in letting go of the "Boskop race" and race typology, he, Singer and De Villiers were a new generation of physical anthropologists in South Africa. They expanded their thinking as a result of exposure to genetics, which had not been an area of attention for their predecessors. All three were introduced to "The New Physical Anthropology" of the 1950s during their travels in the United States and Europe. This new approach was post-typological and relied more heavily on statistics, genetics and population dynamics to explain variation rather than on type. Of the three, Tobias was by far the most reluctant to let go of the typology paradigm.

This new physical anthropology had grown out of new thinking that was often referred to as "the synthesis", as it represented the bridging of a gap between genetics and palaeontology. In the 1930s and 40s, internationally, there had been a growing alienation between the two disciplines. In 1944, palaeontologist George Gaylord Simpson of the American Museum of Natural History, had described it as follows: "Not long ago, paleontologists felt that a geneticist was a person who shut himself in a room, pulled down the shades, watched small flies disporting themselves in milk bottles, and thought that he was studying nature ... On the other hand, the geneticists said that paleontology had no further contributions to make to biology ... and that it was a subject too purely descriptive to merit the name 'science.'" Another way that scientists viewed the two approaches was that genetics was focused on lower order molecules and genes and that palaeontologists were looking at higher order populations and species.[231]

It was in the mid-1940s that a "synthesis" began to form around evolutionary theory, bringing the two areas of science together by integrating the Darwinian idea of natural selection with the changing regularity of genes in the gene pool. The synthesis accepted that evolution was a gradual, long-term process of small genetic mutations which over time accumulated to create larger changes that were passed on from generation to generation. The synthesis also embraced the ideas that a species consisted of unique individuals and populations and that there was no such thing as an ideal type.

One of the evolutionary biologists most responsible for the new synthesis was Ernst Mayr, a German-American scientist based at the American Museum of Natural History. At a breakthrough conference in the US in 1950, Mayr made the point that the process of classifying hominid fossils had been different from the process in other areas of zoology. He claimed that all hominid fossils that had been found to date should be lumped into the genus *Homo*. This new thinking related to the process of classifying hominid fossils and argued for more lumping and much less splitting.[232]

This thinking arrived too late in South Africa to have any impact on the earlier generation of physical anthropologists, Dart included, but the typologists had left their legacy and it did shape some of the next generation. This did not mean, however, that they all got along.

In 1958, Ronald Singer applied for Raymond Dart's post at the department of anatomy at Wits. He was one of three candidates shortlisted for the position but he lost out to Tobias. This was after Singer had applied for the chair of anatomy at UCT in 1955, but the post had been given to Lawrence Wells. After that disappointment, in 1956 Singer applied to be the chair of anatomy at Stellenbosch University's medical school. The selection committee chose him as a leading candidate but the university senate did not confirm his appointment because they did not want a Jewish man whose first language was English to hold such a position at an Afrikaans university. They gave the chair to a member of the Broederbond. Singer then applied for the chair at Wits and was in competition with Tobias and Joe Weiner, before Tobias was chosen for the job. This series of events contributed to Singer's decision to take a position at the University of Chicago, and move to the United States.[233]

By the time Tobias took over from Dart, his relationship with De Villiers had soured. It was a tragedy, thought De Villiers, when Dart retired in 1958. She would have been happier with either of the other candidates – Ronald Singer or Joe Weiner. Dart must have supported Tobias as his successor, but De Villiers quoted Dart as once saying that "Tobias [had] a razor sharp mind to cut out and steal other people's ideas". She herself often described him as a "nasty man". Although it is unclear what contributed to their falling out, one possible clue is that in 1955 De Villiers wrote a letter to Tobias while he was in England, asking for his advice about her research with the Dart skeleton collection and she signed the letter, "Love Hertha". Uncharacteristically, it took Tobias months to write back and when he did, he signed his note merely "Sincerely Phillip".[234]

On another occasion, Tobias wrote an internal departmental note to De Villiers: "I noticed the extraordinarily interesting sacrum of the skeleton you had brought up for Professor Howells. I think it was the skeleton of Keri-Keri, (sic) or of one of the other 'Bush' specimens." Tobias asked De Villiers to bring the "abnormal" sacrum to one of his students to study. Two days later, De Villiers responded in a memo, telling Tobias that the sacrum could not have belonged to /Keri-/Keri because her skeleton was articulated.[235]

At that time /Keri-/Keri's family was living in a township outside Upington. The apartheid classification system had assigned them a "Coloured" identity, with no origins, no history and no past.

10

An Indifferent Kindergarten

In August 1959 Louis and Mary Leakey presented Mary's fossil find from Olduvai Gorge at the Pan African Congress on Prehistory in Kinshasa. They had been working in Olduvai for 30 years, unearthing many stone tools, but without finding a significant hominid fossil, and so this was a major breakthrough. At the congress it was numbered OH5, Olduvai Hominid 5. The Leakeys had decided to place Mary's new fossil in its own genus and species – *Zinjanthropus boisei* – and they nick-named it "Dear Boy" or "Zinj". There was great excitement and debate about whether the Leakeys had done the right thing to be splitters or whether they had been too hasty in this assignment. Some people thought the fossil should be placed in the same genus and species as the South African *Australopithecus africanus* specimens such as the Taung child skull and Mrs Ples. In each case, in addition to the discussion about the categories for genus and species, these nicknames personified the fossils and portrayed them as having a strong connection to the corresponding scientists – Dart, Broom, and now Leakey. Mary Leakey was often left out of the equation. One man, one fossil.

Phillip Tobias, who had just taken over from Raymond Dart at the department of anatomy at Wits, chaired a session after the Leakeys' presentation and fielded questions from the floor. At the end of the session, a small number of physical anthropologists were invited to join the Leakeys for a special viewing of the fossil. They set up a table in the garden, which was surrounded by the lush forest of the Congo. Including the Leakeys, there were about ten people around the table, Raymond Dart, Lawrence Wells, and Phillip Tobias among them. Louis Leakey showed the group all the separated pieces of the fossil bones arranged on the table. "I'd like to know what you think of them," he said. Everyone was picking up the bones and examining them. Then one of the group, Sherwood Washburn, turned to him and said: "Well, this is an extraordinary piece of good luck. But must you give it a new genera?" Leakey began to go red in the face and a few beads of sweat appeared on his forehead. Another viewer said, "I don't think so." As the fossil went around the table and each scientist had something to say, the tension was palpable.

When it was his turn to comment, Raymond Dart picked up the fossil and said, "This is a remarkable find." He paused for effect, seriously reviewed the bones, and looked up with a sly smile. "I wonder what would have happened if Dear Boy had met Mrs Ples one dark night?" The entire group around the table fell into laughter. Tobias later reflected that Dart's comment was brilliant, not only because he diffused the tension with humour, but also because he was suggesting that if Dear Boy and Mrs Ples were in a different genera, they would not be able to procreate; but if they were in the same genera and species, they would have been able to create little hominid children.[236]

After the laughter, the group dispersed but the Leakeys asked Phillip Tobias to stay behind. They put the pieces of fossil back into its box and then took a walk together in the forest. The Leakeys asked Tobias if he would like to undertake the scientific description of the specimen. Tobias recalled how his mouth went dry and his knees went weak. "It was simply a marvellous opportunity because nothing like that had ever happened to me." Up until that point, Tobias had confined his research to the study of the living people of Africa, including the people of the Kalahari and the Tonga Valley, whom he had been studying for almost a decade. "Louis and Mary brought me into the field of paleo-anthropology," he said. "That was a turning point in my career."[237]

One of the reasons the Leakeys chose Tobias was because he had already done Louis Leakey a favour. During Tobias's time in England, he had examined one of Leakey's hominid fossil specimens. This was the Kanam jaw fragment that Leakey had found in Kenya in the early 1930s. Tobias was particularly interested in the controversy over the Kanam mandible that was said to have a bony chin. Unlike *Homo sapiens*, most hominid fossils that had been found by then illustrated a receding chin – except for this Kanam piece. Back in the 1930s, Leakey had claimed that it was an ancient fossil, but it exhibited this modern trait. He had been severely critiqued. When Tobias examined the Kanam jaw 20 years later, he thought that the chin was in the wrong place and slightly off centre. He wondered whether it wasn't a "chin" at all but rather a pathological growth of bone. After seeing that there was a fracture, he concluded that there had been an infection and resulting bone swelling. Tobias presented his findings at the congress in Kinshasa in 1959, helping to clear Leakey's scientific reputation. Leakey was grateful.[238]

The association with the Leakeys would shape Tobias's career for decades to come. Although they took Dear Boy back to Kenya with them, not long afterwards they allowed Tobias to take it to South Africa, where he kept it while he examined, photographed and X-rayed it. He compared it with other South African fossils and with fossil casts from other parts of the world.

Earlier that same year, in March, Louis Leakey had written to John Robinson, who was still at the Transvaal Museum, to invite him to visit Olduvai and to look at the archaeological site there. They had been in correspondence since 1952 discussing their views on the field of human origins. The two often disagreed but their exchanges were collegial. Robinson didn't immediately accept the invitation, but he wasn't surprised when Leakey wrote to him again in July, saying he had found an early hominid tooth at Olduvai Gorge. He sent Robinson photographs of the molar tooth and asked for his opinion. Soon thereafter, Mary Leakey made her discovery, which they announced in *Nature* as a new species, *Zinjanthropus*. Robinson was certain the Leakey find was in the same species and genus as the South African fossils. If John Robinson had been standing around that table in Kinshasa, no doubt he would have added to the tension. He strongly agreed with Sherwood Washburn's question: why did the Leakeys have to put their fossil in a new genus and species?[239]

Just before the congress in Kinshasa the Leakeys had made a trip to South Africa. They went to Johannesburg first to visit with Raymond Dart and Philip Tobias. They examined the Taung child skull and the growing number of australopithecine fossils from Makapansgat. Both Tobias and Dart welcomed the Leakeys and their fossil with excitement. Dart told Leakey: "I am so glad that this has happened to you of all people."

Hertha de Villiers remembers seeing Mary and Louis Leakey when they visited the department at Wits. She described them as being treated like "God and Goddess". She remembers Louis Leakey sitting on a bench examining a fossil and proclaiming his views loudly to those around him. While De Villiers could not remember exactly what he said, she did recall that he had a "whacking great hole in his sock".[240]

The Leakeys then went to Pretoria to see the Transvaal Museum's collection but their meetings with Robinson were strained. From the start, he was irritated that they wanted to present *Zinjanthropus* as a new species and he tried to convince them it was the same as the South African australopithecines.

Dr Bob Brain, who was Robinson's student at the time, vividly remembers the tension between the Leakeys and Robinson. He recalls that the Leakeys recognised that the East African and the South African fossils could belong to one species, but because they had already described theirs as a distinct species in *Nature*, they didn't want to change their position. As a result of Louis Leakey's energy and excitement, he was well received by audiences, more so than Robinson, who had a shy and serious demeanour. Leakey travelled to London with a cast of *Zinjanthropus*. Then he was offered support from the Wenner Gren Foundation to conduct a lecture tour in the US at 17 universities. While Leakey was becoming a celebrity, Robinson felt his work and earlier travels were being overlooked.

While he was on his US tour, Leakey met the president of the National Geographic Society, who recognised that he had the personality and appeal that could promote an understanding of science. The society decided to award Leakey a large grant to continue his work at Olduvai and, in 1960, they asked him to write an article for their magazine about the finding of *Zinjanthropus*.

Two years earlier, in January 1958, Robinson had received a letter from *National Geographic* asking him to write an article about the "Man-Ape Finds of South Africa". They sent him ten rolls of film so that he could take photographs to accompany his piece. Robinson's focus throughtout 1958, however, was on other writing and research projects, and so he did not write that more popular article for *National Geographic*. By late 1958, although he had sent off the developed photographs of the fossils and the fossil sites, he had still not completed a draft article. Finally, in March 1959, he submitted a draft, but it maintained a scientific, rather than a popular tone, and included large amounts of facts and data. The editor came back to him asking for significant changes in the tone and style. Again Robinson delayed on the project.[241]

Then, in January 1960, Robinson received a letter from *National Geographic* returning the photographs he had sent to the magazine over a year earlier. The letter stated: "As you may know, the story has not yet been scheduled for publication because of the wealth of other unusual and timely material that has been acquired and because of other archaeological subjects that have been brought to light recently." In June 1960, they informed Robinson that instead of his work, they had sent an article by Leakey to press. Leakey's article, "Finding the World's Earliest Man", was published in *National Geographic* in September.[242]

One of the projects that had distracted Robinson in 1959 was his exploration of the role parapsychology might play in science. While Broom was still alive, Robinson had learned a great deal from him as a mentor, not only in the sciences and evolutionary research, but also on a personal level. Robinson had been raised a devout Methodist, but while he was at university he began to question his religion and inclined towards the view that the scientific method was the best way to expand his knowledge of the world around him. It was Broom who led Robinson back to religion. On Broom's suggestion, he began attending lectures at the Theosophical Society in Pretoria and to study the philosophy of theosophy. Within five years Robinson had begun giving talks on the topic of death and reincarnation in which he presented death as the transition from one form of existence to another, and therefore as nothing to fear. He also began to question whether the scientific method was able to explain all that exists in the world. He saw science as explaining the material world, but he looked to his spiritual side to explore non-material aspects of the universe.[243]

Through his membership in the Theosophical Society, Robinson heard about the work of Geoffrey Hodson, a well-known clairvoyant from New Zealand.[244] He was impressed with Hodson's systematic work in a range of fields, including forensics, the power of music, and also communicating with the dead. Robinson posted several fossils to Hodson and asked him if he could see into the past lives of the australopithecines. He asked him to record on tape the description of the mental images that came to mind when he held the fossils in his hands. Robinson was intrigued by Hodson's descriptions so they made an arrangement to meet in South Africa and carry out further experiments together.

In March 1960, Robinson drove from Pretoria to Sterkfontein with Hodson, who was then in his 70s. He had looked forward to this moment for months. Both men wore formal suits with shirts and ties and had carefully shined shoes. Robinson did not tell Hodson where they were going or exactly what they were going to do. He didn't want too much information to shape Hodson's thinking.

They arrived at the Sterkfontein field site before dark and parked the car. Robinson chose a location near a short stone wall. He took a box of fossil specimens out of his boot and set it down on the wall. Then he laid a blanket and pillow on the ground, set up a tape-recorder and took out his notepad and pen. As instructed, Hodson lay down on the blanket and closed his eyes. Robinson placed a fossil specimen on his forehead and waited. Hodson said he did not know what it was but that he felt it was something like a stone. Robinson told him it was the fossil of an "ape-man" but nothing more. Concentrating intensely, Robinson held all the information he could in his mind about this "ape-man". He wanted Hodson to give him as much information about this creature as he could. After some minutes, Hodson said: "I am very sorry, I can see no ape-man at all. I am stuck with an ape!" Robinson asked him to persist and for some time they continued in silence. Hodson continued to feel the fossil as an ape rather than as an ape-man. Robinson eventually told him that the fossil was "from an ape" and although Hodson later described this as "a trick and laying a trap for the poor investigator", he was pleased with Robinson's technique to determine whether his information was valid.

Robinson continued the session by setting another fossil on Hodson's forehead and asking him questions about that particular creature. Hodson would speak at length and Robinson would take notes. They didn't discuss any of Hodson's analysis until after the series of tests were completed. At times, Robinson would ask questions such as "What did this creature eat?" or "What kind of teeth did the creature have?" Robinson later wrote that Hodson "never misidentified a specimen or gave conflicting statements about a specimen that had been presented more than once. As far as I could determine, his information was always accurate and he gave me a strong impression of complete reliability." There was one occasion in which Hodson's response puzzled Robinson. This was when Hodson repeatedly told him that he saw the small-canine creature eating after having climbed a tree. Robinson had earlier concluded that this particular creature – *Paranthropus* – lived on the ground. It was only several years later, after the collection of more fossil material, that he found that certain features of the skeleton did suggest that *Paranthropus* might well have lived in trees, as well as on the ground.

Robinson and Hodson made a number of trips to Sterkfontein together and carried out numerous sessions over a period of weeks. Robinson offered up a range of fossils to the clairvoyant, many of which were over two million years old, and he recorded everything and wrote up a report. His view was that these experiments opened up a compelling world of hidden forces and energies. He believed clairvoyant powers could be of great value to science.

Meanwhile, the Leakeys and East Africa had burst anew onto the international scientific scene. Not only was the scientific community impressed, but the duo attracted public attention as well. And having chosen Phillip Tobias to help them describe their famous fossil, he, too, was now at the centre of attention. As a result, John Robinson felt personally isolated, believing that once again the South African australopithecines were not being fully recognised. Robinson was methodical, meticulous and cautious about his scientific analysis, whereas Leakey was exuberant and ready to move quickly to his own conclusions. Leakey's personality was more akin to Raymond Dart's. Robinson didn't easily catch the attention of the public imagination.

At that time as well, the South African government was moving ever more strongly into the implementation of apartheid legislation. Attention to East African discoveries suited the international audience, which had begun to take notice of the political developments in South Africa.

Politically, the early 1960s was a tumultuous time in the country. Apartheid was no longer in a period of consolidation as it had been in the 1950s but had reached full force. On 21 March 1960, police fired more than a thousand rounds of live ammunition against peaceful protestors refusing to carry passes – the hated identity document – in their march on a police station in Sharpeville, near Johannesburg. Sixty-nine people were killed that day and the deaths made international headlines. Chief Albert Luthuli, the president of the African National Congress, published his autobiography *Let My People Go* in 1962, writing, "We Africans are depersonalised by the whites ... Our humanity and dignity is reduced in their imagination to a minimum. We are 'boys,' 'girls,' 'Kaffirs,' 'good natives' and 'bad natives'. But we are not, to them, really quite people scarcely more than units of their labour force and parts of a 'native' problem."[245]

In 1963, Nelson Mandela made his speech from the dock saying that he had dedicated his life to the struggle of the African people. "I have fought against white domination, and I have fought against black domination. I have cherished the ideal of a democratic and free society in which all persons live together in harmony and with equal opportunities," he proclaimed. It was an ideal for which, he said, he was prepared to die. After the Rivonia Treason Trial much of the leadership of the liberation movement was incarcerated or had been driven into exile.

It was in this context that Tobias increasingly began to spend less time in South Africa, and more time in East Africa with the Leakeys. Just at the point when South Africa might have become the focus of international popular imagination regarding fossils and human origins, the country's politics intervened. The world became more interested in South Africa for its political developments under apartheid and less for its scientific efforts. Scientists were more comfortable focusing on East Africa.

Fortuitously, the Leakeys had found more hominid fossils. In 1960 they had announced another find from the Olduvai Gorge in Tanzania. While *Zinjanthropus* had been similar to the australopithecines of South Africa, the Leakeys were convinced that this new fossil, found by their son Jonathan, was the first to represent early *Homo*. Tobias did not agree with them at first, believing that the new fossil should also be described as *Australopithecus*. "One specimen does not a species make," he said. He thought it was possible that one fossil might represent "one freakish individual" and that it was important to find more specimens. The Leakeys went back to Olduvai to look for more fossils. They found more specimens there, which indicated that they were indeed *Homo*. The newly found specimen seemed to have a larger brain than *Australopithecus*.[246]

Over the next four and a half years, Tobias spent time with the Leakeys and studied first the one fossil, and then the additional similar fossils they found. It took years, but eventually he changed his mind and agreed with Leakey that the new fossils should be placed in the genus *Homo*. Some people thought that Tobias had been convinced by Louis Leakey's force of personality, but Tobias insisted that he had arrived at his conclusion "due to the sheer weight of the evidence". The fossils had teeth that were smaller than *Australopithecus africanus* and more like human teeth. The new fossils also had a larger brain size. In January 1964 Tobias wrote to Louis Leakey to say "I am now convinced that we have a new species." He was also convinced that the new species belonged to the genus *Homo*. Leakey was thrilled.[247]

After they had reached common ground, the Leakeys and Tobias felt the new species needed a name. Tobias decided to ask Raymond Dart for his ideas since he had a way with words. He told him the new species had a larger brain and smaller teeth than *A. africanus* and also that it was likely it had been responsible for making the Oldowan tools that Mary Leakey had found near the fossils. Tobias was somewhat concerned that he had asked the man who invented words like Aus-tra-lo-pith-e-cus and Os-te-o-dont-o-ker-a-tic for assistance, but this time Dart came back with a name that was short and sweet. He suggested they call it *habilis* – a Latin word meaning manually dextrous. Tobias liked the name, recognising in it the English word "ability" and both he and the Leakeys agreed with Dart's suggestion. They named their fossils *Homo habilis*.[248]

Once again, as with *Zinjanthropus boisei* in 1959, Mary and Louis Leakey asked Tobias to describe their *Homo habilis* fossils in detail. Tobias decided that given time constraints, he would have to concentrate on the crania, the brain, the teeth and the jaw. He suggested they invite John Napier from London to describe the hands and feet. Leakey, Tobias and Napier published their paper jointly in *Nature* in 1964.

"All hell broke loose," recalled Tobias. John Robinson was one of the first to critique the new species. As Tobias remembers it: "Almost immediately John Robinson sprang to the attack." He argued that Tobias, Napier and Leakey didn't understand the principles of taxonomy. "It was a bit rude," said Tobias, "but I suppose he was entitled to say that." Tobias responded to Robinson in *Nature* defending his stance. Robinson published a second critique, to which Tobias responded. Robinson came back at him a third time. "And then I got sick and tired of this going back and forth," said Tobias, "because it was beginning to be a little bit nit-picking and a little bit personal. So I abandoned it." Tobias felt that many other scientists took their position based on Robinson's opposition so that very few people accepted *Homo habilis* initially. Bob Brain remembered a time when he brought Tobias and Robinson together for lunch and urged them to work out their differences in person rather than in the pages of scientific journals. According to Brain, his efforts were not terribly successful.[249]

Tobias felt that, with *Homo habilis*, he spent 16 years in the wilderness without acceptance. He truly was convinced the fossils represented a new species but many others were not. Questions continued in terms of where and how early *Homo* had separated from its precursors. How did *Australopithecus* differ from *Homo*? Which one was ancestral to *Homo sapiens*? What were the various factors that contributed to the evolution of humankind, including physical traits such as bipedalism and brain size? What about the role of language? And how did cultural factors relate, such as the making of stone tools and the first use of fire?

At the same time as these questions about human evolution were being explored, internationally another question was on the table. The question concerned the origin of race. Although the focus on race typology had begun to wane, a new set of questions had arisen about the relationship between human evolution and race. Carleton Coon, a physical anthropologist from the University of Pennsylvania, published *The Origin of Races* in October 1962. He believed that the first evolutionary step from *Australopithecus* to *Homo* happened not in Africa but in Java in south-east Asia with the formation of *Homo erectus*. "If Africa was the cradle of mankind, it was only an indifferent kindergarten. Europe and Asia were our principal schools."[250]

Coon took on several aspects of Franz Weidenreich's theory that different racial groups evolved separately around the world. He argued that over 500 000 years ago, the pre-cursor of humans existed as a single species, *Homo erectus*, and that they existed in five different geographic regions. Each of the five groups of *Homo erectus* then evolved separately "by one genetic process or another" at different rates into five different groups of *Homo sapiens* representing five different race groups that he considered subspecies: the Caucasoid, Mongoloid, Australoid, Congoid and Capoid. (Capoid was Coon's word for the peoples of the Cape of Good Hope.) Coon's book was applauded for its comprehensive and meticulous review of the hominid fossil record. However, his overall premise of separate evolution of different races was hugely controversial. He wrote: "Each major race had followed a pathway of its own through the labyrinth of time. Each had been molded in a different fashion to meet the needs of different environments, and each had reached its own level on the evolutionary scale."[251]

Phillip Tobias remembered his first meeting with Carleton Coon in 1955 at the Pan African Congress on Pre-History in Livingstone, Zambia. He remembered Coon as a brash but jovial and hard-hitting personality. The next year, when Tobias was in the US, he met Coon again, this time at an American Association of Physical Anthropologists conference in Chicago. Coon was in the audience when Tobias delivered a paper about the "evolution of the Bushmen", where he suggested the "Bushmen constituted a peculiarly African line of human evolution". He said that if the earlier fossil record with heavy brow ridges and a large jaw had changed to the smaller, delicate and "virtually infantile characteristics of the San", then perhaps the remodelling took place first, followed by the change in size. As soon as Tobias finished speaking, Carleton Coon leapt up to contribute during discussion time. According to Tobias, Coon was excited to think that if these changes had happened rapidly, then scientists might have to revise their thinking about human evolution.[252]

In preparation for writing *The Origin of Races*, Coon continued to communicate with Tobias and many other palaeoanthropologists during his phase of worldwide research in the late 1950s. Coon was coming to the conclusion that the relationship between race and human evolution was clear – "human evolution was racial evolution".[253] As Coon's theory entered the debate in the early 1960s, Tobias began his own exploration of the concept of race.

11

The Meaning of Race

In 1960, after Tobias first met the Leakeys in Kenya and before he wrote his *Nature* paper on *Homo habilis*, the Union of Jewish Women in Johannesburg asked Tobias to give a lecture on the topic of race. They advertised the event widely and there was a huge crowd at the lecture theatre in the medical school. People from the Institute of Race Relations (IRR) attended the lecture and came up to Tobias afterward suggesting that they publish it, which they did in 1961 with the title "The Meaning of Race". The Institute had been established in 1929 by white academics, as a liberal organisation focused on advocacy and research. Their publications were written by political analyst, economists and sociologists and focused on the social and economic conditions of black people in South Africa and the impact of government policies. They produced the first, of what would become an annual, Survey of Race Relations in 1947. Thirty years after their founding, apartheid weighed heavily on the country and Tobias's lecture appealed to the Institute as a way to encourage a discussion about race.

This was the first time Tobias spoke publicly on the issue and published for a broad audience. The debate that had begun with the UNESCO Statements on Race had continued, so Tobias's article was of interest both locally and internationally. Tobias argued that the idea of race had become a "national neurosis" in South Africa. "Every single aspect of our life has come to be dominated by the thought: to what group or race does *that* man belong?"[254] That was a question Tobias and Dart had been asking within the department of anatomy over the previous 40 years: evidenced by the skeletons, the skulls, the face masks, and the expeditions to measure and describe living people of southern Africa. But now Tobias was describing the political, cultural and social reality in South Africa.

Tobias wondered whether a scientist like himself, "whose special field of research and study is *race* is likely to be accused of meddling in politics merely by talking about the scientific aspects of race". He stated that it was the duty of a scientist like himself, who had made race his special study, to speak out about "the scientific truth about race".[255]

In offering a definition, Tobias turned to biology. He explained that mammals, birds and plants were divided into species and that species were often divided further. He also reverted to the concept of a type. "The 'typical' representatives of the different races are simply those which are most different from the others. But members of one race can and do interbreed with members of another race, producing intermediate forms."[256]

Tobias told his white South African audience that race was a valid "biological concept which helps us to bring order out of the otherwise meaningless range of human variation". He explained that "confronted with an enormous diversity of different-looking people, anatomists and physical anthropologists have classified the varieties into a number of races." It was generally agreed, he said, that the first draft of the UNESCO Statement on Race in 1950 had gone too far in claiming that race was a fallacy. He lamented the fact that not a single physical anthropologist or geneticist had been included in the seven teams that were created by the minister of the Interior to classify races in South Africa. But he did not suggest what he would have contributed had he been invited to join one of those teams.[257]

In direct reference to Dart, Tobias wrote that scientists in the past had tried to relate living peoples "to idealised racial types" and that "some anthropologists (had) in a measure, been responsible for the over-emphasis on race purity". And he spoke to Ronald Singer's critique of the "Boskop race": "Let us in South Africa beware lest we become race-purity-drunk and go mad in search of the entirely mythical El Dorado of race purity."

Appearing to get somewhat confused when he discussed "racial hybridisation", Tobias quoted Raymond Dart's work, suggesting "that living Bushmen of the Southern Kalahari included strange foreign facial types, Mediterranean, Armenoid and Mongoloid". Then, referring to the African face masks, he said that recent studies at the department of anatomy at Wits had revealed that there was a "variety of racial elements woven into the faces of various South African groups". In keeping with what he had written in 1953, Tobias said "miscegenation (had) created a large hybrid population, as in the Coloured people of South Africa" and suggested that the "the Coloured population (was) emerging as a new race".

Tobias acknowledged that many scientists had worked hard to show that there was a hierarchy of race. He explained that the human line split off from other primates possibly 10 million years ago, and he said that "the modern races of man developed only 25 or 50, 000 years ago". His conclusion was that it was "palpably meaningless to compare living races of man with living apes in order to arrange them in some sort of hierarchical order ...We must conclude that there is no scientific basis for arranging the living races of mankind in a series from the more ape-like to the more god-like."[258]

It would have been very helpful for Tobias to spread this important message far and wide, because many South Africans continued to hold serious misconceptions for decades. The fact that human evolution was not taught in schools contributed to the problem.

While Tobias's predecessors had argued that biological race determined culture, Tobias distanced himself from this belief in his paper but he still used the language of diffusion. He argued "race and culture are totally separate concepts, that in time past cultures have diffused from race to race". In closing, Tobias said that the term "race" was heavily charged emotionally and politically. "It is in the name of race that millions of people have been murdered and millions of others are being held in degradation. *That* is why *you* cannot afford to remain ignorant about race."[259]

In the early 1960s, Tobias had already taken a stand against apartheid. However, overall, this lecture illustrated that he remained quite conventional in his thinking. He had not yet broken from race typology completely, and the lecture showed that he was not ready to reject the relevance of racial classification.

In the same year that Tobias's "The Meaning of Race" was published by the SA Institute of Race Relations, Hertha de Villiers published a paper in the *South African Journal of Science* entitled "The Tablier and Steatopygia in Kalahari Bushwomen". This was a topic that she continued to research and speak about throughout the 1960s. The Latin word "tablier" was defined as "an elongation of the labia minora" and steatopygia as "the localized accumulation of fat over the buttocks". In the paper, De Villiers wrote that she participated with Tobias in the Nuffield-Witwatersrand University expedition to the Kalahari in 1958. This was when she had taken most of the measurements – from 54 females – for her paper. De Villiers wrote that during the Nuffield expedition of 1959, Professor Phillip Tobias "measured the tablier on a further twelve individuals ... These included six adolescents."[260] Tobias, writing in as late as 1975, recalled how "we were able to make studies of this fascinating phenomenon and to see it being taken out from the withdrawn position. In its full extent, it is greatly elongated and hangs well below the labia majora."[261]

There was an extreme contradiction between Tobias's opposition to apartheid in support of human rights and his scientific practices which offered no dignity to African women. De Villiers and Tobias had continued the centuries-old preoccupation with Khoisan women's most private parts. They were not at European hospitals doing comparative research with young white adolescent girls.

In 1961 Prime Minister Verwoerd withdrew South Africa from the Commonwealth and this withdrawal had an impact on Tobias's funding for his Kalahari Research Committee. The Nuffield Foundation, which had funded the committee in the past, confined its support to countries in the Commonwealth and as a result South Africans were no longer eligible for funds. After a long period of negotiation, however, Nuffield agreed to continue funding Tobias's work and Wits led 14 more expeditions, involving more physical measurement, and blood and hair samples. Their goal, wrote Tobias, was not only to understand the biology of the San, but also to look, still, for their potential link with early man.

The paradox of Tobias's opposition to apartheid and his scientific practices came through in another instance, when he led another expedition, this time to Campbell in the northern Cape – not far from Douglas, where Robert Broom used to live. The purpose of this expedition was to exhume skeletons from the graves of people of Griqua ancestry. It was motivated by Tobias's continued efforts to collect other "types" of skeletons, such as the Griqua, for the Raymond Dart collection at Wits. When this new group of skeletons was added to the collection, contextual information and biography were not included. Scientist researchers labelled them merely "Griqua skeletons".

It was 18 April 1961. Tobias's interest in this particular exhumation was because they were looking for the skeleton of a known personality who had died 103 years earlier, in 1858. Captain Cornelius Kok II was a Griqua leader whose direct descendants were members of the Griqua community living in Campbell. Specifically, Tobias was especially interested in whether the Griqua were descendants of the Khoisan, the European settlers or the Bantu-speaking peoples, or whether they were an "admixture". The Griqua and their distinct sense of identity had been largely ignored by the South African government. However, anatomists and physical anthropologists, including Tobias, were interested in what their physical attributes might reveal about their biological ancestors, similar to the way Aleš Hrdlička had investigated skeletons in the US decades earlier, and also similar to the way Dart had been interested in /Keri-/Keri in the 1930s. According to a local newspaper article at the time of Tobias's 1961 expedition, upon seeing a member of the Griqua community at the exhumation – the 70-year-old Niels Watermond – Tobias described him as "a wonderful link with the primitive".[262]

Abraham Kok and his father Adam Kok IV were also at the graveside that day. Cornelius Kok II was Adam's grandfather and Abraham's great-grandfather. Adam Kok IV didn't actively participate in the excavation process because he was 72 years old but 29-year-old Abraham stood by, interested in how the scientists from Johannesburg were going to address the job.

Abraham Kok at the excavation site at Campbell in 1961. Courtesy of AJB Humphreys.

Tobias had originally heard about the grave site from his colleague Dr Gerhard J Fock from the McGregor Museum in Kimberley. Fock had been alerted to the graves by a local farmer, Basil Humphreys, whose farm was just outside Campbell. The initial intention of Basil Humphreys and Adam Kok IV was to move the remains of Cornelius Kok II to a more accessible place near the mission church.

Humphreys' family had lived in Campbell for several generations and Basil had taken an interest in the town's history. According to written records, the history of the settlement of Campbell dates back to 1805 when a group of Griqua people travelled from Klaarwater, which is now known as Griquatown, into territory that had been settled by Tswana people near the town of Kuruman. After a visit by the Reverend John Campbell in 1813, the town's name was changed to Campbell. Cornelius Kok II was among the first Griqua people to live in the area and in 1816 he was declared the Captain of Campbell.

By 1961, when the exhumation took place, the Griqua population had gone from being reasonably prosperous to near abject poverty. In addition, their identity was a clear challenge to the apartheid ideology's emphasis on racial classification into one group or another. As early as 1953, government officials realised that this would prove difficult in many cases, especially when trying to distinguish between African people and "Coloured" people. The 1957–58 survey of the Institute for Race Relations reported that the Griquas of the Northern Cape were a mix of "White, Hottentot, Bush and a little African".[263] The report concluded that the Griqua were a defined group with a distinct appearance and language. These statements were part of what had led Tobias as a physical anthropologist to become interested in the Griqua.

It is important to point out the tragic irony that government classification

137

did not allow people described as "Hottentot" or "Bushman", whose ancestors had lived in southern Africa for thousands of years, to be described or classified within the racial category "African". This fact has continued to affect the language used throughout the country today, more than 50 years later, and more than 20 years after official racial classification ceased.

In 1955, an apartheid Population Registration official visited Kimberley and classified Griquas as African. From that point onward, they had to carry reference books, to obey curfews, to pay poll tax and to draw pensions at a lower rate than people who were classified as "Coloured". If someone told a government official that he or she was Griqua, they would be classified as African. However, many people also qualified for classification as "Coloured" and carried both sets of documents, one requiring them to pay poll tax, and the other exempting them from it. Some people stopped saying verbally to government officials that they were Griqua because it would result in them being subjected to legal penalties. While it was in Griqua people's practical interest to be classified as "Coloured", to do so also subsumed the Griqua identity so that it no longer remained distinct. By the 1960s most people in Campbell who had previously called themselves Griqua were described by others as coloured.[264]

Basil Humphreys recalled that not only was Tobias excited about the possibility of uncovering a 100-year-old well-preserved Griqua skeleton, but also that he was pleased the descendants of the deceased would be "at hand for comparative study".[265]

Tobias had decided to exhume three graves, and to explore one of the graves first that was not that of Cornelius Kok II. He and his assistant removed the pile of stones on the surface and began digging with spades and trowels, in one uniform rectangular area. They found that this first skeleton had been buried with what was called the "chamber" method. With this method, an initial grave is dug to a certain depth. Then along one side of the grave a "chamber" or tunnel is dug more deeply. The body is then lowered into the grave and moved sideways into the "chamber" or natural coffin, which is then closed with a layer of stones. The entire grave is then filled with soil. The remains in the second grave also turned out to be buried using the "chamber" method.[266]

Tobias carefully brushed the soil off the first skeleton, photographed it, and then packed it for its journey to Wits University in Johannesburg.

Basil Humphreys and Dr Fock from the McGregor Museum took the lead in investigating the third grave, which was believed to be that of Cornelius Kok II. Their first job was to remove a thorny tree or *kareeboom* at the foot of the grave, which took them several hours. As they began digging, they found the edge of a coffin. The group saw this as a good sign because if there was a coffin, the different method of burial might indicate that the grave was in fact that of the leader of the community, Cornelius Kok II.

According to Humphreys, at this stage in the exhumation "Captain Adam Kok IV rose from the large drum of plaster-of-Paris on which he had sat most of the afternoon". He stood at the grave's edge amongst amateur and professional photographers, and "seemingly half the population of Campbell"[267] while Humphreys and Fock continued to dig. They removed more soil until finally the collapsed and disintegrated top of a coffin was revealed, through which they could see a skull. Excitedly, Tobias joined Humphreys and Fock. Getting down on his hands and knees to take a good look at the skull, Tobias proclaimed that the skeleton was of a man who was approximately 80 years old at the time of death, which was in keeping with the age of Cornelius Kok II when he died. Much of the coffin had decomposed and for another four hours the group carefully removed the soil from amongst the bones inside it. After some time, they found a button and then another. Not long afterwards they found a copper ring around one of the finger bones on the right hand. It would be more than three decades before DNA testing of bones would be available as a tool of identification, but they were almost certain that they had found the correct grave.[268]

Basil Humphreys wrote about his experience for the Kimberley newspaper, the *Diamond Fields Advertiser*, in June 1961, explaining that Professor Tobias was preparing to deliver a paper in Port Elizabeth at the Congress of the South African Association for the Advancement of Science about the method of burial of Cornelius Kok II. Humphreys wrote it would be "fully two years before experiments and investigations will be completed and the remains of Cornelius Kok can be reinterred in Campbell".[269]

In fact it would be 46 more years before the skeletons were returned to Campbell. Neither member of the Kok family who was at the excavation that day would see the skeletons again. Adam Kok IV passed away in 1978 and his son Abraham died in 1991.

The story of Tobias's work in Campbell is not told in his autobiography, nor in any of the papers in his collection. It is not in the book of interviews where he reflects on his life either. The reason that his work in Campbell, which began in 1961, came to light publicly, beyond the knowledge of interested academic researchers, was because in 1996, after the end of apartheid, the descendants of the dead demanded that the remains of their family members be returned to Campbell.

In 1962, Ronald Singer, who had "stung" Tobias with his critique of the Boskop race concept, addressed the South African Archaeological Society. He spoke about the damage being done by researchers in the fields of archaeology and physical anthropology in South Africa. In his speech, he referred to "grave-digging", which he said scientists fondly called "excavating". Singer suggested that these researchers might have produced skeletons but they might also have destroyed any evidence that could have provided more information about the period and the environment in which the skeletons were found. Referring to the growing skeleton collections in South Africa at the Bloemfontein Museum, the University of Cape Town, the Transvaal Museum and the Dart collection at Wits, he suggested "we have large collections of skeletal remains and artefacts, but we have gained little precise knowledge of the sequence of events concerning ancestral and indigenous populations". Singer was critiquing archaeology, physical anthropology and palaeoanthropology. These three disciplines were related in terms of the desire that scientists had to try to understand race and human biology and its relationship to human evolution.[270]

In the same address, Singer said that some physical anthropologists tended to be very sensitive and concluded that any disagreement with their view "may be a personal attack on themselves". He said that because of the small number of people working in these fields "it has become easy for some people to adopt the guise of the expert especially on the subject of race".[271]

By 1963, the major impact of Carleton Coon's book *The Origin of Races* was clear, with a ripple effect felt by palaeoanthropologists around the world. The book, which argued that the different races of humans had evolved separately, had revived the scientific defence of the old typological approach. In the US it aggravated the long-standing conflict between the anthropology departments at Harvard and Columbia universities.[272] Coon had been trained at Harvard by Earnest Hooton and his methodology relied on taking measurements and looking at morphological characteristics to classify human beings. Coon had become president of the American Association of Physical Anthropologists. In contrast, all of the anthropologists at Columbia trained by Franz Boas had shifted their focus from the study of race to a focus on culture, two concepts they saw as completely distinct from one another. Although it has not often been mentioned, Boas also had begun his career at the turn of the twentieth century, like Hrdlička, pillaging graves for human skeletons in order to better understand race. But Boas left his focus on race behind and gained a reputation for his emphasis on cultural anthropology, so that by the 1960s students at Columbia disagreed with Coon.

Coon's book also fed into the racist political context in the US at the time. For several years before the publication of *The Origin of Races*, he had been in discussion and correspondence with his cousin, Carleton Putnam, who was a major supporter of racial segregation in the American South. Putnam believed the core problem with integration was the racial inferiority of the "Negro". He questioned why the truth about racial inferiority was widely denied and he blamed anthropologists in general and Franz Boas in particular for this situation, writing, "Two generations of Americans have been victimized by a pseudo-scientific hoax in this field."[273] The cousins corresponded regularly and they shared their concerns about cultural anthropology and the influence of Franz Boas and his students. Quietly, Coon helped Putnam shape his arguments against both cultural anthropology and against integration. These conversations had informed Putnam's own book, *Race and Reason*, which had been published in March 1961. In it he'd argued that desegregation would lead to the destruction of white civilisation. "I must ask the Northern integrationist, by what authority he claims the right to gamble with the white civilization of the South against the will of its people, while he personally sits with his children in all white schools." The state of Louisiana made *Race and Reason* required reading for all high school students.[274]

When Carleton Coon's book came out the following year, arguing that different races had evolved separately, Putnam wrote: "When ... the president of the American Association of Physical Anthropologists, a magna cum laude graduate of Harvard and a native of New England, states that recent discoveries indicate that the Negro is 200,000 years behind the White race on the ladder of evolution, this ends the integration argument."[275]

Anthropologists were already engaging with the debate about how anthropology sat within the political context of the United States. In November 1962, Margaret Mead wrote: "The use that is being made of Carleton Coon's book by racists is very disturbing to all of us." She went on to say, "... in dealing with this new development, especially the campaign waged by Putnam, author of *Race and Reason*, in which he makes heavy use of Coon's speculations, we have tried to steer a course between adding to the publicity by attacking the racists, and yet making quite clear where anthropologists stand."[276]

Two widely known white supremacists wrote a letter to the *New York Times* just after Coon's book came out. They repeated the concern about the "cult" of Boas and they quoted Coon in *The Origin of Races*: "It is a fair inference ... that the subspecies which crossed the evolutionary threshold into the category of *Homo sapiens* the earliest have evolved the most." Coon drafted a letter to the editor in response. Apparently, he never sent it, but it remains in his papers.

The letter expressed no argument against the segregationists who were using his findings for their own purposes, but rather conveyed his feeling that he was "discouraged that a work of substance which took me five years to write and which covers the racial history of all mankind should be dismissed as a mere prop for domestic, partisan argument".[277]

The Origin of Races did not, however, only influence thinking in the United States. The International Association for the Advancement of Ethnology and Eugenics (IAAEE) had been founded in 1959, and Carleton Putnam had attended its first meeting. The IAAEE established their journal *Mankind Quarterly* in 1960 as a right-wing counter-balance to the anti-racism of post-war anthropology in the United States and Britain. Coon's work, which had already had an impact in the US, now had a mechanism through which to influence scientists around the world.[278]

One member of the IAAEE, the Scotsman Robert Gayre, asked Coon to join the board. Coon turned down the invitation but said he was "very glad to get your monographs and also your magazine ... to accept membership on your board would be the kiss of death, here in the so-called land of the free and home of the brave".[279] Gayre, however, who edited *Mankind Quarterly*, was invited by JDJ Hofmeyr, another member of the IAAEE, to attend the South African Genetic Society conference in 1962. Hofmeyr (not to be confused with South African politician Jan Hofmeyr, who died in 1948) was a professor at the University of Pretoria. He was one of the first members of *Mankind Quarterly*'s international advisory board, and the founding president of the South African Genetic Society. He became a part of the network of scientists who were opposing desegregation in the US and upholding white supremacy in South Africa. Gayre's trip was paid for by the University of Pretoria and he gave several lectures at the second congress of the South African Genetic Society. The lectures were so inflammatory they caused a public response. John Robinson, who was still at the Transvaal Museum, claimed that Gayre was "misusing scientific evidence to bolster up ideological concepts" and the *Sunday Times* ran an article on 23 September 1962 with the headline "Pro-Apartheid Scot's Visit Causes Stir" and called him a "fascist geneticist".[280] When Gayre sued the *Sunday Times*, the newspaper turned to Phillip Tobias, privately asking him for advice on whether Gayre's views as an anthropologist were fascist. The judge ruled that Gayre's reputation had been damaged so the newspaper had to reach a settlement with Gayre and apologise. Despite the insult, Gayre continued to return to South Africa throughout the 1960s. Hofmeyr went on to make the case that there was genetic evidence to support the concept of the Bantu homelands, which were then under development as a component of South African apartheid in the 1960s. Hofmeyr was part of a worldwide effort to oppose UNESCO's anti-racism campaign and to strengthen the scientific support for apartheid.[281]

Phillip Tobias was wary of Hofmeyr's thinking on genetics. He also rejected Carleton Coon's premise in *The Origin of Races*. In the 1960s Tobias was still thinking about how the different races of humans developed and he was greatly interested in human variation. However, he rejected Coon's statement that each race "had reached its own level on the evolution scale".[282]

Soon after the publication of *The Origin of Races*, Sherwood Washburn, the physical anthropologist who had met with Tobias in Nairobi in 1959 and had encouraged the new synthesis, gave his presidential address to the American Anthropological Association. He, like Coon, had been educated and trained at Harvard by Earnest Hooton. "If we look back at the time when I was educated, races were regarded as types. We were taught to go to a population and divide it into a series of types and to re-create history out of the artificial arrangement ... This kind of anthropology is still alive, amazingly ... Genetics shows us that typology must be completely removed from our thinking if we are to progress." He said physical anthropologists were so preoccupied by the divisions amongst humans that they forgot evolution worked on a species level; and he made the point that it was critical to integrate genetics into anthropology. Washburn closed by proclaiming: "A ghetto of hatred kills more surely than a concentration camp, because it kills by accepted custom and kills every day in the year."[283]

The dispute between Coon and other scientists continued for several years. While the debate about segregation in the US carried on, the US Congress passed the Civil Rights Act in 1964. In South Africa, however, under Verwoerd, the development of the homeland system was fully under way. Events in the 1960s further solidified apartheid, and the measurement of living people by physical anthropologists continued.

The apartheid environment in the early 1960s also had an impact on the work at the Transvaal Museum. Frustrated by his inability to shift from the museum to a university post, John Robinson was planning to leave South Africa permanently in 1963. He was offered a position at the University of Wisconsin in the United States, where he began teaching, but he also continued his research on australopithecines. In September 1962, before Robinson left, the museum's board asked him to modernise the palaeontology exhibits, especially in terms of human evolution. These hadn't been updated since Broom's time in the late 1940s, or since the church had protested in 1952. Robinson began the task and then handed it over to his successor, Jurgens Meester, who had previously worked in the mammology department at the museum. With the help of carpenters, artists, taxidermists and scientists, the new displays went up. Not only did they show fossils, but also artistic impressions of what various forms of pre-humans looked like. The exhibit was called "The Descent of Man According to the Theory of Evolution".[284]

Even before the new exhibit was complete, the public morals commission of the Dutch Reformed Church sent a memorandum to the board of trustees at the Transvaal Museum, asking that the exhibit be changed so as to eliminate any links between the animal kingdom and man. The memorandum stated that "man was created in the image of God and there could be no question of a gradual change from animals to man". The letter was read to the board of trustees on 28 May 1963. They had also received a verbal message from the minister of education, arts and science, Senator Johannes de Klerk, asking them to consider the DRC's request. The board asked the minister to put his views in writing, which he subsequently did.[285] Referring to the exhibits on evolution as "contentious" and "offensive", he asked that they be removed from public view.

"EXHIBIT ON EVOLUTION MUST GO, SAYS DE KLERK" read the headline in the *Pretoria News* on 24 July.

The term of the board of trustees had expired on 30 June and the museum's director, Vivian Fitzsimons, wanted to wait until the new board was confirmed so that it was the new board who would consider the minister's letter and respond. Meester, who had worked with John Robinson to design the exhibit, said: "Should an order be transmitted to me, I regret to say that I would probably feel obliged to disregard it. We have made every effort in preparing these displays to avoid errors of fact or interpretation."[286]

The apartheid government's Christian National Education syllabus, first implemented in 1958, was taking effect. As had been the case with the protest against the evolution display at the museum 11 years earlier, there was still concern about how public education related to the concept of evolution. Tobias entered into the debate by telling the *Pretoria News* that he had examined the exhibits himself and that he found them to be a "cautious and scientific portrayal of the lines of human development, supported by a wealth of fossil evidence from all over the world". He reminded the public that the Transvaal Museum was one of the world's greatest repositories of fossils concerning "our human ancestry" and suggested that the South African government and the people of Pretoria did not appreciate that they were "one of the five greatest centres in the world for fossil man [alongside collections at Wits, London, Utrecht and Paris]". Fitzsimons said it would be a great pity if the exhibit had to be dismantled.[287]

Before the matter was resolved, in late August Dr Jurgens (Waldo) Meester, who had by then taken over from Robinson, decided to leave the Transvaal Museum to take another job in King William's Town. This was an opportunity for the media to review the controversy and Meester was quoted in the *Sunday Times* as saying: "The Minister's objections are not of a scientific nature. To request the removal or modification of the displays in question on any other grounds represents an attempt at curtailing our academic freedom, which I feel it my duty to resist." It was reported that while Meester had not resigned in protest, he did fire a "parting shot" by writing a letter to an [un-named] Afrikaans newspaper. "We certainly do not want everybody to agree with us that our exhibition is accurate," he wrote. "What we do ask is that we should be allowed to have the exhibition so that those who wish to do so can look at it and decide for themselves whether we are right or not."[288]

Minister De Klerk (who was the father of FW de Klerk, who would later become South Africa's president) announced his intention to convene a symposium of scientists in the field of evolution in order to gain clarity on the contentious issues. He also confirmed the reappointment of the old board of trustees of the Transvaal Museum, along with three new appointments. The board announced they would meet in late October to discuss the dispute. The debate carried on.[289]

On the morning of 29 October 1963, the *Pretoria News* announced a "NEW TURN IN MUSEUM QUARREL OVER EVOLUTION" and reported that the board was taking a firm stand against removing or changing the display on human evolution. By 15 November, Minister De Klerk had not officially spoken on the matter. "The decision taken two weeks ago continues to remain a closely guarded secret," declared the *Rand Daily Mail*.

Then someone broke the silence and leaked to the *Pretoria News* that the trustees had decided not to give in to the pressure. "MAN-APE SHOW LIKELY TO STAY AT MUSEUM" was their headline. According to their source, only two of the 19 board members advocated removing the exhibit or toning it down.[290] Finally, on 18 December, the *Pretoria News* reported that the evolution exhibit would not be changed or removed. Not revealing what, if anything, had happened behind the scenes, they quoted the minister as stating that he had been convinced that the exhibit did not offend and that the debate should be considered closed.[291] On Christmas Eve, the Reverend de Beer, the Public

Morals secretary of the Dutch Reformed Church, said the church still objected to the displays and he accused scientists of holding onto their "pet theory". While many people continued to hold their personal views, after more than six months, the public row, however, was over.[292]

12

"There is Reason to be Suspicious"

The original Campbell Griqua Expedition, as it was referred to by Phillip Tobias, led to growing scientific interest in the Griqua skeletons throughout the 1960s. Tobias led three further trips to Campbell for excavations, in 1963, 1967 and 1971. He returned with his students to the graveyard where there were about 70 graves identified by the presence of headstones, footstones and piles of stones which appeared in rows. Tobias's scientific interest went far beyond the initial intentions of Basil Humphreys and Adam Kok IV to move the remains of Cornelius Kok II to a more accessible place near the mission church and to investigate for a period of two years. Over the years, he exhumed a total of 35 skeletons. He brought them all back to the Raymond Dart skeleton collection so that they could be available for further research.

As a result of Tobias's initial interest, other scientists from Wits University began to travel to Campbell as well. Hertha de Villiers was one of them. She went there in 1967 and subsequently delivered a paper entitled "The Morphology and Incidence of the Tablier in Bushmen, Griqua and Negro Females" at an anthropology conference in Japan in 1968. De Villiers said that the question of whether "the elongation of the labia is a natural, inherited or artificially produced phenomenon" had long been discussed in the literature. In her research "all women examined insist that the growth is a natural phenomenon." She based her research on 103 Bushman individuals – 12 from Tobias's study in 1959, 45 from her own research in 1961 and 46 from their Wits colleague Trefor Jenkins' research in 1964. It is hard to believe that all three scientists would have gathered data of this kind. De Villiers also compared 36 Griqua women from a previous study as well as 97 women she had examined that she described as "Negro".[293]

In carrying out her study, De Villiers had gone to Baragwanath Hospital in Soweto, south west of Johannesburg. The women she met there were either ill and in wards, or waiting for attention in the out-patient department or at the ante-natal clinic. All of these women were living in oppressive conditions in apartheid South Africa. They had to carry passes. Many of them had to travel long distances every day to work in the homes of white people as domestic servants. There is no documentation of how these women reacted when a scientist asked to measure their labia, nor is it recorded whether they were paid to participate or coerced. De Villiers asked each woman to stand up with her legs slightly apart. She placed a sliding caliper gently on the labia minora at its point of emergence. Keeping the caliper in position, she shifted the movable point of the caliper down to the lowest point of the labia. "With this technique, it is clear that any undue pressure exerted on the labium majus will constitute a source of error in the length of the labium minus," wrote De Villiers.[294]

In Tokyo, De Villiers explained to the anthropologists gathered there that "Amongst Bushman women, it is fairly common practice to conceal the labia minora within the vagina, resulting in a swollen and puffed appearance of the labia majora pudenda. In measuring, it is necessary, therefore, to ensure that the labia have been completely extracted and are hanging freely." De Villiers concluded that "Morphology, language and culture serve to separate present day Khoisan and Bantu speaking populations. However, genetical surveys over the past ten years have brought to light a number of serological and other genetical resemblances between these two populations groups. The genetic resemblances point to a common ancestry, while the morphological and cultural differences point to a lengthy period of differentiation in relative isolation ... It seems possible that the tablier which is found in the Khoisan, the Griqua and in the South African Bantu-speaking Negro populations is yet another African gene-determined morphological trait."[295]

In the early 1970s, George Nurse and Trefor Jenkins began genetic research of the Griqua people in Campbell.

Trefor Jenkins first met Tobias when he joined the department of anatomy in 1963. He was hired as a demonstrator, what was often called a table doctor, and worked in the department for two and a half years while in his early 30s. Of Tobias, Jenkins said, "He was very autocratic and authoritarian and we invariably had fights." During those years, Tobias was not doing any field work because he was focused on his writing and analysis of the fossils in Kenya. "A lot of people resented that he wasn't around and that he was neglecting the department," Jenkins said. "But it was good for me. I got into South African field work without close supervision." Tobias sent Jenkins on his first expedition, with Stan Blecher, on a trip to the Kalahari.[296] He wanted him to do chromosome studies with the San people.

Jenkins was originally from a coal mining family in Wales. He had studied medicine in London and then worked in a mining hospital in Wankie, Rhodesia. There he had been introduced to sickle cell anemia, a genetic mutation in red blood cells that results in a condition where blood cells are unable to carry oxygen throughout the body. Jenkins began to do sickle cell screening. He would drive into the valley to hold clinics and collect blood samples. This experience led to Jenkins' ongoing work in genetics, and it was one of the reasons Tobias thought he was the right person to do the work in the Kalahari with the San people. At the time of the trip in 1963, Jenkins said, "there were still murmurs and rumblings in the German literature that the San people were a different species from *Homo sapiens*. Tobias wanted to confirm that they had 46 chromosomes." Needing the blood samples to be fresh, Jenkins took them from people at the airport at the end of the trip.

Tobias had been concerned about this particular issue for several years. In 1961, the same year he published "The Meaning of Race", he spoke at the International Human Genetics meeting in Rome. Addressing the same accusation that the San people were not members of the human species, Tobias delivered a paper on what he called "Bush-Caucasoid hybrids". He showed a photograph of a man named Jimmy Morris who was married to a San woman who had four or five children to prove that the San people could reproduce with white people. "They lived in Ghanzi in the Kalahari," recalled Jenkins. "I met Jimmy and his wife and his children. They were very nice people."[297]

At the Johannesburg City Hall in 1963, Tobias put together an exhibition called the "Man in Africa Exhibition", sponsored by the Institute for the Study of Man in Africa (ISMA). Tobias had established ISMA in honour of Raymond Dart in 1959. Jenkins helped Tobias with the displays. It was a very popular exhibit and well attended. "I held forth on the subject of race to all the visitors," said Jenkins.

Tobias's hosting of the exhibit, his lecture on race, and his work in genetics resulted in an increasing number of visitors to the department of anatomy asking for assistance with racial classification. Visitors were often accompanied by their lawyer and often they were trying to make the case to be allowed to live in a certain area or attend a certain school. Race Classification Appeal Boards had been established by the minister of the Interior and by the early 1960s, the appeal procedure had resulted in growing numbers of people turning to physical anthropologists such as Tobias as expert witnesses.[298] Tobias and Jenkins both became involved in this process, often hoping to help people deal with difficult circumstances, to enable someone to marry a person who was classified as a different race, or to assist one spouse to be reclassified in order for a married couple to live together in the same place.

Jenkins remembered that before he left the department, he observed Tobias taking this task very seriously. "He was meticulous," said Jenkins. "He would measure. On one occasion when I was in the department, a nurse brought in this child and Tobias called me in. 'Those eyes are a bit dark,' he said. 'And that nose is a bit flat.'"

Given Tobias's opposition to apartheid legislation and his public stance on race, one might have expected that he would protest the process of racial classification and refuse to participate. In fact, Tobias offered assistance to various institutions regarding racial classification. On one occasion in 1966, it was the Princess Alice Adoption Home that asked Tobias for help.

On 29 March 1966, at about 2:15pm, Marie Arendt, a social worker and the Adoption secretary from the Child Welfare Society of Johannesburg, accompanied a nurse and a baby from the Princess Alice Adoption Home to the Wits Medical School to visit Professor Tobias. They needed his assistance to help classify the baby's race before they could place the baby for adoption. According to Arendt, when Tobias first saw the baby, he "did not query the appearance". Then she told him that the baby had an Australian mother and that the father's surname was Amamoo. She told him that the adoption home had been concerned about the baby's profile and her hair. According to Arendt, a long discussion followed. Tobias pulled out slides of Aboriginal people from Australia and they examined them at great length in comparison to the baby. "There is reason to be suspicious," he concluded. It was the big upper lip that gave him cause for concern. He could not confirm the baby was white.[299]

It had been on 16 February 1966 that a young blonde Australian woman named Suzanne Alcock gave birth to the baby girl at the Queen Victoria Hospital in Hillbrow, Johannesburg. She named the baby Tandy Jane Alcock, and five days later she signed a consent form requesting that her child be admitted to the Princess Alice Adoption Home so that she could be placed with suitable adoptive parents.

At the time, Suzanne, or Suzie as she was often called, stated the name of the father of the baby as Samuel Amamoo. Marie Arendt met with Suzie and pointed out to her that it would be essential to have more details regarding the background of the baby's father. Arendt expressed concern that the name Amamoo might be Indian, which Suzie did not confirm or deny. Arendt explained that there would be difficulties with the placement of a child because of the population registration laws in South Africa. Arendt noted in her case report that Suzie "appeared to fully realise the implications and promised to cooperate".

Arendt and the Child Welfare Society made attempts to find out more about the surname Amamoo. They contacted the Greek Consulate and were told that it was definitely not a Greek name, but that it could be a Lebanese name. The Lebanese Consulate assured Marie Arendt that the name was not Lebanese but said it could be Arabian. Arendt contacted the Australian Consulate as well but was told that they had never heard of such a name before. "In spite of further enquires," wrote Arendt, "I could not ascertain what race this name belonged to." She went on to note that "it is the profile and the black almost silky hair of this baby that causes concern".

Realising that she might have run into a problem, Suzie Alcock returned to the Child Welfare office and said that she had previously given incorrect information. The surname of the father was not Amamoo but North. It was John Brian North, a 26-year-old Australian man she had met in the Royal Adelaide Hospital where she was training. He had left Adelaide for London in July 1965 before she realised that she was pregnant. She wrote to him to tell him that she was pregnant but offered to "settle the matter without his assistance". Miss Alcock said that the reason that she gave incorrect information earlier was because she did not want Mr North to be involved. She had thought up a name, she said, not thinking that the adoption agency would be so concerned about it. When Marie Arendt asked Alcock why she'd used such an unusual name and whether she knew anyone by that name, she said "No." She had heard the name mentioned by her brother, she said, who knew a fellow student named Amamoo. Alcock said that she had never met the student and didn't know what he looked like.

The Child Welfare Society and the Princess Alice Adoption Home were not convinced by the change of name. Alcock had made a sworn statement in court that the name of the father was Samuel Amamoo, and they were still concerned about the child's appearance. Dr Wunsch from the adoption home was concerned about Marie Arendt's investigations about the baby and wanted to clarify the child's race. It was Wunsch who suggested that they take the baby to see Phillip Tobias.

By the end of March, more than a month after Tandi Jane was born, the committee at the Princess Alice Adoption Home decided, based on Tobias's examination, that they would not be able to place the baby. Marie Arendt told Suzie Alcock that they could not be responsible for the adoption. She asked her if it was possible that she could take the baby back to Australia. Suzie Alcock said that was definitely not possible. In addition, Miss Alcock had arranged a vacation with two Australian friends to drive to Mozambique. Marie Arendt asked that Miss Alcock shorten her holiday as they could not keep the child indefinitely.

It was on 2 May that Miss Alcock next came to see Marie Arendt. She again said that it was impossible for her to take the child back to Australia. At this point the department of social welfare and the department of immigration got involved with the case. They decided that Tandi could not be repatriated to Australia because Suzie Alcock had a permanent residence permit at the time of Tandi's birth. Her case remained in limbo for several months.

The adoption home took the baby back to Tobias to be examined a second time. This time Tobias called Hertha de Villiers in to assist with the examination, looking at the baby's hair, eyes and head shape. "There might be some Aboriginal features here," conjectured Tobias. "Don't be silly," responded De Villiers. "She's a beautiful child. I'd be happy to adopt a child like this." Just at that moment the phone rang and Tobias sat down to take the call; it was international. He glanced up at De Villiers and said, "You handle this," and proceeded to talk on the phone.

Irritated, De Villiers turned to the nurse. "I'm afraid I can't help you," she told her. But the nurse hadn't taken De Villiers's passing comment as a joke. "Why don't you adopt the baby?" she said. "That's absurd," said De Villiers. "I'm 42 years old. I'm not looking for a baby." In that moment, the nurse decided to take a chance. She looked down at the baby and then turned back to De Villiers in her white lab coat. "You know, if you don't take her, she'll end up in a coloured orphanage."

On 5 August 1966 Hertha de Villiers wrote a letter declaring that after close examination of the baby's features, she had reached the conclusion that "all her features fall definitely within the range of variation of the features encountered in Europeans of European national origin" and that they showed a blend of features most commonly encountered in "the Mediterranean branch of the European Caucasoid race". De Villiers carried on to comment specifically on the baby's hair, eyes and the colour of her skin which, she said, fell within the "range of variation of the degree of pigmentation encountered in members of the Alpine, the Armenoid and especially the Mediterranean branches of the European or Caucasoid populations". Under South Africa's Population Registration Act, Tandi Jane Alcock was registered as a European child.

As a result of De Villiers' letter, the department of social welfare declared that "Re-classification of this child as a Coloured, and placement in an institution for coloureds, or with coloured adoptive parents would be unjust not only because of the child's appearance, but also because the child's origin is in no way related to any negroid strain at all". They concluded that it would be in the child's interest to be adopted as soon as possible by European parents and that according to Dr de Villiers' report, European society was "in no way likely to suffer from such placement".

Hertha de Villiers discussed the matter with her husband, her son, and her foster son, and they agreed that they would adopt the baby. De Villiers perjured herself and declared that she was 40 years old, the age limit for adoptive mothers. To her son Adrian, who was 18 at the time, De Villiers remarked on the baby's beautiful dark violet eyes. "Classification is nonsensical," she said. On 8 September 1966, when she was seven months old, baby Tandy Jane was brought home to her new family. They renamed her Phillippa. And so began a uniquely South African story.

13

Sterkfontein Revisited

After spending seven years focusing on East African fossils, Tobias decided it was time to bring some attention back to Sterkfontein in South Africa. It had been largely ignored for almost 20 years since Robert Broom found Mrs Ples there in 1947. In 1966, the same year he had examined Tandi Jane Alcock, Tobias decided to open new excavations at Sterkfontein on 30 November, the centenary of the birth of Robert Broom. He also prepared to open a small museum at Sterkfontein in Broom's name.

Five organisations – the South African Association for the Advancement of Science, the Institute for the Study of Man in Africa, the Transvaal Museum, the University of the Witwatersrand and the South African Archeological Society – came together to organise the 100th anniversary of Broom's birth. To start the proceedings, Professor Lawrence Wells, the head of the department of anatomy at UCT, gave a memorial lecture in the Wits Great Hall. Wells had been a student of Dart's at Wits and he had written a thesis entitled "The Foot of the South African Native". In his lecture, Wells applauded Broom as a zoologist, palaeontologist, anthropologist, thinker and teacher. And he emphasised that Broom had resented the belief that London was the "centre of the scientific universe".[300]

Wells introduced his talk about Broom by saying that he had travelled to the University of Edinburgh in February 1951. The human skulls that Broom had sent there attracted Wells' attention so he wrote a letter to Broom about them. Broom responded to Wells days before he died saying, "I sent I think four skulls of 'Hottentots' from Port Nolloth to Turner in 1897 ... I cut their heads off and boiled them in paraffin tins on the kitchen stove and sent them to Turner."

Several years later, in 1972, Raymond Dart wrote the preface for George Findlay's biography of Broom in which he said, "In both scientific and academic circles the mention of his name evoked unflattering comments", but suggested that Broom was brought back into the fold at the Transvaal Museum because of those "whose respect for his scientific work outweighed their disapproval of his past misdeeds".[301] Broom's "misdeeds" as assessed by Dart and others in the 1960s and '70s was not his treatment of human heads and skeletons, but rather his selling of Karoo fossils to the American Museum of Natural History in 1912.

Robert and Mary Broom's son Norman spoke at the opening of the Sterkfontein museum on 1 December 1966. Norman had become the Consul to Panama and made the trip up from Cape Town for the ceremony. Earlier that same year he had been in correspondence with Findlay, who hadn't yet finished his father's biography. In September he sent a letter to Findlay asking if he might see him at the opening of the small museum at the Sterkfontein Caves which was going to be opened "believe it or not by yours truly". Norman Broom hadn't had a close relationship with his father and he was nervous about the event but felt that he couldn't refuse. "What on earth I shall say on that day only the good Lord knows ... Probably some stories/incidents taken from my tender years at home for as you know I left home at the comparatively early age of sixteen."[302] He ended up speaking of Robert Broom's poor driving skills and his flair at chess. "He seemed a severe unsmiling type of person always dressed in forbidding dark suits and always peering. Every object would be brought to within an inch or two of his eyes and up would go his glasses to rest on his forehead." Norman told the people gathered that day that "it was not uncommon for a human skull or some other horror to be placed on the stove to cook merrily alongside whatever was being prepared for the next meal. Mother never took kindly to this and neither did the servants."[303]

It was at this occasion that the University of the Witwatersrand placed a bust of Robert Broom at a clearing near the opening of the Sterkfontein Caves. The bust was a replica of the one that had been unveiled at the Transvaal Museum by Jan Smuts 20 years earlier. The one at Sterkfontein is now seen by thousands of people every year as they pass by it when they complete their tour of the caves.

Concluding the proceedings, Tobias said he was certain that Broom would be pleased to see that Wits University was embarking upon a "major long-term excavation of the Sterkfontein fossil site. We may confidently expect that in the next ten years this fossil treasure house will yield many new secrets bearing on the evolution of man."

Tobias was correct in his predictions but not exactly on target with the time-frame. Alun Hughes, who would lead the excavations in the field at Sterkfontein for the next 25 years, would produce many findings. However, it would take over 30 years, until 1998, before Ron Clarke would unearth a full skeleton at Sterkfontein. In the 1960s, Clarke, a young English technician, was working with the Leakeys in Kenya, cleaning fossils and making casts. He and Tobias first met in Nairobi where Clarke helped Tobias with his studies of *Homo habilis*. Clarke made casts for Tobias and helped him to do brain volume measurements. "I was very young and looked up to him as my elder," said Clarke.[304]

In 1969, after seven years with the Leakeys, Clarke made a trip to South Africa to study fossils there. He had been working on a reconstruction of a crushed fossil hominid from Olduvai Gorge in Tanzania and Leakey thought that comparing it to some South African fossils would help. While Clarke was working at Swartkrans, near Sterkfontein, he found a cranium, later called SK47. "It launched me into the forefront of paleo-anthropology. I knew that it was a very important fossil and I thought it was early *Homo*." Clarke shared his find with his colleague Clarke Howell and with Bob Brain, who had taken over the palaeontology department at the Transvaal Museum in the wake of Robinson's departure. The three of them wrote a paper that was announced on the cover of *Nature*. Clarke compared the cranium to a piece of maxilla that had been described as *Homo* by John Robinson. Then he realised that both fossils were from the same individual. Clarke developed a knack for finding separately classified specimens that fit together as a single individual.

Clarke recalls that in the late 1960s there were ten black labourers who worked at Sterkfontein. All of them lived in a compound on the premises and needed permits from the department of Bantu affairs to stay there. Every day their job was to move rocks and breccia, to shovel sand and to pick through the mining rubble looking for bones. Very little has been documented about these men, their lives and the contribution they made to the work at Sterkfontein.[305]

In the late 1960s, both the department of anatomy at Wits and the Transvaal Museum were receiving support from the Council for Scientific and Industrial Research (CSIR). The amounts of funding were not large, but they were enough to keep the modest excavations going at Sterkfontein and the adjacent Swartkrans.

In one CSIR publication in 1968, Tobias wrote an article entitled "The Emergence of Man in Africa". He wrote that "by studying the very beginnings of some living races, scientists may learn something of the dynamics of race formation and the environmental conditions affecting race differentiation". He also spoke about the human diversity in South Africa and how that diversity related to the major racial groups into which physical anthropologists classified the world's population. The language that Tobias used at the time was uncomfortably close to the way the South African government was classifying the entire South African population into racial categories. Interestingly, the article, which was predominantly about human evolution in Africa, extensively considered the concept of race.[306] To Tobias, the scientific community and the broader public reading the article, the two concepts were still inextricably intertwined.

By 1970, the Leakeys, Tobias, Clarke, and Bob Brain focused on the ancient species of *Australopithecus*, *Homo habilis* and *Homo erectus*. There was no consensus internationally, however, on the role Africa played in the more recent development of *Homo sapiens*, both in terms of their physical traits as well as their cultural traits such as art and language. At that time, the role of Africa was still seen as unimportant in terms of later human devlopments, and the scientific emphasis was still on other regions, including Asia and Europe. Over the following two decades, many palaeoanthropologists and archaeologists would, however, begin to focus on Africa as the place of origin for these human developments. The 1970s and '80s would be a time of great change in the thinking about human origins and the role that Africa played in the story.

However, there was one more major influence that Raymond Dart would have in the 1960s on the way that the world viewed the australopithecines and their influence on human nature. His work in Makapansgat would be publicised by the actor and scriptwriter Robert Ardrey in his hugely popular books, and even influence the film director Stanley Kubrick; Dart's ideas would explode across the globe.

14

Clubs and Daggers

In 1961, Robert Ardrey, an American playwright and screenwriter, published a book called *African Genesis: A Personal Investigation into the Animal Origins and Nature of Man*. Based on Raymond Dart's research at Makapansgat in the 1940s and '50s, the book popularised Dart's theory that humans were naturally violent. Dart had based his theory on the fact that pre-humans had acquired a more upright posture and that their somewhat increased brain capacity allowed them to use an aggressive striking movement that was not possible for living apes. After spending time with Dart, John Robinson and the Leakeys, Ardrey was enthralled by the idea that one species of *Australopithecus* – *africanus* – was the aggressive carnivore and the other species – *robustus* – was the vegetarian. In a letter to Dart in late 1960 before his book was published, Ardrey wrote: "One is an armed hunter with a killer's mentality and demands. The other is non-aggressive. One is Cain, one is Abel ... Sooner or later their paths crossed. And Cain slew Abel."[307]

While it was Ardrey's book that catapulted Dart's theory around the world, it had been circulating in scientific circles without much support for years. After five years of research at Makapansgat, Dart published an article in 1953 entitled "The Predatory Transition from Ape to Man". He explained that there were no stone tools found at the limeworks at Makapansgat. Instead there was a specific collection of certain kinds of bones and Dart believed that these bones were used as weapons. He saw fractures in skulls as proof that antelope humerus bones were used as clubs and that other bones were used as daggers. He argued that the *Australopithecus* bones found in Taung, Sterkfontein and Makapansgat told a consistent story of the carnivorous habits of *Australopithecus africanus*. Dart used extreme language to make his point. He said that the *Australopithecus* likely "tore the battered bodies of their quarries apart limb by limb and slaked their thirst with blood, consuming the flesh raw like every other carnivorous beast". The editor of the journal that published Dart's article wrote in a foreword that "of course, they [the australopithecines] were only the ancestors of the modern Bushman and Negro, and of *nobody else*".[308] This note revealed the bizarre assumption that evolution of the australopithecines related to black people only and did not apply to white people. The editor's comment seemed to illustrate the association between the claim that early humans were dangerously violent, and a prejudice that black people were inherently dangerous. This view was in keeping with the concept of "*swart gevaar*" (black danger) in South Africa, a slogan that played on the fears of white South Africans.

Dart took things further. From the bones at Makapansgat and his conclusions about a carnivorous *Australopithecus*, he concluded that humankind was unable to avoid its deeply ingrained carnivorous habit and that the predatory transition had doomed humans to a violent existence. "The blood-bespattered, slaughter-gutted archives of human history from the earlier Egyptian and Sumerian records to the most recent atrocities of the Second World War accord with early universal cannibalism, with animal and human sacrificial practices of their substitutes in formalised religions and with the world-wide scalping, head-hunting, body mutilating and necrophilic practices of mankind in proclaiming this common bloodlust differentiator, this predaceous habit, this mark of Cain that separated man dietetically from his anthropoidal relatives and allied him rather with the deadliest of Carnivora."

"What impressed me about Dart," said Bob Brain, the palaeontologist who would become the director of the Transvaal Museum, "was his amazing concept of the predatory transition from ape to man. He used very dramatic language to describe our ancestors: 'Slake their ravenous thirst,' 'Devour their writhing flesh'. I asked him once, 'As a serious scientist, why do you use such dramatic language?' 'That will get them talking,' was his response."[309]

When Dart published his memoir in 1959, he included a chapter about a historic clash at Makapansgat between Afrikaner *Voortrekkers* and Chief Mokopane, which might give us another clue to Dart's view of violence. Mokopane (referred to as Makapan in Dart's era) was the chief of the Kekana people who had occupied the entire eastern part of the Waterberg in the Transvaal for generations. In 1835, Louis Trichardt led a trek north, of farmers from the Cape Colony, who rejected British rule there. This movement became known as The Great Trek north and across the Transvaal. Thousands of Boer families settled on large farms in the area and Trichardt's descendants declared the Transvaal their own republic in 1852. Conflict between Mokopane and the settlers began in 1854 when there were disputes over trade routes. In

September, Mokopane attacked a group of the settlers. As soon as news of the attack spread, a large commando, an informal milita, assembled and rode towards Mokopane to seek revenge.

Historical reports of what happened next differ. Pretorius and Kruger wrote at the time that Chief Mokopane moved close to 2 000 men, women and children into the immense dolomitic cave – now known as Makapansgat (Makapan's Cave). However, it is now unclear if the numbers were that large. Dart wrote that Mokopane had stocked the cave with plenty of food but didn't have adequate water supplies. After a week, the Boers at the entrance showed no signs of leaving. The blockade lasted 25 days, after which the commandos stormed in. Differing reports say that anywhere between 300 and 3 000 people died inside the cave.[310]

According to Dart, and more recent archaeological excavations, the skeletons of some of the defenders remained in the cave for many years, and then gradually dispersed as tourists took bones home with them from the cave as mementos. In 1936, the cave was declared a historic monument. Dart seemed to interpret these modern events at Makapansgat as giving weight to his hypothesis that humans were naturally violent and murderous.

Robert Ardrey first came upon Dart's work at Makapansgat when he heard about it from a colleague at Yale and read Dart's obscure article about the predatory transition from ape to man. Ardrey was struck immediately by the implications of the theory. In 1955, Ardrey wrote: "In Johannesburg, city of violence and nerves, of ugliness and stomach ulcers and corrosive distrust, a story is being written more explosive than apartheid. It is a story translated in old caves from ancient bones. Its thesis is all too simple: that the earliest human assertion was murder."[311]

At the time, the scientific community was still undecided as to whether it should shift its attention in the search for human origins from Asia to Africa. As the fossil record grew, thanks to the work of Dart, Broom, Robinson, and the Leakeys, there was increasing evidence making the case. Ardrey believed scientific opinion was beginning to shift toward African origins and that some scientists were reclassifying australopithecines from anthropoid to hominid. Dart's theory that australopithecines used bones as weapons suggested to Ardrey that they were more human than ape.

Ardrey published his impressions of South Africa in the 1950s in *The Reporter*. He had an affinity for the Afrikaner; he described the Boer as an African who had lived in Africa as long as Americans had lived in America, but who "had no Indians to fight". To Ardrey, the "Bushman [was] that yellow-skinned Paleolithic left-over". He accepted the myth of the empty land.

The Boers had retreated from the British in the Cape, he wrote, into a "land that nobody wanted, not even the Bantu". For the article, which was entitled "South Africa: A Personal Report", Ardrey interviewed a spokesman from the department of Native affairs, a Stellenbosch cleric, the Anglican Bishop, anti-apartheid activist Trevor Huddleston, and the author Alan Paton, but he did not quote one single black person. "The traveler in Africa should be warned to buy his native curios in the West Coast ... Visit the Basuto in his protectorate; you will find straw hats. Visit the Zulu in Natal: you will find walking sticks and a few childish, ingratiating wooden animals."[312]

Ardrey travelled to Johannesburg to meet Raymond Dart who, he said, was "regarded by many of his fellow townsmen as slightly mad and by much of international science as ... remarkable and gifted". Ardrey asked Dart what would happen to the premise of man's innate goodness if his interpretation was correct. Dart responded, "You are the first layman who's ever come to me who cared." Then, Ardrey recalled, Dart began opening drawers in his laboratory and "rolling out skulls like apples". He showed Ardrey many baboon skulls that had been crushed. Dart claimed that this was not as a result of rock falls, but of weapons made of bone. He showed Ardrey many of the 3 500 bones that he had catalogued. He also showed him a fossil jawbone Alun Hughes had found in 1948, the mandible of an adolescent *Australopithecus* who had perished by a heavy blow to the head. The four front teeth were missing. One side of the jaw was cracked and the other side had been broken through. After Dart and Ardrey discussed all the possible causes of death, Ardrey "dismissed accident as the mother of injury. The youthful creature had died of purposeful assault."[313]

"When are you going to present all this?" asked Ardrey. "When I'm ready," answered Dart. "I've made myself a bit of trouble on more than one occasion – this thing of speaking too soon."

Dart was ready in 1957. He named the use of bones, teeth and horn as weapons as the "Osteodontokeratic (ODK) Culture" and published a detailed monograph entitled *The Osteodontokeratic Culture of the Australopithecus Prometheus*. In preparation for Dart's monograph, workers had sorted 15 tonnes of bone-bearing breccia from the hundreds of tonnes of dumped lime at Makapansgat over a ten-year period. While Dr Bernard Price had passed away in 1955 and his financial support came to an end, new financial support for the Makapansgat work was taken up by a $3 000 grant from the Wenner Gren Foundation and another $9 000 grant from the Wilkie Foundation. In that one year alone, teams of African workers were on site every day. It was the painstaking hammer and chisel work by Dart's research team that made it possible for him to put together his conclusions in the 1957 publication. Dart claimed that the large tusk of a warthog or the mandible of a hyena could become "a slashing, ripping or tearing weapon". He argued that the transformation of bones into tools and weapons had contributed to mental development. Arm and leg bones became clubs and daggers. Individual horns were used as tools after they were broken off the skull. Dart concluded that "it was his unique capacity to kill with a weapon that set proto-man apart from his fellow animals. The greater brain came later – perhaps only a little later – to satisfy the complex demands of the confirmed and specialized killer." Ardrey looked at all of this and decided "weapons had produced man, not man weapons".[314]

Ardrey's *African Genesis* became an international best-seller and was translated into dozens of languages. "Not in innocence and not in Asia, was mankind born" was its opening line. Ardrey told his readers that "hierarchy (was) an institution among all social animals and the drive to dominate one's fellows an instinct three or four hundred million years old".[315] He wrote that the australopithecines were the ancestors of man and the "authors of man's constant companion, the lethal weapon".[316] He contended that human history reveals that weapons are a *Homo sapiens*' largest, single preoccupation.

He argued that "with an instinct as true as a meadow-lark's song", humans turn to weapons as their "most significant cultural endowment".[317]

In 1964, the British science fiction writer Arthur C Clarke read *African Genesis* and saw there were similarities between Dart's theory (and Ardrey's exposition of it) and the plot-line he had been working on with Stanley Kubrick for the film *2001: A Space Odyssey*. Clarke took the book to Kubrick and they used *African Genesis* as the inspiration for the opening scene of the movie – the "Dawn of Man" sequence.[318] A group of apes gather around a water-hole on the sparse African plain. When a second group of rival apes arrives, their only defence is grunting and screeching and motioning for the others to go away. In a subsequent scene, one individual in the group of apes receives inspiration, a new idea. He sits next to the dried skeleton of a bovid, looking at the skull clearly laid out and the prominent bones of the ribcage. The ape character looks at the skeleton and then cocks his head slightly as if that new idea had come to him for the first time. The ape breaks off a large bone, a humerus. In slow motion, he lifts his arm to the sky and then down to hit the skeleton with the bone, causing other bones to fly. He recognises his new power. The man-ape realises he can use the bone as a weapon. Over and over again, he uses the bone to crush the skull and break apart the bones of the skeleton on the dry ground.[319]

In the next scene, the apes have gathered again at the water-hole, and confront the second rival group. This time, they use their new technology. The bone has become a weapon and they use it to destroy an enemy ape and frighten the opposing group. After the destruction of the enemy ape is complete, the camera turns to the victorious ape. He slowly throws the bone weapon up into the clear blue sky. The large bone turns over and over in the air, and transforms into a satellite orbiting the earth in space. A slow series of single trumpet notes accompanies the scene followed by an explosive orchestra of sound.

In a documentary that helps to explain the film, Arthur C Clarke says that the bone turns into an "orbiting space bomb, a weapon in space". He laments that this transition wasn't made perfectly clear and that some people assumed it was some sort of space vehicle. The scene was a "three million year jump cut" from the bone weapon to the orbiting space weapon.[320]

2001: A Space Odyssey was one of the first big-budget science fiction movies. The movie begins with the evolution of humans from apes in the past and moves forward to the evolution of human technology in the future. Weapons and the natural instinct of humans as violent is the common characteristic across deep time. With this theme, the film secured a place in popular culture for Dart's scientific theory of innate aggression. One biographer of Stanley Kubrick said that Kubrick took his cue from Ardrey for every film he made after *2001*. The film linked Dart and Ardrey and Kubrick to the anti-war protests of the 1960s and '70s as well as the political debates about the arms race and the Cold War.

Ardrey was condemned by many for advancing "a fascist doctrine of original sin that justified totalitarian rule".[321] In February 1972 a letter appeared in *The New York Times* critiquing Kubrick's films as being "anti-liberal" with "fascist overtones". Kubrick wrote a letter in response, quoting Robert Ardrey at length. "It is, I am convinced, more optimistic to accept Ardrey's view that '... we were born of risen apes, not fallen angels, and the apes were armed killers besides ... The miracle of man is not how far he has sunk but how magnificently he has risen.'"[322]

African Genesis told readers that the development of a superior weapon was the central human dream of the human species, that there had been more energy focused on weapons development than any other endeavour and that it was the "most profoundly absorbing of human experiences". Ardrey argued that without this driving force, *Homo sapiens* would not have advanced. "Let us not be too hasty in our dismissal of war as an unblemished evil ... Across history, war has defended Christianity, the rule of law, the value of individual worth and laid the foundation for the Industrial Revolution." Ardrey argued that he was able to write *African Genesis* because he was a citizen of a country that "obtained freedom for its citizens through war, and that has successfully defended my freedom, by the same means, on all occasions since." He asked the reader: "Do you care about freedom? Dreams may have inspired it, and wishes promoted it, but only war and weapons have made it yours."[323]

After their first meeting Ardrey and Dart maintained a correspondence that spanned 25 years. In August 1963, Ardrey was so excited about the success of *African Genesis* that he wrote to Dart saying that he planned to take the next ten years to work on a study of man's evolutionary nature and write five more books in a series. After *African Genesis*, in 1966 Ardrey published *The Territorial Imperative: A Personal Inquiry into the Animal Origins of Property and Nations*, which also became hugely popular. His opening pages again reviewed his first meeting with Dart and acknowledged Dart's research as the inspiration for his books. This second book in the series investigated one aspect of human behaviour, what he describes as the innate tendency for humans to defend territory and its implications for private property and nation building. Ardrey argued that conflict is often constructive. For example, he wrote that for 80 generations, Jewish people had survived difficult conditions which resulted in their becoming tough and resourceful. The new state of Israel benefited from being surrounded by Arab states, he said, and Israelis should dread the day that the Arab League appeared at their doorstep with an olive branch. The defence of territory, said Ardrey, was what made Israel strong.

Ardrey proclaimed on South Africa as well. "The pariah state South Africa is attaining peaks of affluence, order, security and internal solidarity rivaled by few long-established nations" in part because it was surrounded by countries hostile to it. He agreed that "a degree of tyranny" contributed to South Africa's success but applauded the "natural alchemy" that had "transmuted a divided, unstable, near-bankrupt state on the verge of racial explosion into a stable, united, incredibly prospering nation in which the threat of racial explosion is almost nonexistent." By contrast, he said the black African countries throughout the continent "stagger along on one side or the other of the narrow line between order and chaos, solvency and bankruptcy, peace and blood".[324]

Carleton Coon reviewed *The Territorial Imperative* in the *Boston Globe*. "For over thirty years brainwashers have been trying to persuade us that man has no instincts ... All this documentation of territoriality in man and other species makes excellent sense and the burden of proof is on those who deny it. It is a powerful, courageous book."[325]

Ardrey's books led the way for other authors to join in the study of animals to show that humans are naturally aggressive. Konrad Lorenz published *On Aggression* in 1963 and it was translated into English in 1966. In 1967 Desmond Morris's *The Naked Ape* was published and it, too, gained a hugely popular following.

In 1968, anthropologist Ashley Montagu, who had been part of the original team to draft the 1950 UNESCO Statement on Race, and had also been one of the prominent opponents of Carleton Coon's *The Origin of Races*, brought together 14 scientists to write a volume of essays called *Man and Aggression*, in response to Ardey's work. In 1970, UNESCO hosted an International Aggression Debate that was attended by over 20 scientists. They were unanimous in rejecting the view that aggression was innate, inevitable and even beneficial. Later that year, Dart wrote a letter to George Price at the University College London, in which he said that it would be up to future scientists to determine "whether human beings are genetically disposed towards killing, breaking down and cruelty rather than saving life, building up and loving". Price later became well known for his studies of altruism in human behaviour, biology and genetics.[326]

With his controversial claims, Ardrey had pushed the boundary between academic science and popular science. His books and the debate around them exerted a large cultural influence and resulted in an increased interest in palaeoanthropology and human origins. They raised the question of who was best suited to interpret science for the broader public, and how. These questions would arise again in the 2000s, with Wits University palaeoanthropologist Lee Berger's approach to the field, and his efforts to share the academic science with a broader audience through popular books and social media. As a playwright, Ardrey had used drama and a theatrical style in his writing. His books utilised line drawings of australopithecines and other animals produced by his wife Berdine Ardrey, a South African. Ardrey wrote to his editor that the artwork "extends the essentially poetic and imaginative approach to the material." The role of artistic impressions of ancient hominids was often controversial, and would become so again after the announcement in South Africa of *Homo naledi* in 2015.

Ardrey produced a third book, *The Social Contract*, in 1970, and a final book in the series, *The Hunting Hypothesis*, was published in 1976. Berdine Ardrey wrote to Marjorie and Raymond Dart in October 1975 saying, "While I am glad that he has written a last short book on the old theme, I am hoping he will be inspired to start something new quite soon. When the inspiration happens, I will surely let you know." In 1977 the Ardreys moved to Cape

Town and Ardrey passed away in January 1980.

In the *New York Times* obituary, Phillip Tobias was quoted as saying, "He [Ardrey] has made an incalculable contribution to the science of human evolution. Thousands of people around the world, especially in the United States, were made aware [through his writing] of the fascination and the importance of studies on man's place in nature."[327]

Raymond Dart's theory of a predatory transition from ape to man spread further and had a greater impact on the general public than did his analysis of the Taung child skull. For many, the theory promoted the conclusion that genes had a greater influence on behaviour than did culture and learning. It advanced the idea of biological determinism in the famous phrase "nature over nurture". Given apartheid South Africa's need to defend its apartheid system, and the global realities of the Cold War, the narrative of human origins had been adapted to suit the needs of the day.

15

A Rather Odd Detective Story

Bob Brain first heard Raymond Dart speak about the blood-soaked archives of human history as a young man, in July 1955, at the Third Pan-African Congress of Prehistory in Livingstone, Zambia. Brain was intrigued by the concept that the transition from ape to man had been a violent, predatory one. He was taken by Dart's conclusion that nearly every piece of bone in the vast collection at Makapansgat had been used as a tool or a weapon. He mulled over Dart's dramatic theory for ten years before embarking on his own long-term project designed to throw new light on the subject. In 1965 he began his work at Swartkrans, an area not far from Sterkfontein, to investigate whether he could replicate Dart's findings.

John Robinson had been Brain's supervisor on his PhD thesis, "The Transvaal Ape-man bearing Cave Deposits based on geological work at Sterkfontein, Kromdraai and Swartkrans". Brain had almost completed his thesis when he realised he hadn't gathered any stone tools in the area. He thought that was strange so he started looking specifically for stone implements at Sterkfontein and he found them – beautiful artefacts made of quartzite. Then he looked at Swartkrans and was excited to find them there as well.

Robinson was away for a few months overseas so Brain sent him a long telegram telling him about the stone tools. Robinson replied saying that he would look at them when he got back. When he did, Brain showed him the tools and told him that he was so excited that he was going to write about them in his thesis. Robinson's reaction was not what Brain expected. "I forbid you to write anything about stone artefacts from the ape-man sites in your thesis because I am going to publish them," he said. Brain couldn't believe it. "Stuff you" was his response, and he went ahead and put the stone tools in his thesis. Brain remembers Robinson saying that he would ensure that Brain never had a position in palaeontology while he was active in the field. Brain was pleased Robinson left South Africa in 1963. "There's a lot of jealousy in the academic field, especially in this hominid field where there is a lot of public attention," Brain recalled.[328]

When Brain finished his PhD in 1957, there was no post available for him at the Transvaal Museum in the palaeontology department, but there was one in the reptile section, so Brain took the job. He spent many months counting the scales on every known species of snake in southern Africa. He also spent four years working as the head of zoology at the Queen Victoria Museum in Rhodesia, where he began to develop a narrative concept of museum display, something he would return to at the Transvaal Museum years later.

While he was in Rhodesia, Brain continued to speculate about the nature of early man. He examined the conditions of living primates in order to help him answer questions about extinct hominids. He questioned why early human ancestors had left a secure source of food in green forests to make a new life on the open savannah which had limited sources of food. Brain wondered whether this change had an impact on their behaviour. He investigated the behaviour of two species of monkeys in southern Africa, the samango and the vervet. Samangos live in forests. Their predators are few and their sources of food are plentiful. Brain discovered that samangos had a very relaxed social structure and showed little aggression towards one another. Vervets, on the other hand, lived on open savannah where predators were more numerous and food supplies uncertain. Vervets have a social structure that is much more hierarchical than the samangos and they maintain dominance in the group with frequent displays of aggression. Brain wondered if something similar had happened to the social structure of our pre-human ancestors when they had moved from the forest into the open. Brain's multidisciplinary interests in geology, zoology, and palaeontology served him well because these enabled him to draw from one discipline to inform another and resulted in his interest in mixing the study of fossils with the study of animal behaviour.

Robert Ardrey visited Brain's family in Rhodesia on a regular basis. When he was working on his second book, *The Territorial Imperative*, Ardrey asked if he could spend a week with Brain talking through his work. They drove Brain's Land Rover to the Gorongosa game reserve where they could talk undisturbed. Brain recalls that "Bob Ardrey's capacity for conversation through day and night was matched only by his capacity for Portuguese brandy. I took several days to recover from the experience."[329]

When Brain's work in Rhodesia was finished, he applied for a position back at the palaeontology department at the Transvaal Museum. John Robinson had emigrated to the US, and Jurgen Meester had moved to King William's Town so the position was open. Bob and Laura Brain and their four children moved back to Pretoria. Brain had one goal in mind. While Raymond Dart had investigated bones from Makapansgat to reach his conclusion about the violent origins of humans, Brain wondered what could be discovered from close investigations of the collections at Sterkfontein, Swartkrans and Kromdraai. Brain thought "it would be fun to find out".

The existing fossil collections at the Transvaal Museum from Sterkfontein and Kromdraai were quite extensive so Brain decided to expand on the Swartkrans collection with further field work. He began in April 1965.

Soon thereafter, the museum was looking for a new Director. The Chairman of the Board Fritz Eloff, was an Afrikaner zoologist. He was a member of the Broederbond, the secret male society devoted to the advancement of Afrikaner interests in South Africa. Brain remembers that Eloff was "sufficiently good as a zoologist to realise that evolution was a possibility."[330]

Brain made it onto the shortlist for the position of director of the museum, but Eloff received instructions from a government official to give the job to another candidate, William Steyn, an Afrikaner who was from Windhoek. After the protests in 1952 and 1963, perhaps the museum board members wanted to avoid another clash with the church and the government about human evolution.

After about a year, Steyn developed cancer, became seriously ill and died. The job was readvertised. Brain remembers that he almost didn't apply because of what had happened the first time. But he did and was appointed director in 1968.

It took Brain seven years to clear the rubble from the site at Swartkrans. He cleared out the remains of the mining at the site and the rubble from work that John Robinson and Robert Broom had done in the late 1940s and early 1950s. According to Brain, Broom and Robinson were only looking for hominids so the rest of the fossils were tossed aside. A large dump on the lower hillside contained thousands of tons of rock that had been blasted from the site so Brain and his team meticulously worked through that as well. The cave itself was also filled with rubble. Clearing and ordering all this rock was tedious, but Brain's quiet, patient nature helped him carry on with the task. It was frustrating, none the less, because all of the fossils that he knocked out of the breccia were not *in situ* so they weren't exactly sure which part of the cave they came from.[331]

Nevertheless, those first seven years of work, from 1965 to 1972, resulted in a much better understanding of the structure and the stratigraphy of the cave. Only then could Brain begin the second phase of the project. He told Phillip Tobias that he was looking at the whole faunal assemblage at the site and that he was trying to reconstruct what sorts of animals had lived there. Brain regarded the hominids as part of the fauna, but not the most interesting part. "Tobias more or less gave up on me," remembered Brain. "I think his view was that anyone who thought the hominids were just part of the fauna must have been a bloody fool."[332]

It took another seven years to clear the natural rock and soil that had obscured a lower older deposit. This work was done by Brain's field team and supervised by George Moenda. A seemingly interminable search began for this lower bank. Brain felt that it took on some of the qualities of a religious quest. Colleagues would ask him, "Have you found it yet?" and Brain would reply, "No, but I know it's there!"[333]

Once the concrete-like breccia was removed from the site, much of the time-consuming process of removing fossils from the breccia was done by Laura Brain and others in a work area outside the Brains' house in Irene, Pretoria. "We dissolved the specimens in acetic acid in my back yard because I didn't have facilities at the museum," remembers Brain. "The whole family was involved. Our daughter Virginia did amazing analysis of the composition of the Swartkrans bone accumulations in terms of the skeletal parts and the animals represented. We knew exactly where every scrap of bone came from within a 10cm spit in our excavation. I had over 240 000 pieces of fossil bone." When asked how they kept track of all the bones, Brain said, "With great difficulty." The door from the house to the back yard was off the kitchen. The family kept track of thousands of fossil bones inside and outside the house amidst school projects, dinner preparations and three exuberant dogs.[334]

"This is a detective story, but rather an odd one," Brain writes in the opening line of *The Hunters or the Hunted?*, the book he published in 1981, 16 years after he began his work at Swartkrans. "The clues are bones, and the aim of the investigation is to establish causes of death, but the evidence is ancient and no witnesses survive to relate their experience." The book provided an alternative view to the hypothesis originally presented by Raymond Dart. "It may seem ridiculous for anyone to devote sixteen years of his life to the study of a single cave," Brain wrote, "yet this is how long it has taken for me to understand something of the complexity of the cave's form and its filling, as well as to gain some insight into the ways that the bones found their way there."[335]

The analysis Dart had undertaken at Makapansgat and which Brain continued at Swartkrans became known as the field of taphonomy, the study of what happens to decaying organisms over time. In this new field, Brain investigated modern animal behaviour to inform his analysis of ancient fossil remains. Survival of different parts of skeletons was not haphazard but related to the qualities of different parts of the bone. Brain's investigations led him to conclude that the unchewable sections of the skeleton survived the best.

Dart's analysis of predation was based on the remains of many antelopes in Makapansgat. Mandibles survived most often and spinal vertebrae survived least often, a pattern that Brain saw in the Sterkfontein Valley as well. Another similarity was that in Sterkfontein and Swartkrans there were fewer primate remains than bovid remains, a pattern Dart had also found in Makapansgat. Bovids are horned animals, including antelope, impala and springbok. Brain saw that primate skulls remained behind most often and that the rest of the skeleton was rarely found. Dart had looked at these accumulations of bones and concluded that the australopithecines were headhunters. But Brain had another, very different theory. In the hope that it might shed light on the way the sabre-toothed cats of ancient times treated bones, he had begun to investigate the way cheetahs damaged the skeletons of their prey. With the help of the Natal Parks Board, Brain was able to observe the feeding behaviour of six cheetahs who were about to be released into the park. They were given an antelope as food. Brain observed very little damage to the skeleton.[336]

He repeated his study of the food remains of cheetahs a few years later and the results were the same. Then Brain gave the cheetahs a male baboon to eat and he was astonished by the results. In the process of eating, the cheetahs chewed away the ribcage. They also chewed up and swallowed the spinal column. This suggested to Brain that a primate backbone is much easier to chew than that of an antelope (or any kind of bovid). On another occasion, when they were equally hungry, the same cheetahs were given a sheep, which they proceeded to eat, but the vertebral column was left intact.

Brain was not suggesting that cheetahs were contributing to the bone assemblages in Swartkrans and Makapansgat. Instead, he thought the experiments might be suggesting that carnivores in the past behaved similarly and might have left similar bone remains from their prey. Perhaps this could explain why, in their field work, they were finding more abundant and complete remains of bovids and fewer remains of primates. The reason was in the different ways in which carnivores could chew up the remains. The bones that were very dense survived and the ones that were fragile were eaten. Unlike Dart's conclusion that different bones survived because they were selected as weapons, Brain concluded that certain bones survived because they weren't easily chewed.

Many scientists agreed with Brain. A 1977 *Time* magazine article said: "Most anthropologists reject the notion popularized by Robert Ardrey (*The*

Terrorist Imperative) and others that man is inherently aggressive and that his murderous instincts derive from his apelike origins." The article went on to say that humans were likely peaceful hunter-gatherers until the invention of agriculture around 10 000 years ago which resulted in property and antagonisms between people.[337] This was also the time-frame in which many anthropologists began to develop the myth of the isolated, peaceful San people. Instead of being portrayed as savages as they had previously, "Bushmen" were being described as being in perfect harmony with nature.[338] Also, at this stage, in terms of the storyline of human origins, the anti-war movement of the 1970s was in full force and instead of the concept of a Killer Ape, the idea of co-operation and peace was more popular.

Brain was so excited by his conclusion about the use of bones that he took a large sample over to Wits to Raymond Dart's office. He laid the bones out on a wooden table for Dart to take a look. Brain showed him the survival proportions and shared his theory. For about ten minutes, Dart was completely quiet. He didn't say anything. Then suddenly his eyes lit up and he said "Brain, this is remarkable. At last we are getting closer to the truth."

"I was very impressed by that," recalls Brain. "The fact that Dart's much loved concept could be shown to have an alternative explanation. The fact that he was big enough, accommodating enough, to accept that. And to be enthusiastic about how the work should go on. That's not something you find with many other researchers."[339]

In contrast, Robert Ardrey was not impressed with Brain's work. He did not accept Brain's evidence that Dart's conclusions needed to be reinterpreted. He held onto Dart's concept of the powerful hunter and the killer instinct. "He never gave it up," said Brain. "And he never came back to see us. He just stopped all contact."[340]

Just as Piltdown Man had led scientists astray for 40 years in the first half of the twentieth century, Raymond Dart's osteodontokeratic culture led the public astray for more than 20 years in the second half. In fact, Ardrey's books are still circulated and debated. The theory of a violent human ancestor met with Robert Ardrey's expectations, and his popular books helped promote a theory that cemented his assumptions, even after they had been proven to be false. Sometimes it is not the factual information about human origins that is the most popular, but rather the epic manner in which a story is told.

Brain has described another scene with Dart from the 1980s. When Brain started to explore the role of stone tools at Swartkrans, Raymond and Marjorie Dart visited his house one afternoon. Dart was over 90 years old and almost blind. He felt the texture of the stones and felt the point of each stone carefully. Then he said "Brain, what have these stones been used for?" Brain said, "I think they've been used for digging in the ground." Dart's jaw fell open and he said, "Brain, that is the most unromantic explanation I've heard in my life." He took the stone tool and stuck it right between Brain's ribs and said, "I could run you through with this."[341]

16

It's Not Only About the Bones

In November 1977, a young Richard Leakey was on the cover of *Time* magazine with the title "How Man Became Man". Mary and Louis Leakey's second son, Richard, had followed his parents into the field of archaeology and he continued his father's work after the latter's death in 1972. By the time Richard turned 32 in 1977, he had spent ten years working in the field, and had built his own reputation gathering over 300 fossils of human ancestors on the banks of Lake Turkana. The coverage in *Time* magazine was an indication that not only was scientific opinion giving credence to the idea that human ancestors came from Africa, but also that broader public opinion and interest was shifting towards Africa as well. The article focused on Leakey's recent finds in Kenya, but it also devoted space to Raymond Dart and the Taung skull as well as Robert Broom's finds of australopithecines at Sterkfontein in 1936, 1938 and 1947. But the discovery of *Homo habilis* by the Leakeys and its description by Tobias and Napier in 1964 had raised a new problem. At first scientists thought *Homo habilis* might have evolved from its South African cousin, but then evidence emerged that both species lived at about the same time. If both species of pre-humans lived at the same time, which one truly led to the development of *Homo sapiens*?[342]

There were many discoveries that were starting to put together pieces of the vast, complex jigsaw puzzle. In 1974, palaeoanthropologist Don Johanson's team found something remarkable in the Afar region of Ethiopia – not a tooth here and a mandible there, but large portions of an entire skeleton of an *Australopithecus*. The team gave it the name Lucy after the Beatles song "Lucy in the Sky With Diamonds" which was playing the night they celebrated the find. Once again, a fossil had been given a name to personify and humanise it. Lucy was about three and a half feet tall. She had short legs but was definitely bipedal and walked upright, supporting Dart's theory that the australopithecines were bipedal. Initially, the team placed Lucy in the same species as the Taung child skull and Mrs Ples, *Australopithecus africanus*, but after there were more fossil finds with similar morphology, they decided the bones represented a new species, which they named *Australopithecus afarensis*.[343]

The *Time* article described Leakey's find of a skull around Lake Rudolf (now Lake Turkana) in 1972. Leakey concluded that it was probably a specimen of *Homo habilis* and that it was more than two million years old, making a strong case that *Homo habilis* and *Australopithecus* had co-existed, "thus weakening arguments that the latter was man's direct ancestor". Leakey and his team made a policy decision not to name a species, but rather to refer to a fossil by its catalogue number (1470 at the National Museum of Kenya), and then study it more fully. They also believed that the specimens were so fragmentary at that stage that it was hard to come to firm conclusions about where the fossil fit into the human family tree.[344] In 1978, a Russian scientist, Valery Alekseyev, proposed that the fossil be called *Homo rudolfensis*, named after the lake on which it was found.[345]

In the same article, Phillip Tobias commented on the changing thinking about human evolution. "We can no longer talk of a great chain of being in the 19th century sense from which there is a missing link." Instead, said Tobias, "we should think rather of multiple strands forming a network of evolving populations, diverging and converging, some strands disappearing, others giving rise to further evolutionary development." The human family tree with one main trunk was developing multiple branches of different sizes and turning into a bush.[346]

Scientists were beginning to ask questions about the diet and behaviour of these creatures. Some people believed that *Australopithecus africanus* began to eat meat when it was struggling to increase its food supply, probably scavenging from other predators. *Australopithecus robustus*, however, remained vegetarian, eating seeds and nuts and leaves. Stone tools entered the equation as well. Bob Brain and others started to argue that hunting brought about co-operation and encouraged hominids to be more social. It was no longer only about the bones, but also about the geology and the stratigraphy of the surroundings, as well as what other clues there were about the environment. The field of cave taphonomy, developed by Brain, had begun to play an important role in understanding the context of the fossils.

Leakey closed the article by saying that "racial differences, as they are commonly perceived, are a superficial and recent development, having arisen only about 15,000 years ago". Leakey said: "I am aghast that people think they are different from each other. We all share a tremendous heritage, an exciting bond. We are all the same."[347] While the theory of Carleton Coon had been squarely rejected by most scientists, there was still uncertainty in the public domain. Coon's public lectures had caused many people to question the role Africa played in contributing to human evolution in other parts of the globe, and Coon's work still left some people convinced that people of different races evolved separately.

In the same 1977 issue of *Time*, there was another article about the fact that the United Nations was preparing to adopt a mandatory arms embargo against South Africa in response to the South African government's latest wave of arrests and bannings. A voluntary arms embargo had been in place since 1963 but more recent events had resulted in the UN wanting to make it mandatory. In June 1976, the violent police response to the Soweto uprising had resulted in hundreds of deaths, at least 23 on the first day. Just two months before the article appeared, in September 1977 Steve Biko, one of the founders and chief proponent of the Black Consciousness Movement, had been killed while in detention. A new wave of arrests and the banning of organisations and newspapers had resulted in the UN wanting to take further action.[348]

Richard Leakey was much more aware than his parents had been of the political implications of working with a South African in the apartheid era, and the relationship between Tobias and the Leakeys had begun to change. Throughout the 1960s and '70s, Tobias had never had trouble travelling to Kenya or Tanzania because he travelled on a British passport, as a result of his father having been born in England. Given the political developments in South Africa, however, it became increasingly difficult for Tobias to travel, even with his British passport. Also, after Louis Leakey's death in 1972, Richard discouraged his mother from making any more trips to apartheid South Africa.

Throughout the apartheid years, the concept of evolution was not regularly taught in schools in South Africa. As one example of the impact, in 1976, just after the Soweto uprising, the Wits Council of Churches provided support to black students who had not been able to finish matric. One teacher who dared to mention evolution to the group was shocked to find that the students thought that they would be taught the horrible falsehood that black people had evolved from baboons. This myth had grown from the fact that many white people used the term "baboon" as an insult. In many schools, black children were being taught that they were culturally inferior to white people because of their worship of their ancestors instead of the Bible. Ironically, at the same time, scientists were searching for hominid ancestors. In fact, there have often been misunderstandings about the term "ancestor". Many African people use the word to refer to their relatives who lived a century ago, whereas paleoanthropologists use the same term to refer to hominids that lived one or two million years ago.[349]

Shortly after Steve Biko's death in September 1977, Tobias, who had recently become the dean of the Wits Medical School, called a mass meeting of the students and faculty to protest Biko's death and the shocking way doctors had mishandled the situation. Biko had been badly beaten before he died. Despite his injuries, the doctors who examined him minimised their impact. Dr Ivor Lang recorded that he found nothing wrong with Biko. Dr Benjamin Tucker examined him and recommended that Biko be taken to the hospital, but after the security police argued with him, he changed his mind and said it wasn't necessary. Dr Lang allowed the police to drive Biko 700 kilometres to Pretoria. Biko made the trip naked in the back of a van where he was injured further – reminiscent of /Keri-/Keri's body making the trip in the back of a bakkie from Oudtshoorn to Johannesburg. Biko died the next night.[350]

Tobias co-ordinated submissions on Biko's death to the Medical and Dental Council but they were ignored. A group of people from the medical schools at Wits, Cape Town and Durban, including Tobias and his colleague Trefor Jenkins, took the Medical and Dental Council to court. They took the case to the Pretoria Supreme Court, and to their surprise obtained a unanimous verdict invalidating the decision of the Medical and Dental Council that there was no case. On 30 January 1985, the judgment ordered the council to hold a formal disciplinary hearing for the two doctors involved, and to pay the costs for the case. The council did hold the hearing and Dr Lang and Dr Tucker were found guilty but given very light sentences. Several months later, however, the council took the decision to increase the severity of the sentences and both doctors were struck off the roll.

Four of the people who took the SA Medical and Dental Council to court over the treatment of Steve Biko by the medical doctors in Port Elizabeth. From left to right, Phillip Tobias, Frances Ames, Joe Veriava and Trefor Jenkins. Courtesy of Tobias in Conversation published by Wits University Press.

During this period of increasing protest and opposition to the apartheid regime, scientific work continued. South African government institutions shied away from supporting any research that looked at the importance of African culture, and funding for the research at Mapungubwe suffered during this period. Phillip Tobias and Bob Brain, whose work focused on bones rather than culture, continued to receive funding for their work at Sterkfontein and Swartkrans. The CSIR continued to provide a modest amount of funding for Tobias at Wits and for Brain at the Transvaal Museum. The New York-based Wenner Gren Foundation also offered targeted funding for palaeoanthropology.

One month before Steve Biko's death, Trefor Jenkins, who later became involved with taking the Medical Council to court, gave his inaugural lecture at Wits. The title was "From Generation to Generation", and it referred to the history of human genetics, which was Jenkins' field.[351] Jenkins began by explaining that the early history of human genetics was greatly influenced by eugenics and its founder, Francis Galton, who advocated the importance of "selective breeding" in order to improve the future generations of humans. Jenkins described Galton's efforts, in 1851, to measure the buttocks of a young woman in South Africa from a distance with his sextant. Harking back to his own unseemly research in the 1960s, Jenkins said that he had not been able to find Galton's original measurements "so I have not been able to compare them with my own, obtained, I must confess, by much bolder and less discreet techniques."[352]

Jenkins went on to discuss the influence of Eugen Fischer, the German physical anthropologist who conducted research in South-West Africa among the "Rehoboth Basters", in 1908. Jenkins concluded that Fischer himself was a Nazi and suggested that the field of eugenics had dragged anthropology and human genetics down with it. Jenkins said that human genetics continued to be a field in which "our science can easily be misapplied by politicians and because of its vulnerability we must constantly be on our guard."[353] Turning to race classification in South Africa, Jenkins said that when a society imposed discriminatory legislation, "definitions" became bizarrely necessary, as had been the case in Germany in terms of how to define a Jewish person. Jenkins gave his speech in August 1977, but he was still referring back to problems with the approach of typology that looked for a particular archetype for each race. He applauded the shift to population genetics instead that saw each individual as "unique, with a unique genome, which ensures that he is different from every other person." Despite his support for this positive development in scientific thinking, Jenkins proceeded to share with his audience that his research estimated the genetic composition of the "Johannesburg Coloured population" to be "33 percent Caucasoid, 33 percent Khoikhoi, 25 percent Negro and 9 percent San". Jenkins argued that this kind of research was important because some people had incorrectly argued that "the Caucasoid contribution to the genetic constitution of the Coloured people of South Africa (was) minimal". It appeared that Jenkins had not followed his own advice, and had dropped his guard.

At the same time, Bob Brain was working at Swartkrans. He also led the development of a narrative display sequence for the Transvaal Museum called "Life's Genesis". The plan was to tell the story of the gradual evolution of life on earth. Brain was convinced that visitors to the museum would benefit from seeing a narrative display sequence rather than in an uncoordinated group of displays. His team spent five years putting together the first part of the exhibit, which covered the story of life's development from the smallest protozoa to fish to reptiles. The exhibit was opened in November 1978 and in the first three months it was visited by 17 000 people. "… this narrative type display sequence has been well-received," said Brain. "I have no reservations about the wisdom of proceeding with the second part of this particular project."[354] The plan for the second part of the exhibit, "Life's Genesis II", was to display the story of the development of life from mammal-like reptiles to mammals to humans. Brain was taking a chance that a new exhibit would not result in a third protest at the Transvaal Museum against the concept of human evolution.

Elizabeth Vrba was another South African scientist at the Transvaal Museum who made important contributions to palaeontology in the 1970s. Vrba received her PhD from the University of Cape Town in zoology before she began teaching high school science at St Alban's High School for Boys in Pretoria. She approached Bob Brain, asking him for suggestions on how she could pursue her studies. Brain showed Vrba a storeroom filled with unstudied bovid fossils from the caves in the Sterkfontein Valley. Like Brain, Vrba was interested not only in hominid fossils but also in what their surroundings could reveal. Brain remembers that the two of them "had several fierce encounters as each thought the other to be intruding on what should have been an exclusive intellectual hunting territory", but these disputes taught him that "the subject matter in natural history is always more interesting than is the position of one's ego relative to it; that the significance of a discovery lies not, first and foremost, in who made it."[355]

As Vrba began to study the antelope fossils, she noticed there had been dramatic evolutionary change in the bovids about 2.5 million years ago. In a sudden burst, many new species emerged and many older ones died out. Vrba began looking for the cause. She found that one species of impala that had survived unchanged for three million years did so because it was able to adjust to different kinds of vegetation available, but that other antelopes, those with more specialised diets, when faced with changing patterns of vegetation, had to move to other areas or face extinction.[356]

Vrba concluded that environmental change, and in particular, massive change in the climate, was causing this major evolutionary change. Based on

Vrba's observations, she developed a theory of how climate change had affected evolution. Her view was that abrupt changes in the climate had produced bursts of new species in Africa, including hominids. When there was no climate change, there was little evolutionary change. Vrba called her theory the "turnover pulse hypothesis". After looking at what happened to the bovids around 2.5 million years ago, she then saw that the major cooling of the world at that time caused a major change in the African environment. When she noticed that it was around 2.5 million years ago that *Australopithecus robustus* first appeared and that *Homo* made its first appearance, she realised there had to be a connection.

Since the 1950s, the predominant theory of evolution internationally, which most scientists agreed with, was called "the evolutionary synthesis" or "the modern synthesis". This was the idea that evolution consisted of the slow, steady accumulation of small changes over long periods of time. This was the theory Ernst Mayr had put forward in the early 1940s. But in 1972, Stephen Jay Gould and Niles Eldredge, who was from the American Museum of National History, presented a counter-theory. From his study of an ancient species of land snail, Gould had noticed that millions of years went by with little or no evolutionary change. Eldredge had noticed a similar trend in his study of ancient marine arthropod fossils. They argued that if evolution was the result of gradual and continuous change over time, the fossil record should reflect that – but it didn't. The fossil record included great gaps. For a long time, scientists attributed the gaps to the need to find more fossils to fill in those gaps. But even with the increase of fossils found over time, the gaps remained. Gould and Eldridge both agreed that evolution didn't happen gradually over time, but rather in bursts of great change, which they called "punctuated equilibrium". They proposed that these sporadic periods of great change were followed by long periods where there was little or no change at all. Gould and Eldridge's critics called their theory "evolution by jerks". Gould retaliated, saying his critics promoted "evolution by creeps".[357]

The theory of "punctuated equilibrium" fit well with Elisabeth Vrba's "turnover pulse hypothesis". However, they were different because the Gould/ Eldridge theory was the result of internal, genetic changes within a species and Vrba's theory was the result of external influences of climate change. "Speciation," said Vrba "does not occur unless forced by changes in the physical environment."[358] Her work on the rate of evolutionary change was recognised internationally. She and Gould wrote a paper together.

The debate continued between the jerks and the creeps, and between the jumpers and the gradualists. The jumpers pointed out that no one had ever found an intermediate form between one species and the next. The existence of big leaps in evolution meant that the exact point of human origin could be a fantasy. The concept of a missing link began to make no sense. Not only was there no missing link in the chain, there was no chain.

Vrba's work suggested that when you find a major change in the hominid lineage, there must have been massive environmental pressure to account for it. Also, these hominids survived because they were resourceful and able to adjust to changing conditions. Creatures that were reliant on one set of environmental conditions would die out.

One major leap in hominid development was the changing of the pelvis and the foot so that apes could become bipedal. This happened with the australopiths. The next big major change was the growth of the brain, which likely happened about four million years later, possibly with *Homo habilis*. Both of these massive changes seemed to have occurred alongside major changes in climate. When the rainforest shrank and vast areas of land in Africa became drier, apes that had been in the trees were forced onto open grassland.

"Man," said Vrba to the author Bruce Chatwin, "was born in adversity. Adversity in this case, is aridity." "You mean that man was born in the desert?" he asked her. "Yes," she said. "The desert. Or at least the semi-desert." "Where the sources of water were always undependable?" "Yes." "But there were plenty of beasts about?" "A carnivore doesn't care where he lives so long as he gets his meat. It must have been terrible!"[359]

It made no sense, Vrba said, to study the emergence of man in a vacuum, without pondering the fate of other species over the same time scale. About four million years after the australopiths became bipedal, around 2.5 million years ago, hominids took another jump, at the same time that there was a "tremendous churning over of species. All hell broke loose among the antelopes."[360] Again there was a massive drought. To survive, any species had to adapt or move.

Just as scientists began to look more closely at the surrounding environment of the ancient fossils, so too did physical anthropologists begin to look at modern skeletons within the context of where each individual had lived, and the influences while that person had been alive. The typology approach was falling away and a contextual approach was gaining ground.

One of the scientists who emphasised the contextual approach was a young man from Canada, Alan Morris. Morris came to South Africa to study with Phillip Tobias and he decided to focus his PhD on a study of the Griqua skeletons from Campbell. While he did not focus on ancient hominid fossils, Morris did carry forward Tobias's work in physical anthropology and the examination of more modern human skeletons.

181

Morris had read about Tobias's work in the Kalahari and wrote to him, asking if he could study with him. Before his arrival in South Africa, Morris completed a project on the thigh bones of the Iroquois people, and another study on the human remains and pelvic anatomy of the Ossossane people, one of the First Nations of Canada. He arrived at Wits Medical School in Johannesburg in January 1975. At the time, he admits, he didn't give much thought to the politics of the country. What struck him most on arrival was that there was no television. The South African government believed that television would introduce wrong thinking about the interaction between races; it did, however, introduce it the following year. In the meantime Morris had to adjust to radio.[361]

Tobias assigned Alun Hughes, the often bad-tempered but generally kind chief technician at the department of anatomy, to help Morris settle in. Morris recalls that Hughes constantly complained about working for Tobias, who he thought was authoritarian and demanding. Hughes had worked for Raymond Dart in the late 1940s and 50s as a technician in Makapansgat and had stayed on to work for Tobias when he took over in 1959. He had worked for Tobias for over 15 years and offered the professor his respect, but Hughes always longed for the "good old days" with Dart.

Alan Morris, Phillip Tobias, Ron Clarke, Ivan Suzman, Raymond Dart and Fred Grine, July 1976. Courtesy of Tobias in Conversation published by Wits University Press. Copyright: School of Anatomical Sciences, University of the Witwatersrand.

Morris went on one of the annual field trips with Phillip Tobias out to Sterkfontein and Swartkrans. He described it as the one time in the year when Tobias allowed himself to relax his status slightly and socialise with the students. Tobias was about to turn 50 later in 1975 and treasured the respect he had gained as a professor, and the iconic status he had attained among the white liberal community in Johannesburg.

Tobias asked all of the staff – both technical and academic, black and white – to come together each day for a morning tea break. In the 1970s, most of the academic staff called the African technicians by their first names but expected to be called by their title and last names in return. This struck Morris as unusual (it wasn't the norm in Canada) and he encouraged the black staff to call him by his first name. He was impressed when Tobias invited Desmond Tutu to speak with the medical students at the cadaver dedication ceremony in March 1975. At about the same time, Morris was putting together a group of what he described as "Zulu skeletons" from the Raymond Dart collection so that he could conduct research on the thigh bone in order to deliver a paper at a conference. No one was studying the distinction between the Afrikaner thigh bone and the English thigh bone, however.

Over time, Morris clarified the area of his PhD thesis. He would examine four sets of skeletons from the northern Cape and the Western Orange Free State. One of the sets of skeletons would be those that Tobias had exhumed in Campbell. Because Tobias had not yet written or published about the Griqua skeletons from Campbell, he gave all of his notes, excavation data, photographs and documentation to Morris. This included notebooks from each of the four exhumations in 1961, 1963, 1967 and 1971. Since Tobias had not conducted any research on the skeletons throughout the 1960s or '70s, it is not clear what his plan was in terms of "serving science" and then returning the skeletons to the people of Campbell. The two-year expectation from Basil Humphreys had come and gone. In fact, ill health had forced Humphreys to leave his farm outside Campbell in 1965 and move to Kimberley. He passed away in 1971. There was no one at that time in a position to advocate that Tobias return the skeletons. Adam Kok IV was not in a position to do so, given his age and the apartheid dynamics at the time. In 1978, at the age of 89, he too passed away. It was not until Alan Morris settled on his PhD topic in 1977 that any focused scientific attention was given to the skeletons from Campbell.

Morris had complete access to the 35 skeletons, all of which were stored in wooden boxes. By March 1979, he had measured all the bones and made all the comparative observations he needed for his research. He took a range of cranial measurements, analysing the shape of each brain cast, face and mandible. Morris compared the measurements of each skeleton amongst the group and then compared them with three other groups of skeletons from graves along the Orange River. He followed Hertha de Villiers' methodology and, unlike Dart and Tobias, used multivariate statistics in his analysis. Morris was not looking for an elusive pure racial type; rather he wanted to see how specific individuals compared to a range of variation. He wanted to examine the archaeological and historical context as well.

Morris was one of a number of international students who travelled to Johannesburg to do research. In the late 1970s, Bernard Wood, a young British palaeoanthropologist who had been studying with Richard Leakey, visited the department of anatomy at Wits. When he couldn't find a place to stay, Hertha de Villiers offered that he could stay with her family in Halfway House, midway from Johannesburg to Pretoria. Wood befriended De Villiers' daughter Phillippa, who by that time was 11 years old, and they often chatted about human evolution. As a gift, he gave her a book he had recently published called *The Evolution of Early Man*. The book inspired Phillippa and fired her imagination. She loved the chapter on "*Australopithecus* – Ancestor or Rival", knowing that these hominids lived millions of years ago in South Africa, not far from her home. At school, Phillippa was surprised to find that some people didn't believe in evolution. How could you protest against something that was so obvious, she thought.[362] She also liked the page in the book that showed "Races and racial types" from across the world – all the different faces and expressions. The caption said: "There is no evidence that any one modern race is the direct and sole descendant of a single ancient population which evolved at a different rate from others; for example man in Europe would not have become modern long before this happened in Africa or Asia." Fifteen years after Carleton Coon's book, Bernard Wood's text was speaking out against Coon's theory.[363]

At about the same time as Wood's book came out, Hertha de Villiers had completed important research of her own. She had been examining human skeletal remains at the Bushman Rock Shelter in the Eastern Transvaal and at Border Cave on the border with Swaziland and concluded that the ancestors of "Bantu-speaking people" could have made their home in South Africa as long as 80 000 years ago. *The Star* reported that if De Villiers could confirm her theory, the findings would not only support studies that refuted "the popular dogma that negroids arrived in southern Africa only 300 or 400 years ago", but in addition would "push that time span back beyond the presently-believed 2000 year date". These research findings were in direct opposition to the views that had been held by Dart and others on Mapungubwe and Great Zimbabwe and the myth of the empty land.[364]

The Star said that De Villiers' work would "help piece together the jigsaw puzzle of mankind's origins". Tobias explained that the thinking at the time was that the "Khoisan" lived in southern Africa and that the "Negroids" lived in Central Africa and only moved south 2 000 years ago. He also said that the popular theory was that all sub-Saharan Africans had a common ancestor and that the two lineages diverged about 20 000 to 30 000 years ago. "If Professor De Villiers is correct in her theory," said Tobias, "it means that two kinds of man were co-existing in southern Africa and this raises many questions such as how did they survive without merging? We might have to change our thinking."[365]

In 1978, Tobias published a book that brought together his own research findings from 15 years of leading the Kalahari Research Committee. It was called *The Bushmen: San Hunters and Herders of Southern Africa*. In it, Tobias reviewed recent efforts to clarify the terms used to describe the biology, language and economies of different groups of people. At that stage, the term "San" was used to describe the race or physical type of people who were hunter-gatherers, and Tobias explained that "Bushman" referred to their collection of languages. The term "Hottentot" referred to the language of people who had been cattle owners and he described their race as "Khoikhoi". Tobias wrote that the term "Negro" described the race of Bantu-language speakers.[366] Raymond Dart wrote the foreword for the book and in it he referred to his initial expedition to the Kalahari back in 1936. He congratulated Tobias for the 21 expeditions Wits had led to the Kalahari since. "Forty years ago there were few if any more than forty professional anthropologists in the United States; today there are thousands," wrote Dart.[367]

In Tobias's opening paragraph, he explained that the Bushmen had long lived a life of subsistence by hunting and gathering, a way of life "followed by Man's fossil ancestors, ever since the first appearance in Africa of *Homo habilis*". He suggested that *Australopithecus africanus*, "the probable ancestors of all later forms of humanity", might have lived "a way of life that foreshadowed that of the Bushmen."[368] In his chapter entitled "The San: An Evolutionary Perspective", Tobias wrote: "In general, it would not be accurate to speak of them as 'living fossils', since most of them do not represent the survival of an otherwise extinct kind of man. It is rather their *way of existence* that is a living fossil for it represents the persistence of a lifestyle that has all but vanished from earth."[369] Either way, it was preposterous that Tobias still referred to living people as "living fossils". At the end of his chapter, Tobias returned to this term that had first been used 50 years earlier in the 1920s by Dart and Smuts.

Tobias wrote that the Khoisan and the Negro had a common ancestor as recently as between 15 000 and 25 000 years ago. "Once the dichotomy was established, there followed a fairly lengthy period in which the Khoisan line and the Negro line were geographically isolated from each other." Tobias closed by saying, "In the sense that their gene-set may well provide us with some telling clues as to the gene make-up of ancient Africa, we may regard the San as being living fossils, not only in their hunter-gatherer ways of life, but perhaps even in their biological heritage."[370]

17

"Apartheid Fossils Not Wanted"

On 2 February 1984, on a warm, sunny day in the Sterkfontein Valley, Bob Brain was working at Swartkrans with George Moenda, his experienced excavator and foreman. Moenda was digging out small areas of dirt and passing fragments of bone on to Brain for further inspection, something he had been doing for 14 years. The British author and explorer, Bruce Chatwin, had joined them that day. Standing above the excavation site, Chatwin could see across the grassy hills towards Pretoria. In the other direction, he could see the roof of the Sterkfontein tea room and museum, and off in the distance was the Krugersdorp Mine. During Chatwin's short visit, he wrote to a friend: "I've had perhaps the most stimulating discussions in my life."[371] Brain showed Chatwin the lower level of the cave where two million years before the australopiths were dragged there and eaten by carnivores. Then he pointed to the upper layer where the carnivores had disappeared and man's ancestors took control. How had things changed? "The only way to inhabit a cave which is inhabited by predators," said Brain, "is to deter them with fire."

There were very few trees in the valley so in the heat of the day, the three men rested in the hut. Half an hour later, they were back at the excavation site. Moving forward another inch with the search in the upper layer of the cave, Moenda found a piece of blackened bone. He paused to look at it with great interest and surprise, and then walked over to show it to Brain. Brain turned the bone over in his hands and said, "That bone is remarkably suggestive!"[372] Not wanting to jump to conclusions, Brain took another four years to confirm that the bone had been burnt, not by chance, but by controlled fire. Inspired by Brain's persistence over decades and the scope of his work, Chatwin sent a postcard to a friend saying Brain "should be given a Nobel Prize on the spot".

Bob Brain deserved wide recognition for his never-ending patience and the important achievements in his career. So did George Moenda. Yet there is very little information available on Moenda's life and work. According to the librarian at the Ditsong (formerly Transvaal) Museum, there is currently no record of his employment. In a book that Brain edited in 2004, he mentions that Moenda had been working at the museum for 34 years. Brain wrote that Moenda "helped with all the escavations and built the extensive stone walls and steps that now make the site accessible".[373] Despite Moenda's relatively recent presence in the field, there is less information about him in the public record than there is about Saul Sithole and Daniel Mosehle.

The third person on site at Swartkrans on that fateful day, Bruce Chatwin, wrote a book. Twenty-seven years after Robert Ardrey's *African Genesis,* Bruce Chatwin's *The Songlines* was another long meditation on the origins of humans, but this time Chatwin concluded that early humans had a nomadic nature that was unaggressive and respectful of the land. In the book, Chatwin learned to "reject out of hand all arguments for the nastiness of human nature. The idea of returning to an 'original simplicity' was not naïve or unscientific or out of touch with reality." Chatwin believed that it was a source of hope for the future.[374]

The Songlines was about Chatwin's trip to Australia to research Aboriginal song and its meaning. But in the second half of the book, Chatwin explored his ideas about early humans and their nomadic lifestyle. It is in this section of the book that he refers to his time with Brain. He didn't write about finding fire because he respected Brain and wanted him to publish first. His focus instead was on Brain's new interpretation of the fossil evidence. "All those antelope bones which Dart claimed were bludgeons and daggers and so forth were precisely those parts of the skeleton which a big cat would leave after its meal." Chatwin asks "Could it be that *Dinofelis* [a pre-historic cat] was Our Beast? A Beast set aside from all the other Avatars of Hell? The Arch-Enemy who stalked us, stealthily and cunningly, wherever we went? But whom, in the end, we got the better of?"[375]

Chatwin was fascinated by the research to find out when and how ancient humans developed a larger brain. "'I know this may sound far-fetched,' Chatwin asked Elizabeth Vrba, 'but if I were asked, 'What is the big brain for? I would be tempted to say, 'For singing our way through the wilderness.' She looked a bit startled. Then, reaching for a drawer in her desk, she brought out a painting, an artist's impression of a family of hominids tramping in single file across an empty waste. She smiled and said, 'I also think the hominids migrated.'"[376]

It was around the same time of Chatwin's visit that Bob Brain opened the "Life's Genesis II" exhibit at the Transvaal Museum, which focused on the development of mammals and primates and humans. Brain carefully avoided the use of the word "evolution". He did not want South African government ministers or members of the church to protest the exhibit, as had happened in the past. He bravely placed a wooden box at the end of the displays with a slot in it, in which people could place pieces of paper with comments. Brain received endless abuse and insults on those slips of paper. The propaganda of the apartheid government had worked well, and convinced many people that "evolution" was not only incorrect but also morally wrong.[377]

In an effort to break South Africa's isolation, Wits and the Transvaal Museum were preparing, in the early 1980s, to host the Harvard palaeontologist Stephen Jay Gould. Tobias wrote to Gould saying, "The Taung skull sits in my laboratory waiting for your visit!"[378] Gould had proposed that he would speak about "scientific racism and theories of intelligence" as well as the theory of evolution. However, a week before the trip was scheduled to proceed, in July 1982, Gould was diagnosed with cancer and had to cancel.

Two years later, Gould was able to reschedule the trip. In addition to speaking about human evolution and race, he was interested to learn more about a South African woman he had first heard of in Paris, Sarah Baartman. In early 1982, Yves Coppens, a professor at the Musée de L'Homme in Paris, had taken Gould on a tour of the museum's back rooms. They had been looking for the preserved brain of Paul Broca, a scientist from the nineteenth century. They found it on a shelf in a jar amongst the brains of other scientists from the same era. As Gould said, they were "all white and all male".

On the shelf above Broca's brain, Gould saw three smaller jars. They provided a "chilling insight into nineteenth century mentality and the history of racism": in each jar were the dissected genitalia of what Gould called "Third World" women. He didn't see the brains of any women, nor Broca's penis or male genitalia in jars. The three jars were labelled "une Negresse", "une Peruvienne," and "la Venus Hottentotte".

Later that year, Gould published an essay in *Natural History* – something he had done every month for 20 years – with the title "The Hottentot Venus". It told the story of Sarah Baartman and how she had been taken first to London in 1810 and then to Paris for exhibition. Gould wrote: "On the racist ladder of human progress, Bushmen and Hottentots vied with Australian aborigines for the lowest rung, just above chimps and orangs … In this system, Saartjie exerted a grim fascination, not as a missing link in a later evolutionary sense, but as a creature who straddled that dreaded boundary between human and animal." Gould described the fascination that European men held with Baartman because of two features of her anatomy – the accumulation of large amounts of fat on her buttocks and the long labia of her genitalia.[379]

Gould's essay brought Baartman's fate to light again 167 years after her death and 27 years after Tobias had seen her remains. While other scientists had visited her skeleton and her body cast in Paris over the years, few had written publicly about them. Gould's interest in Baartman reflected one of his greatest concerns, what he called "scientific racism", a topic he had written about forcefully in his 1981 book *The Mismeasure of Man*.

On arrival in South Africa in 1984, Gould stayed with his colleague Elisabeth Vrba in Pretoria. He was surprised when one of her domestic workers called him "Master". He kept a journal throughout his stay and in it wrote that his visit to the Transvaal Museum was exciting because "we looked at the original australopithecine material ... behind a large bank vault door. I then toured the museum's 'genesis' exhibit and was disappointed to find 'evolution' omitted as a word but well treated as a concept."[380]

Tobias took Gould out to Sterkfontein. "I don't see why they don't search here more assiduously," wrote Gould. "Alun Hughes is the paradigm cautious man who will not proceed below until he disarticulates every block of breccia above." Gould said perhaps the details of the site "demands patience and nonimagination in a way ... It is always a great thrill to stand on such a celebrated spot and to see concretely what has been so well known but only imagined for so many years." While Tobias had been supportive of the excavations at Sterkfontein since he had re-opened the site in 1966, in the 1970s and '80s he did not excavate there himself. He left the work to Alun Hughes and his team. Wits University and the CSIR provided enough funds for the work to continue. While more and more hominid fossils were found there, Tobias had not described them in detail, nor published about them in scientific journals. Instead, his focus was on completing his conclusive description of *Homo habilis*, which he published in 1990.

After a weekend at Makapansgat, Gould was interested to see the collection of fossils Dart had collected. When he got back to Johannesburg, he made a visit to the Bernard Price Institute (BPI) where Dart spent his years of retirement and where many of the bone tools he had brought back from Makapansgat were stored. It was during that visit that Gould had an epiphany. "I also finally found the little puzzle with big implications – the coordinating theme that will make sense of this visit and give it an intellectual focus ... Here I saw (at the Bernard Price Museum) the famous Osteodontokeratic tools of Makapansgat and suddenly realised how profoundly Dart's "Killer ape" made no sense and was based on nothing. And now I must find out why Dart developed it. Was it just a personal idiosyncrasy, or did it arise from South African politics? Tobias says Dart was apolitical and would participate in nothing, citing his Australian origins. Or did he just have an [illegible]* pessimistic view of human nature?" Several days later, when Gould went out to Swartkrans and spent time with Bob Brain, he asked this question about Dart. According to Gould, Brain could provide "little enlightenment" on the source of Dart's tragic view of human nature.[381]

Stephen Jay Gould in front of the bust of Robert Broom on the occasion of giving the first Robert Broom Memorial Lecture at the Transvaal Museum in Pretoria on 31 July 1984. Courtesy of the Ditsong National Museum of Natural History.

Gould gave a lecture in the Wits Great Hall focused on intelligence testing and racism, and a seminar on punctuated equilibrium, his theory that evolution was not slow and gradual, but occurred in bursts of change. He spoke of Sarah Baartman as well as the "biological absurdity of South Africa's racial classification". Gould also gave the first Broom Memorial Lecture at the Transvaal Museum, which Gould thought was his best. He began by referring to the introduction Jan Smuts had written to Broom's *Australopithecus* monograph in 1946 about the idea of progress. He thought that referring to Smuts was "a delicious irony since Smuts now is the hero of neither camp (too sucking up to the Brits for the Afrikaners, too racist for the liberals)."[382]

Gould's overriding impression of South Africa was the impact of its growing isolation. He saw that Phillip Tobias ran Sterkfontein, Bob Brain worked at Swartkrans and Elisabeth Vrba was in charge of Kromdraai. He saw that they would all benefit from a greater flow of ideas and information. He described Tobias as a "very self conscious Professor. He is a good example of South African isolation (and courage). He is very conscious of rank, placing all titles on stationary, all degrees and citations framed on the wall, and called "Professor" by all – even Alun Hughes, a man of his age who has excavated Sterkfontein with him for 20 years. Yet he is immensely warm and kind, energetic and efficient ... I got the feeling that his sense of self importance would diminish if he were one among many in a flow of free information rather than the best in a domain of restriction."[383]

In honour of the 60th anniversary of Dart's publication in *Nature*

---

* The writing here is illegible. The illegible word could possibly be "essentially" but it is very unclear.

describing the Taung child skull, *National Geographic* put a holograph of the fossil on the front cover of its November 1985 edition. A holograph provided the impression of depth on a two-dimensional page, thereby allowing the viewer to look deep into the Taung child's eyes. The image first appeared to blend into the silvery grey background, just as the skull did into the breccia all those decades ago. But when the reader shifted to another angle, the fossil turned yellow, then red, green, and blue – a rainbow of colour appearing on the page.[384]

At the height of this celebration of the Taung skull, both in South Africa, and around the world, especially since *National Geographic* was distributed to millions of readers, global dismay over apartheid was also peaking. Given the pariah status of South Africa at the time, the fossil did not represent a collective effort to understand human evolution. The Taung child skull was seen by many as more of a mascot for white supremacy.

In April 1984, Tobias travelled to New York with the original Taung skull to participate in a study session at the American Museum of Natural History. He carried the skull in a specially made "carrycot" across the Atlantic. At JFK, they were met by security guards, whisked through customs, and into a stretch limousine with tinted windows. They rushed through the streets of New York City and into a side entrance of the museum. With armed guards at their side, Tobias and the carrycot entered a strong-room where the fossils were unpacked.[385] Not only the Taung skull, but 40 fossils from 21 institutions in nine different countries were flown in and treated like precious diamonds. On the eve of the opening, the Tanzanian government withdrew, objecting to the South African specimens in the exhibit.[386] The exhibit was called "Ancestors".

For a week, scientists from around the world gathered to investigate more than 50 original fossil specimens. "Never has there been such an assembly," Walter Sullivan wrote in *The New York Times*. Next to the Taung skull was the original Neanderthal skull found in 1856. The fossils, spanning millions of years of human evolution, were brought together for the first time. Raymond Dart, who was then 91 years old, also participated in the discussions. He had only limited peripheral vision, having lost much of his central eyesight. Many of the visiting scientists appeared more excited to see Raymond Dart than they were the fossils.[387] Four years later, in November 1988, Dart passed away and his ashes were scattered at Makapansgat.

Meanwhile, anti-apartheid groups gathered outside the building, protesting against the South African fossil being included. One of the placards read: "Apartheid fossils not wanted." Tobias recalled: "It was with the most mingled of feelings that I confronted this turn of events. On the one hand, I had every sympathy with the anti-apartheid sentiments of the protesters, and was sorely tempted to cross the road and stand with them – as I had often stood with Wits students and the Black Sash in Johannesburg. On the other hand, the very idea of dubbing the Taung child, who had lived in southern Africa some 2.5 million years before settlers from Europe had even moved in to South Africa, as an 'apartheid fossil' was painful, if not ludicrous."[388]

But anti-apartheid activists disagreed. A coalition of people from across the state of New York had introduced two bills to the City Council, one calling for the museum to remove the South African fossils from the exhibit, and the other calling for the city to cut off its $7 million budget to the American Museum of Natural History unless they removed the South African fossils from the exhibit. Several council members suggested that the museum make a public statement to formally denounce apartheid and place such a statement in the programme. Leaders of the museum refused, saying that by showing a common ancestry for all humans, the museum was already making a strong argument against apartheid. One City councilwoman, Ruth Messinger, said that for the museum to issue a public statement seemed a reasonable thing to do to avoid a showdown. She pointed out that it would be a useful thing for a "world-famous cultural institution to do. And they refused."[389]

Instead, the director of the museum, Thomas Nicholson, said: "To exclude these fossils would be to diminish the crucial impact of Africa in the evolutionary picture. Those fossils existed millions of years before there was a political boundary denoting a country called South Africa." The statement went on to say, "The strongest statement the museum can make in opposition to racism is 'Ancestors' itself. All the fossils and all the participants give evidence in their materials and in their statements of the common ancestry of all humans and of the irrationality of the concept of race and social difference."[390]

"If there's such common ancestry," said Reverend Wendell Foster, the City councilman who had introduced the two bills, "then why doesn't South Africa get rid of apartheid?"[391] The mayor of New York, Ed Koch, said he was against apartheid and that he supported a "change in this abhorrent system of governance". However, he believed "censorship, political or scientific, is not in my judgement, a reasonable or responsible remedy". "… fossils have no nationality," he said. "They are the holy relics of humanity and it is New York City's privilege and right, as a center of intellectual and artistic freedom, to host this exhibition."[392]

This dispute can be compared to the protests against human evolution exhibits at the Transvaal Museum in 1952 and in 1963. In all three cases, a museum was being asked to make a change to their exhibit on human evolution. The protests in South Africa, however, had opposed the concept of evolution; the protests in New York in 1984 agreed with evolution, but opposed the apartheid policies of the South African government. Tobias thought that he could keep politics separate from science, but of course he could not.

In the end, the museum's board of trustees, in a two-hour meeting on 11 June, agreed to put up signs declaring the museum's opposition to apartheid. They agreed that the signs would also make clear that the anthropological conclusion to be drawn from "Ancestors" was that all humans share common ancestry, a position that refuted all forms of racism.[393]

At about the same time as the exhibit, back in South Africa a teenage Phillippa de Villiers was scrambling across the grey rocks of Makapansgat, a place she loved to visit with her mother. The cave always felt like a cathedral to her, huge and overwhelming with its layers and bones. As she came out of the shadows, her eyes squinted against the sun and she sat down next to Alun Hughes, who had worked at Makapansgat for years before taking over the excavations at Sterkfontein. Uncle Alun looked rugged, old and weather-beaten. He had known Phillippa since she was a baby, having spent many Sunday lunches over the years with his colleague Hertha and her family. On this occasion at Makapansgat, Phillippa explored her interest in photography and took her first formal portrait. It was of Hughes, in his grey safari suit, on the rocks. Her mother was getting ready to retire by then but continued to publish, most recently about "new evidence on Negro origins", making the case that Bantu-speaking people had lived in southern Africa for over 100 000 years.[394]

As early as 1980, when South Africans were first having trouble attending international scientific meetings because of anti-apartheid boycotts, Hertha de Villiers had planned to attend the International Congress of Anatomists in Mexico. However, at the last minute, she was informed by the organisers that Mexico and South Africa did not maintain diplomatic relations and that the immigration authorities would not allow her into the country. The president of the congress told her that Tobias was planning to attend with a British passport and he asked if it would be possible for her to do the same. De Villiers wrote a letter to Tobias asking him to raise his voice on behalf of scientists being discriminated against on political grounds. When he responded, he said the situation was "very distressing" and asked her to make a distinction between their scientific colleagues and the refusals from their governments. He said that Pretoria was aware of the situation and that the South African government allowed South Africans to hold a second passport from another country in order to attend such meetings. "This is precisely in accordance with

Pretoria's policy," wrote Tobias, "namely that it is more important to get our scientists to international meetings by any means possible than not to have them attend at all!"[395]

By 1986, the academic boycott against South Africa was in full stride. The organisers of the World Archaeological Congress in Southampton in England were convinced they should ban the 26 academics due to attend from South Africa. The president of the Royal Society in London wrote a letter objecting to what the Southampton organising committee had done. In response to the situation, Tobias said that he found himself in a dilemma. He could see the role that an economic boycott and a diplomatic boycott could play to put pressure on the Pretoria regime. At a push, he would even consider supporting an international sports boycott, but he could not see the benefit of an academic boycott at all and he spoke out against it. Tobias was concerned that the "tactic of isolation" would result in a brain drain from South African universities and academic stagnation. "More than any other single feature of the whole sad business, I fear that the fruits of the boycott will be devastating for the universities of South Africa – and their inevitable decline will be irrevocable, at least for a long time."[396]

De Villiers wasn't pleased that students on the Wits campus had become political. In 1987, at the peak of student anti-apartheid protests, she said, "I feel they have come to learn, not be used by warring political parties. This is a civilized institution and people should behave accordingly. If they can't, they have no place here. I regard all this waving of placards and stirring up of the local population and spouting of Marx and Engels out of context as a cheap form of exhibitionism. My daughter who is studying journalism at Rhodes thinks I'm frightfully bourgeois."[397]

By the 1980s, most scientists had come around to agree with Darwin's early assertion, over a century earlier, that Africa was the place of origin for hominids, creatures that diverged from the apes and chimpanzees to begin the lineage of humans. These were the australopiths. And there was growing consensus that Africa was also the place where the genus *Homo* had originated. For over 15 years, Tobias had been disappointed that other scientists had not fully accepted the distinctive role of *Homo habilis*. But by the late 1970s, this species was referred to more and more by researchers. Tobias believed that a turning point in thinking about *Homo habilis* came in 1980 when Milford Wolpoff wrote in his book *Paleoanthropology* that "*Homo habilis* is a taxon whose time has come."[398]

Scientists, however, were still asking where the species *Homo sapiens* originated. Some scientists believed all *Homo sapiens* developed from a recent, single place of origin. This was called the "Garden of Eden" or the "Noah's Ark" model of evolution, but there was no consensus on where that place of origin was. Some argued Europe, others China or the Near East, and others thought sub-Saharan Africa.

An alternative model of origins for *Homo sapiens*, the Multi-regional model of evolution, argued that *Homo sapiens* evolved in different geographic areas from different groups of *Homo erectus*. This model was based on Weidenreich's work in the 1930s and '40s with a nod to Carleton Coon in the 1960s, and was put forward by Milford Wolpoff and his colleagues in 1984. Like Weidenreich, Wolpoff observed that there was regional continuity in Asia between certain characteristics of ancient fossils and modern humans. Weidenreich argued that "the various racial groups of modern mankind took their origin from ancestors already differentiated".[399] Coon agreed with this idea of regional continuity. Wolpoff was also convinced regional continuity existed, especially when looking at northern Asia and Australasia. He suggested that Coon's theory had been rejected because of his terminology. Coon had written that different races crossed a boundary from *Homo erectus* to *Homo sapiens* at different times. Wolpoff believed that the boundary between the two species was arbitrary, and said, "Coon should never have had to explain how five allopatric subspecies crossed a species boundary but still maintained their identity. It was his own terminology that created the need for an explanation."[400] In support of regional continuity, Wolpoff concluded that "The origins of modern *Homo sapiens* are to be found in the evolutionary process we have described as multiregional evolution, and not in Noah's ark."[401]

Despite the Multi-regional theory, more and more scientists began to be convinced that the common place of origin, the metaphorical Noah's ark, was in Africa. As a result, a major debate flourished in the mid- to late 1980s between Multi-regional evolution and the Out of Africa theory that was also called the Recent African Origin (RAO) hypothesis. In 1988, Chris Stringer from the British Natural History Museum wrote an article in *Science* magazine where he contrasted the two theories and showed that the evidence best supported RAO. In the article, he wrote: "According to the multiregional model, some regional ("racial") features are considered to have preceded the appearance of the *Homo sapiens* morphology and to have been carried over from local *Homo sapiens* ancestors."[402] But Stringer argued that there was a single origin for all humans. In letters and comments to the media, scientists expressed scorn for Stringer's interpretations, "a scorn that seemed to extend to personal abuse at times", he recalled. "I was shocked by some of the vitriolic reactions to that paper."[403]

Tobias supported the idea that early *Homo* first appeared in Africa, but he was not certain which hypothesis he favoured in terms of where and how *Homo sapiens* first appeared. He wanted to take more time to review the evidence. His question was not only about the place of origin, but also about what factors had resulted in the first appearance of *Homo sapiens*. Was it the use of fire? Was it the growth of the brain? Was it the development of language? Tobias agreed with the consensus that the australopiths were not capable of language. However, he had begun to suspect that *Homo habilis* had language ability as far back as 1.8 million years ago. This idea was against the prevailing view that language was only 80 000 years old, and developed first by *Homo sapiens*.

In the 1950s and '60s most scientists believed it was stone tools that distinguished hominids from other animals and made them human. Dart's theory added to the idea that it was violence and the use of weapons that were our first steps towards becoming human. But now, based on Brain's further research and Tobias's theory of early language, along with the fact that multiple hominid species seemed to have existed side by side, there was a new theory that could be a source of hope. Perhaps it was the ability to communicate that made us human.

At the same time as the debate between the advocates of the Multi-regional theory and the Out of Africa theory flared, Tobias wrote an article entitled "Race". It was originally published in *The Social Science Encyclopedia* in 1985. Unlike his earlier treatise, *The Meaning of Race*, Tobias acknowledged that more and more scientists were beginning to question the biological concept and that there was no consensus about the validity of race. He attributed this change to the field of genetics. When scientists look at genetics instead of observable physical traits, he wrote, "they reveal that there are no hard and fast boundaries between races and populations within a species".[404] Looking back on his own career, he said: "As long as one focused on morphological traits alone, it was sometimes not difficult to convince oneself that races were distinctly differentiated, one from another, with clear-cut boundaries between them." But he then agreed that as genetic analysis had proceeded over the previous two decades, the clear-cut boundaries became hypothetical. As he recognised the "blurring of the outlines of each race", he saw the breaking down of the concept of race itself. "The myth of the pure race has been thoroughly disproved. There are no pure (genetically or morphologically homogeneous) human races and, as far as the fossil record goes, there never has been."[405] Tobias made another important point in the paper. This was that there was no scientific basis for the idea that "how one behaves depends entirely or mainly on one's genes" or that "some races are superior and others inferior". In the midst of apartheid oppression, his voice on this point did not travel very far, however.

In terms of race and human evolution, Tobias confirmed that all living human beings were members of a single species *Homo sapiens*. Then he pointed to the growing fossil record that revealed earlier forms of humans existed such as *Homo habilis* and *Homo erectus*, but that it was *Homo sapiens* that made their way across the globe, first across Europe and then into the Americas. According to Tobias, physical anthropology, relying on physical traits, divided *Homo sapiens* into different sub-divisions such as Negroid, Mongoloid and Caucasoid, but there were doubts about whether there was a biological basis for that classification. In fact, there was greater genetic variation within these populations than there was between them. But Tobias continued to ask the question, how long ago did the Negroid diverge from the Caucasoid and the Mongoloid? This question suggests that Tobias thought that the origin of *Homo sapiens* was in Europe or Asia. Tobias agreed that "the formation of the modern races of man is a relatively recent process, extending back in time for probably not much more than 100,000 years".[406] In comparison, he stated that "at least forty times as long a period [four million years] of its hominid ancestry has been spent by each race in common with all other races, as it has spent on its own pathway of differentiation". He was saying that the Taung child skull, Mrs Ples, and Lucy had all existed long before *Homo sapiens* began to differentiate across the globe.

Tobias had been slow in the 1950s and '60s to let go of race typology and, in the 1980s and early '90s, despite the new evidence from genetics, he was reluctant to give up on the concept of race. Despite the lack of scientific consensus on race and the fantasy of a pure race, Tobias stated that the race concept was still of value and that further research was needed. "Time will tell whether we are witnessing 'the potential demise of a concept in physical anthropology' or whether the concept will survive."

In 1985 Tobias made a statement about the role of science in apartheid policies. "No South African physical anthropologist was involved in providing the scientific underpinning for the government's race classification practices," he wrote.[407] Years later, Allan Morris, Tobias's student, wrote that "Tobias was correct in the strict sense that no physical anthropologist submitted proposals to the government, nor did they join in the legislative or administrative process" but that they were passively involved because of their personal views on race as reflected in their professional output. Morris argued that these scientists, especially with their writing before World War II, "provided a fertile growth medium in which the apartheid ideology could flourish".[408] That certainly had been the case during the Dart era in the 1920s through the 1950s, and it was true during Tobias's years as well from the 1960s through the 1980s.

But by the mid-1980s, the fertile growth medium in which apartheid ideology had been flourishing was under threat in South Africa from both internal and external sources. The anti-apartheid struggle inside the country and internationally were having an impact. At the same time, the scientific understanding of race and the theories of human evolution were changing. The field of genetics was about to offer up a new hypothesis about human ancestry that would have a major impact on ideas about human origins – and it all led back to Africa.

Part Three
Searching for Unity

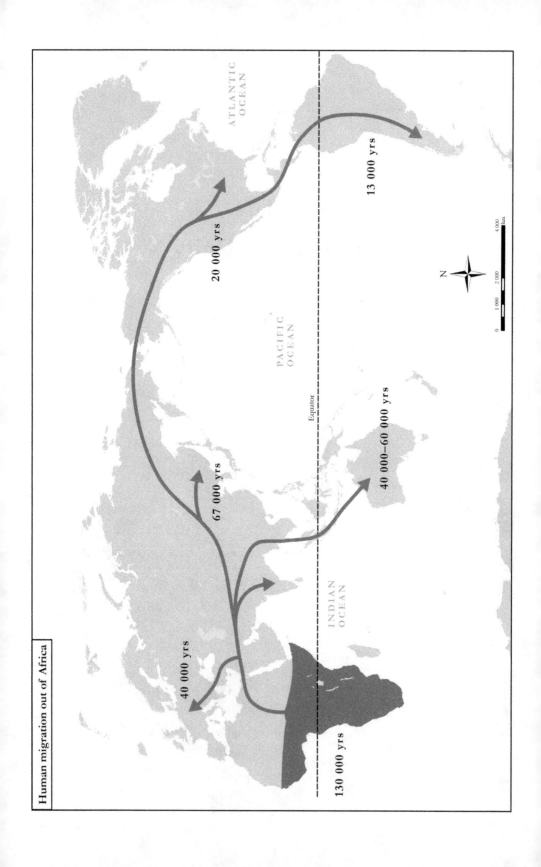

Human migration out of Africa

20 000 yrs

13 000 yrs

67 000 yrs

40 000–60 000 yrs

40 000 yrs

130 000 yrs

ATLANTIC OCEAN

PACIFIC OCEAN

INDIAN OCEAN

Equator

N

0   1 000   2 000   4 000
km

18

Mitochondrial Eve

In 1984, at the University of Durban-Westville, Himla Soodyall held a small bottle of crude oil in her hands. It had been sitting on her professor's windowsill for years. She held it up to the light, shook it a few times, and saw that a few particles of oil floated up away from the rest of the thick fluid. She knew that the oil was breaking apart and wondered how. Her professor in biochemistry thought she should work to find out what caused the reaction and that the answer could have practical implications such as being able to break apart oil spills in the ocean. Soodyall identified some strains of bacteria in the oil that were helping to break down the building blocks of the crude oil chains. Before long she was on her way to the University of the Witwatersrand to do her master's in bio-technology and it was there that she met Professor Trefor Jenkins at the South African Institute for Medical Research.[409]

In the 1970s, Jenkins had worked with Tobias in Campbell and elsewhere in southern Africa on genetics research. By the mid-1980s, Jenkins believed the future of human genetics was in bio-technology research focused on mitochondrial DNA, that exists in cells, that is passed down only through the maternal line. According to Soodyall, it was the legendary Italian geneticist Luigi Luca Cavalli Sforza who visited South Africa and convinced Jenkins of this new research direction. Jenkins had had a terrible heart attack and was in the hospital when Sforza visited him there. "The new era of genetics is going to be in DNA. You should get involved with DNA technology," Sforza said.[410]

Another colleague of Jenkins', Henry Harpending from the US, when doing work in southern Africa, had taken hair samples from San people in the Kalahari. Yet another US scientist, Linda Vigilant, had been able to extract mitochondrial DNA from the roots of that hair and she published a paper in 1989 about the mitochondrial variation in the San. Before Vigilant published, Trefor Jenkins had heard her speak about her work and he wanted his Institute for Medical Research to get involved. He threw that challenge out to his students and Himla Soodyall decided to take it up.

In the 1980s, one of the standard ways to gather mitochondrial DNA was to use tissue from a placenta, and so Soodyall went to a hospital close to her offices to get placenta tissue. At the time, the Human Tissues Act allowed people to gather placentas, with the consent of the donor, for medical research. "There I was waiting with my ice bucket," recalled Soodyall. "The grandfather was pacing up and down waiting for the baby. He got the baby. And I got the afterbirth!" She laughed. "I came back to the lab with the placenta. I had to overcome all of my issues of working with human tissue. I was a religious person at the time so it concerned me."[411]

Soodyall, by nature a shy person, grew up with a conservative religious background and she had not been introduced to the concept of evolution when she was a younger student. "Even in my university science courses, evolution wasn't taught because it was against the religious principles of the person teaching the course." Initially, Soodyall's Hindu religious beliefs clashed with aspects of her laboratory work. She began doing prenatal diagnosis for people as a part of her genetics work because the tools she used could give couples information about whether their unborn child had a genetic disease. "There I was for the first time sitting in the dark room, holding up an autoradiograph to the safety light, to take a quick peek at the result before putting it in the fixer," she said. "It was cystic fibrosis. I was about 23 years old and I wasn't prepared for this. You are now playing the advocate of what we call God in the life of another family. You feel responsible."[412]

But Soodyall enjoyed her work and continued with it. Trefor Jenkins was instrumental in supporting her and broadening her thinking. "*The Peoples of Southern Africa* [by Jenkins and George Nurse] was like a Bible to me," she said. "As I cut my teeth with mitochondrial DNA and seeing these exotic patterns of variation in Africa, I was like a kid in a candy store. So that's how my love of mitochondria started.

"In the late 1980s and early 1990s, we were highly politically fuelled," said Soodyall. There were discussions about political activism. Trefor's involvement in the Steve Biko inquest and the case against the Medical Council was well known. He was very strong on human rights issues and we slowly picked up on those things. At that stage, if you asked me who Nelson Mandela was, I didn't know any of it. I say so unashamedly because I want to position where we came from as a family. My father was a clerk at a bakery from the day he started working until he retired. We were politically neutral." Growing up in the 1970s, Soodyall's family lived in a township set aside for people who were classified as Asian, largely of Indian descent. Throughout her schooling, she never had the opportunity to interact with people from different cultures, but that would change. The first step was when she attended Wits.

Then in August 1990, Soodyall was awarded a fellowship (SHARE – Scholarship for Health Academic Research and Equity) to go to Pennsylvania State University in the United States. "It was a scholarship for people of colour," said Soodyall. Trefor Jenkins encouraged her to apply. It was in the United States where Soodyall first met and talked to people from all over the world, particularly from other parts of Africa and India. "I had never engaged at that level with people of other races ... Every step I took exposed me to new challenges that helped shape who I am today ... It dawned on me that the religion I had been groomed with was actually cultural tradition."

In addition, Penn State had the first Institute for Molecular Evolutionary Genetics in the world. That's where Soodyall met Mark Stoneking, who was starting an Anthropological Genetics Lab. She helped him start the lab and trained his anthropology students in genetics. Meeting Stoneking was fortuitous because he had been one of the three authors of a paper in 1987 in *Nature* entitled "Mitochondrial DNA and Human Evolution". The impact of that article on the field of human evolution and the public understanding of it was dramatic.[413]

The article addressed a debate that had been under way within the field of human genetics throughout the 1980s about classifying human population groups – race. The genetic trees scientists drew were used often to illustrate the passing of genes from one generation to another and to show the time that had passed in this process. These trees were often misinterpreted, in that many people thought that different branches of the trees represented different groups that had evolved away from one another and divided people into different races.

The 1987 article in *Nature* helped discourage this misinterpretation and to break the impasse in the Multi-regional versus Recent Out of Africa debate. It was written by a US scientist, Allan Wilson, together with two of his students, Rebecca Cann and Mark Stoneking, and it was a breakthrough for several reasons. First, a genetic tree was used in the article, but in this case the branches of the tree represented specific individuals (real people), not a hypothetical group or population. It was possible to trace back from each individual to find a common root in the tree, or a common ancestor.

Scientists did not know mitochondria contained DNA until the 1960s. It was only in the late 1970s that it was established that mitochondrial DNA was useful for tracing a family tree because it was inherited only from the mother. It is not a mixture of both parents' genes, like nuclear DNA, so it isn't jumbled together anew every generation. It is passed down from generation to generation as is. The only things that change mitochondrial DNA are mutations: the isolated, random mistakes that occur in copying the genetic code.

Stoneking, Cann and Wilson compared the genes in mitochondrial DNA from many sources and found there were people from different parts of the world with many of the same genes. Relying on their knowledge of the process of genetic mutation over time, the researchers were then able to estimate how many changes had occurred in the mitochondrial DNA over time and how much time would have elapsed for the mothers to have passed down the changed genes to their offspring. Then the researchers worked their way back to estimate how long ago it was that all of the genes in the mitochondrial DNA were the same, or when there might have been one common maternal ancestor for all these different individuals. Looking at 134 people from around the world as the different branches of the tree, they could see that the branches led back to the trunk of the tree, which converged in one single root about 200 000 years ago. What this meant was that the entire modern human species had developed more recently than people had thought.[414]

The article had another dramatic conclusion. For years, there had been different interpretations of the fossil record. A growing number of scientists were in agreement (thanks to Darwin's original hunch, and to Dart, Broom, the Leakeys, and Johanson, among others) that the early ancestors of humans originated in Africa. However, the two schools of thought about how modern humans evolved more recently continued. The Multi-regionalists saw clues in the fossil record that suggested *Homo erectus* had evolved separately in different regions of the world to form different races of *Homo sapiens*. Scientists that promoted the "Out of Africa" or "Recent African Origins" school of thought saw fossil evidence that *Homo sapiens* left Africa about 100 000 years ago and that they replaced the *Home erectus* populations that had developed in different parts of the world. With the Multi-regional approach, a modern European person and a modern Chinese person would have last shared a common ancestor in Africa about one million years ago. By comparison, the "Out of Africa" theory would link the European person and the Chinese person to an African ancestor much more recently. A polarising debate was growing about which scenario was correct.

What the mitochondria gene tree of Wilson, Cann and Stoneking did was suggest that the common ancestor of all modern humans lived in Africa about 200 000 years ago. This fit with the Out of Africa theory but it was greatly criticised by the Multi-regionalists who questioned whether it was accurate. If the theory put forward in "Mitochondrial DNA and Human Evolution" was correct, then it would not have been possible for *Homo sapiens* in different parts of the world to have evolved independently from local groups of *Homo erectus*. There just wouldn't have been enough time. This would mean that the Neanderthals, Peking Man and Java Man were not direct ancestors of *Homo sapiens*. It seemed that the evidence from all the fossils that had been gathered over the past century would have to be looked at all over again and reinterpreted.

While it wasn't mentioned in the original article, scientists started to refer to the common maternal ancestor from Africa as Eve, hence the growing reference to Mitochondrial Eve. *Newsweek* picked up on the implications of the research, and the Biblical allusions, and printed their famous cover story entitled "The Search for Adam and Eve" in January 1988. The artwork on the cover depicted a young, naked, black couple in front of a tree in the Garden of Eden. The tree trunk had a snake curled around it and Eve was holding an apple that she was about to hand to Adam. The new science had made its way into the popular press and would start to shape public opinion.[415]

The fact that *Newsweek* chose cover artwork portraying a scene of Adam and Eve with an apple and a snake illustrated that they felt comfortable perpetuating male and female stereotypes to portray the story of human origins. In the story of Genesis, Adam is alone in the Garden of Eden, and God fashions Eve from a bone – Adam's rib. In contrast, the theory emanating from mitochondrial DNA cut against these stereotypes. Instead of relying on the male line in the family, or the boy child, to carry on the family name, this new theory was dependent on the female, matriarchal line. Instead of being secondary to the man, the woman was the primary carrier of the family line and the continuation of the entire human species.

The article began by saying that "scientists are calling her Eve, but reluctantly" because doing so evoked images of Genesis and Renaissance art. The author clarified that this scientific Eve was not the only woman on earth at the time she was alive, but that she was the most fruitful and that her daughters passed their genes on to every new generation. She wasn't the first woman, but she was our common ancestor. She was, therefore, everyone's "10,000th great grandmother" and has passed down her genes to every single human alive today.[416]

"Eve has provoked a scientific controversy bitter even by the standards of anthropologists, who have few rivals at scholarly sniping," explained the article. Their feuds usually started when someone developed a new grand theory based on the unearthing of a few fossil bones. This time the argument involved a new kind of anthropologist, one of those "who work in air-conditioned American laboratories instead of desiccated African rift valleys".[417]

At the time, Stephen Jay Gould said, "It makes us realise that all human beings, despite differences in external appearances, are really members of a single entity that's had a very recent origin in one place. There is a kind of biological brotherhood that's much more profound than we ever realised."[418] Of course in this case it would have been more appropriate for him to use the word "sisterhood".

After Wilson, Cann and Stoneking's *Nature* article came out, there was a fierce debate at the American Anthropological Association meeting in Chicago later in 1987. Geneticists showed slides of diagrams of DNA. Palaeoanthropologists showed slides of skulls. "What bothers many of us paleontologists," said Fred Smith of the University of Tennessee, "is the perception that this new data from DNA is so precise and scientific and that we paleontologists are just a bunch of bumbling old fools. But if you listen to the geneticists, you realise they're as divided about their genetic data as we are about the bones."[419]

In order to do the genetic research for the article, Rebecca Cann had to request that 147 women donate their placentas. A placenta was the easiest way to gather mitochondrial DNA. Unlike gathering skeletons, often from graves and without consent, Cann had gathered placentas from women with their consent. These women from across the United States had ancestors from Africa, Europe, the Middle East and Asia; she also found Aboriginal women who were willing to donate from Australia and New Guinea. Each placenta was frozen and sent to University of California at Berkeley where the three authors were based. As the *Newsweek* article explained, "The tissues were ground in a souped-up Waring blender, spun in a centrifuge, mixed with a cell-breaking detergent, dyed fluorescent and spun in a centrifuge again. The result was a clear liquid containing pure DNA."[420]

"As a result of speaking about women's genes and human origins for public television," said Rebecca Cann, "I became bombarded with hate mail from both creationists and various political groups. One family member admonished my parents for my behaviour, asking them why I would tell perfect strangers that my ancestors were black when all the family picture albums showed the real truth!"[421] The religious objections to human evolution continued and Cann was the target of much of the venom. Cann's perception was that when the discussion about human evolution takes on "a female face and a black skin" the intensity of the rejection increased dramatically.

The Berkeley researchers cut the samples of DNA into small sections that could be compared with one another. They found that the differences were there, but they were much smaller than they expected. "We're a young species, and there are really very few genetic differences among cultures," said Stoneking. "In terms of our mitochondrial DNA, we're much more closely related [to one another] than almost any other vertebrate or mammalian species."[422] For many scientists, this was a political and ideological relief.

For other scientists, this finding was more problematic. Milford Wolpoff, for example, who had put forward the Multi-regional theory of human evolution, suggested that it was not possible to base a theory of human evolution "using neutrally evolving female genes". There was a backlash against the Recent African Origin theory, and there was even a backlash against the idea that a female held such great importance, whatever her species and wherever her offspring went. The idea that there was a line of female *Homo sapiens* that carried the species forward meant that there could be factors that accounted for success as a species other than male competition and the old concept of a "killer-ape". The overly simplified and misapplied concept of "survival of the fittest" did not necessarily indicate who would survive and why. Some scientists, and many members of the public, appeared to be more comfortable with a patriarchal approach.

And once again, there was a negative reaction to a new theory that humans had origins in Africa. It happened when Darwin put forward his original theory. It happened again in the first half of the twentieth century when Piltdown Man was found, promoting European or British origins and putting off the possibility of African origins. Again, in the 1960s, Carleton Coon promoted a theory that rejected common origins in Africa. Now once again, in the late 1980s, the idea of common human origins in Africa was initially met with disbelief and scorn. In the long term, the science would be proven correct.

Phillip Tobias was initially skeptical of Mitochondrial Eve because he thought aspects of the Multi-regional theory had merit. He believed that forms of *Homo erectus* around the world were ancestors of modern humans and that Neanderthals differed little from *Homo sapiens*. In this regard, Tobias was on to something because in the 2000s it became clear that many modern humans hold a small percentage of Neanderthal DNA. According to Himla Soodyall, Tobias continued to give public lectures that questioned the Out of Africa theory because it suggested a more recent development of modern humans. He also questioned the turn to genetics at the expense of palaeoanthropology and the fossil record.[423]

Himla Soodyall met Mark Stoneking at Penn State three years after his

"Mitochondrial DNA and Human Evolution" paper had been published. They worked together in the new Anthropological Genetics Lab. Soodyall continued her research with mitochondrial DNA and began to look at the relationship between genealogies, history and genetics.

By the time Soodyall returned to South Africa in November 1996, the country had held its first democratic election, implemented a new constitution and was in the midst of significant social and political changes. While she was away, a campaign had developed to call for the return of Sarah Baartman's remains from France to South Africa. Soodyall was asked by the department of arts, culture, science and technology (DACST) to join Phillip Tobias on a reference committee to oversee the process. Soon their paths would cross.

The rise of genetics in the search for human origins in the late 1980s and early 1990s by no means meant the end of the search for fossil evidence. In fact, in the coming years, both fields would come together more effectively in order to better understand human origins, especially as technology has allowed for the testing of ancient DNA gathered from fossils. However, in the meantime, South Africa's transition to democracy and Tobias's retirement brought a period of uncertainty for the protection and management of Sterkfontein, as well as the future of palaeoanthropology in South Africa.

19

"The Human Story is Never Finished"

In New York City in 1988, a "fresh-faced, over-enthusiastic boy" went up to the front of the lecture hall with his father to speak to Phillip Tobias after a presentation he had just given.[424] Lee Berger had travelled all the way from Atlanta, Georgia to meet Tobias, hoping that he might get an opportunity to work with him in South Africa. According to Berger, it was Don Johanson, the palaeoanthropologist who had unearthed the Lucy skeleton, who initially gave him the idea to study in South Africa. Berger had met Johanson on a lecture tour in Georgia and told him it was his lifelong dream to go to Africa to search for hominid fossils. Johanson told Berger that the intense rivalry in East Africa meant that there was not much space for newcomers. "On the other hand," he said, "there's South Africa."[425]

Berger was taken aback by the suggestion. At the time, boycotts against South Africa were in full swing, and the apartheid government was using increasing levels of violence against the anti-apartheid struggle. The US had recently imposed economic sanctions against South Africa and many countries on the African continent refused entry to anyone carrying a South African visa in their passport. "I was very aware," recalled Berger, "that if I went to South Africa, I might never be allowed on Kenyan soil."[426]

Nevertheless Berger made the decision to study in South Africa anyway and he got the opportunity to work with Phillip Tobias, as he had hoped. He remembers his first visit to the hominid fossil lab in the department of anatomy with Phillip Tobias. Tobias took a large key out of his pocket. As he turned the key in the heavy gray metal door of the vault, Berger held his breath. He knew that this hominid fossil vault was the "Aladdin's cave of palaeoanthropology" and the reason that he had travelled halfway across the globe. Since Sterkfontein had been re-opened in 1966, soon after Berger was born, Alun Hughes, under Tobias's supervision, had gathered over 550 catalogued fossil specimens from Sterkfontein alone. The fossils were there, but Tobias had not paid them a great deal of attention. He had spent much of the 1970s and '80s describing East African fossils instead. "Behind this door," felt Berger, "lay not only the largest single collection of undescribed early hominid fossils in the world, but my future."[427]

Tobias suggested Berger focus his PhD research on the australopithecine shoulder. Berger worked with a specimen called STW 431 from Sterkfontein. The fossil included a clavicle (collar bone), a scapula (shoulder blade) and a proximal humerus (the upper arm bone). Tobias had assigned other students to study the *Australopithecus* pelvis, the elbow, the ankle and the hand.

Berger had a father-son relationship with Tobias during those years when he was working on his PhD. As the youngest in the department, still in his 20s, he was often the one asked to retrieve specimens from the Dart collection as Tobias requested. He knew that A43, /Keri-/Keri's skeleton, was considered to be one of the most important and often-studied skeletons in the collection – which suggests that when Berger first arrived in Johannesburg, it hadn't yet disappeared.

Tobias officially reached retirement age in 1990 but the university extended his stay at Wits for another three years through to 1993 and he continued as Berger's supervisor. Tobias was concerned about the future of the Palaeo-anthropology Research Unit (PARU) because as a retiree he was no longer eligible to apply for funding from South Africa's National Research Fund (NRF). Berger went into Tobias's office one day and confronted him. "Professor, we're all anxious to know what's going to happen to the Unit after your retirement," he said. Tobias's response was that it was "in the lap of the gods". He was gloomy and depressed. He knew it would all come down to where they could find the money.[428]

At around the same time, Alun Hughes became ill and wasn't able to continue his work at Sterkfontein. He decided that it was time to retire. Hughes and Tobias decided to contact Ron Clarke, who was working at the American Museum of Natural History in New York, to ask him to come back. Clarke and Tobias had always had a difficult relationship, first when Clarke was Tobias's PhD student in the mid-1970s and then again when they had worked together in the mid-1980s. They didn't seem able to get along for long. But when Hughes called Clarke in 1991 and asked him if he would like to take over the excavations at Sterkfontein, Clarke packed up and moved back to South Africa, reconciled with Tobias, and once again agreed to work for him. He too became concerned about what would happen to PARU in the wake of Tobias's retirement.

According to Berger, Friedel Sellschop, the deputy vice-chancellor for research at Wits, was ready to close down palaeoanthropology and Tobias wasn't sure what could be done about it because he was about to retire.[429] "We had a vault full of unpublished fossils," Berger recalled. "The stuff down here [in South Africa] was seen as a side branch." Berger was to be Tobias's last graduate student. There were changes under way at the Transvaal Museum as well. Elisabeth Vrba had left for Yale in 1986 and Bob Brain retired in 1990. Brain had asked a young South African with a PhD from Yale, Francis Thackeray, to become the head of palaeontology but the museum's future was uncertain as well.

Thackeray remembers when he was a young man interested in palaeontology in 1970 he had visited the Transvaal Museum and met Saul Sithole. Sithole had taught Thackeray how to treat fossils in breccia with acid in order to extract them from the rock. Sithole had started working at the museum back in 1928 and had been by Robert Broom's side at Sterkfontein in 1936. While Sithole left the museum for several years in the 1960s to try to start his own business, he returned in 1970 and carried on as a technician, working for Bob Brain. When Francis Thackeray joined the museum staff in June 1990, he was the one to honour Saul Sithole by giving him a plaque and a cast of Mrs Ples when he retired in December. Sithole lived to see democracy in South Africa but died on 16 October 1997 at the age of 90. Mrs Zondi Sithole Zitha, Sithole's daughter, believes that if her father had had the opportunity, he would have enjoyed further study in science. "He loved his work," she said. In 2015, Zitha continued to proudly display her father's cabinet with momentos from the museum, including the replica of Mrs Ples.[430]

In late 1992, Lee Berger was out at the Rhino and Lion nature reserve with his girlfriend Jackie Smilg. He was down in a cave and Jackie was sitting outside the cave under a tree. At the same time, Mark Read was standing on the border of his property, gazing out into the distance, when across the fence he saw that a tree was shaking quite strongly. He thought there must be a large animal next to the tree so he waited to see what it was. As he got closer, he saw a young woman sitting there so he went over to talk to her. From within the cave, Berger could hear Jackie talking to someone and so he came up out of the cave and joined them. They introduced themselves, and during the conversation that followed Berger told Read that the PARU at Wits was in trouble.[431]

Read mentioned the crisis to his father, who was the well-known art dealer Everard Read, who suggested Berger speak to Gavin Relly, the chairman of Anglo-American, who was already supporting archaeological work in the

Border Cave near Swaziland. Relly was excited about the work that Berger was doing at Gladysvale, another site in the Sterkfontein Valley, and he in turn suggested that Berger talk to Harry Oppenheimer, one of the most wealthy and prominent businessmen in South Africa who had made his fortune in mining. Oppenheimer agreed to meet with Berger and had his driver take them both out to the research site in a huge, black armoured limousine. They came over a hill on a dirt road, the bottom of the car scraping uncomfortably on the rocks. Oppenheimer turned to Berger and asked, "Lee, how much would it cost to support all of this research?" Berger knew that the salaries at Sterkfontein were terrible, and that there weren't many staff members. "Mr Oppenheimer, you could support this science for less than it would cost to buy this car, which we are now destroying on this road."[432]

Mark Read, Gavin Relly and Harry Oppenheimer were willing to provide funding and they helped to bring together a group of interested people. Together, they established a new organisation named PAST, the Palaeo-Anthropological Scientific Trust. PAST was launched at a function at the Anglo-American headquarters in Johannesburg on 12 July 1994 with the original Taung child skull on display. PAST's mission was to raise funds to support research, scholarships and bursaries for students and public education around palaeoanthropology as well as to develop and protect South African hominid sites. The initial grant kept Tobias on board at Wits to run PARU for another three years, from 1994–1996. PARU was saved from closing down.[433]

The new democracy that came into being in April 1994 was exciting for South Africans. Nelson Mandela, the first president of a democratic South Africa, saw it as an era of reconciliation. In his inauguration speech, he said, "Let each know that for each the body, the mind and the soul have been freed to fulfil themselves, Never, never and never again shall it be that this beautiful land will again experience the oppression of one by another and suffer the indignity of being the skunk of the world. Let freedom reign. The sun shall never set on so glorious a human achievement! God bless Africa!"

Despite the euphoria of the time, Tobias was still uncertain about the future of South Africa's ancient past. Writing for the magazine *Optima*, published by Anglo American and De Beers in a special edition to coincide with the first democratic elections, Tobias applauded the 27 years of work at Sterkfontein (the same length of time that Mandela spent in prison), since he had re-opened the caves in 1966. The field team of eight excavators had worked five days a week for 48 weeks a year, and gathered over 650 hominid specimens, from at least three different species of hominids. "There is every scientific reason why palaeo-anthropology – which has long been one of South Africa's most widely known contributions to world science – *should* survive here, well into the 21st century. The question is: *Will* it survive?"[434]

Tobias agreed that South Africa's contributions had not received as much international recognition in recent decades as had those from East Africa, and he lamented the academic boycott. He wondered what level of priority the new South Africa would give to researchers of human evolution. With enormous needs in housing, health and education, he questioned whether the state would be able to offer much support for scientific research at universities and museums. But with this article, Tobias seemed to hit a turning point; he was expressing a new vigour for the idea of shared humanity. He argued that "Whether it be for reasons of pride in one's ancestry and in Africa's priceless gift of mankind to the world, whether it be as a showpiece of the new South Africa, whether for the extraordinary cultural, anthropological, scientific and educational messages that flow from this work, whether it be for the evolutionary underpinning of the concept of brotherhood of humankind – for *Australopithecus* is the great-grandfather of all living human beings – whatever the purposes and motives, there is I believe a cast-iron case for South Africa's palaeo-anthropological impetus to be continued and fostered, and even to flourish as never before."[435] Tobias was reaching out to the new government, but he was also speaking to the private sector who had agreed to float him some support.

As Donna Haraway, the historian of science, wrote, "The human story is never finished." She began to critique palaeoanthropology and primatology in her book *Primate Studies* in 1989. It was as if she had anticipated Tobias's concerns. "The origin in physical anthropological discourse is ever receding, not only because new fossils are found and reconstructed, but also because the origin is precisely what can never really be found; it must remain a virtual point, ever reanimating the desire for the whole." Tobias, Berger and their colleagues in South Africa were hoping for a reanimation of the search for human origins. The new government had already recognised that the field of palaeoanthropology could potentially become "world class" and could contribute to "leading edge global knowledge". The new national department of arts, culture, science and technology saw that South Africa had a geographical advantage in both astronomy and palaeoanthropology. There was hope for reanimation in the New South Africa.

Meanwhile, the scientific debate about the origin of *Homo sapiens* continued. Milford Wolpoff continued to promote the Multi-regional theory of human evolution. In 1994, he wrote, "The evolutionary patterns of three different regions [China, Australasia and Europe] show that the earliest 'modern' humans are not Africans and do not have the complex of features that characterize the Africans of that time or any other ... There is no evidence of specific admixture with Africans at any time, let alone replacement by them ...There is indisputable evidence for the continuity of distinct unique combinations of skeletal features in different regions, connecting the earliest human populations with recent and living peoples."[436]

It was at this point that Tobias decided to weigh in on the debate with a critique of the Mitochondrial Eve hypothesis. He called the 1987 article by Wilson, Stoneking and Cann "astounding" and said that it made a "staggering claim" that women carrying the ancestral mtDNA line lived in Africa as recently as 200 000 years ago. Eight years after the article originally came out Tobias still was reluctant to accept its findings. Tobias did support the theory of migration out of Africa, but in an earlier epoch. "The true placement of the Garden of Eden, in the sense of the cradle of recent humanity, was in Africa, but it is far more likely to have been in ancient, one million years ago Africa, than in recent, 200,000 year old Africa."[437]

Tobias did not support the idea that a second, more recent migration out of Africa had totally replaced the existing hominids in other parts of the world. "This 'total replacement' concept has ruffled the feathers of many palaeo-anthropologists." He argued that the fossil and archaeological evidence pointed to regional continuity. He also suggested that there might be a combination of continuity and replacement. "Why a mitochondrial Eve," he asked, "and not a mitochondrial Adam?" He confessed that the theory of exclusively maternal inheritance of mtDNA had "long troubled" him.[438] "It would be delightful," he said, "if the African origin were corroborated rigorously. Africa already boasts the first emergence of the family *Hominidae* [in terms of the australopiths] and also the earlier appearance of the genus *Homo*. We have thus become used to seeing Africa as the great warm crucible of hominid evolution. Some would find it pleasing if a third major stage were added to that list – the origin of modern humans. But such an attitude would smack of chauvinism and not science. The jury is still out on this question."[439]

In 1994, while Tobias was contemplating the evolutionary theory of modern humans, Ron Clarke decided to review some of the bones that had been blasted out of the Silberberg Grotto from the Sterkfontein cave back in the 1930s. Some of the breccia had been carried out of the cave and fossils had been chiselled out of the rock in 1980 and kept in the storeroom, called the shed, at Sterkfontein ever since. Near the entrance of the shed, there were bookshelves filled with books and knick-knacks. There were old photographs of Robert Broom on the wall that looked as if they had been hung decades before. There were metal shelves filled with cardboard boxes along the back and side walls. From a shelf on the right-hand side of the room, Clarke pulled down an old brown storage box. Out of a box labelled D20, he took several bags of mixed animal bones. In the very first plastic bag he opened onto the table, he found the talus bone of a foot. "Good grief, this is a hominid talus!" Clarke said out loud. "What's that doing here?" Clarke's wife, archaeologist Kathy Kuman, was sitting at another table in the storeroom and Clarke took the bone over to her to show her. He went back to the same bag and found a bone that fit snugly next to the talus bone. They clearly fit together and were definitely from the same individual. Then he found a third bone and a fourth.[440]

At this point, Clarke was so excited by his discovery that he wanted to share the news with Tobias, who was in Philadelphia at the department of anthropology at the University of Pennsylvania. Clarke and Tobias had not yet fully embraced email, so Clarke sent him a telegram, followed by a letter. Tobias asked Clarke to send photographs and casts of the bones as soon as possible. In his initial report, Clarke concluded the bones were from an australopithecine that revealed a tree-climbing capability, but that it was also bipedal, indicating time on the ground. Clarke thought that the australopithecine had fallen down into the underground chamber and died there, along with other animals. Given the small size of the bones, Tobias suggested naming them Little Foot. Clarke sent out a press release in Johannesburg and Tobias sent one out in Washington DC. The *New York Times* covered the story with a big spread including feedback from other scientists, some positive and some critical as usual. Clarke and Tobias published an article together in *Science*.[441] What they didn't know at the time was that there was more to Little Foot.

Three years later, in 1997, Clarke was looking through a cupboard in the hominid vault in the department of anatomy at Wits. He took a look at a box marked "Cercopithecoids" that contained monkey fossils that had been found in breccia from the Silberg Grotto. He opened the box and immediately saw more hominid foot bones and a tibia. He gathered the bones and soon realised that they were a match with the first set that he had found earlier out at Sterkfontein. With twelve bones in total from the foot and tibia bones of a single individual, Clarke reasoned that if he had two feet and a tibia, there might be an entire skeleton down there in the Silberg Grotto. He was excited to begin the excavation.[442]

However, Clarke had a growing conflict with Lee Berger. He was concerned that if he told Berger about his suspicion, he would take over the search and claim any subsequent find for himself. The conflict and tension between the two men was so bad that Clarke decided to keep quiet about the search for the rest of Little Foot.

Tobias's retirement had been delayed twice, so it was at the end of 1996 that the issue of succession began to heat up. Tobias hadn't fully thought through who should take over from him. Three contenders emerged for the position. The first was Ron Clarke. The second candidate was Geoffrey McKee, an American Tobias had recently appointed senior lecturer and who was focusing his research on the Taung site. The third possibility was Lee Berger.

Clarke was the senior person at PARU with many years of experience, significant research and a number of important publications. However, Berger thought that he deserved the position because he had been able to save the future of PARU and secure financial backing for PAST. McKee was relatively quiet and soon realised that the competition was between Clarke and Berger.

While the competition to become Tobias's successor was rising, Clarke and Berger published an article together. They wrote about their theory of what had killed the Taung child. Since the skull had first been described in 1925, scientists had blamed leopards or sabre-toothed cats for the child's death. Clarke and Berger challenged these ideas and claimed the young australopith had likely been killed by a large bird of prey. The skulls and bones of other animals at the modern Taung site in the North West province, they said, showed distinctive evidence of damage caused by the present-day crowned eagle. They proposed that the Taung child, that lived over two million years ago, was also killed by an eagle, or one of its ancient ancestors. The bird of prey hypothesis revealed how vulnerable the australopithecines were in their environment. They were preyed upon by leopards, hyenas, sabre-toothed cats, and even ancient eagles. "Bob Brain was right," Berger later reflected. "We were the hunted, not the hunters."

Despite their joint publication in 1995, Clarke and Berger did not end the following year on a cordial note. It was late in 1996 that Phillip Tobias nominated Berger as his successor at PARU at Wits, a post that he would take on starting in January 1997. At the time, PARU had more than 20 employees, including Ron Clarke, who had been in charge of the field operations at Sterkfontein for five years. Clarke was 20 years older than Berger and was wary of what influence Berger might have on his research. Berger later thought that Tobias had made a mistake. "I was 32 years old and he was asking me to manage people in their fifties. The outcome was predictable."[443]

At the same time that Clarke was wary of Berger's influence, Berger was becoming concerned about Clarke's performance. He was worried that Clarke had been moving very slowly on a skull reconstruction he was working on. He was also concerned Clarke had been keeping fossils, including the Little Foot bones, not in the hominid vault at the department of anatomy, but in his own private safe.

The tension in the department of anatomy at Wits was not confined to Clarke and Berger exclusively. Growing political tension in post-apartheid South Africa was about to spill over onto Phillip Tobias's plate.

After the fall of apartheid, the Griqua National Conference (GNC) had been established and it was encouraging a political voice for Griqua people in a way that had not been previously possible. The GNC approached Tobias and reminded him of the graves of Cornelius Kok II and his family that Tobias had exhumed in the 1960s and '70s. They wanted their skeletons back.

20

Human Remains at Rest

You are cordially invited to attend a unique ceremony at which the skeleton of the Griqua Captain Cornelius Kok II (1778–1858) will be returned by the Wits Department of Anatomical Sciences to the living descendants of the Kok family led by Captain Adam Kok V

The invitation went out from Wits and Phillip Tobias as well as from the organisation "Griquas of Adam Kok V of Campbell". Their invitation stated that, as promised in 1961, when the skeletal remains had "served science" they would be returned for reburial. "This will be the first repatriation ceremony of sensitive remains ever to take place in the new South Africa."[444]

On 20 August 1996, Phillip Tobias and Adam Kok V stood behind a coffin in which the skull and full skeleton of Cornelius Kok II was laid out on white fabric. The 72-year-old professor looked serious, while the 42-year-old Kok looked somewhat bewildered, even sad. Tobias was wearing a suit and tie, not the khaki pants and rolled up shirt-sleeves he'd been wearing in 1961 on the day he'd exhumed the skeleton. There was a strong family resemblance between Adam Kok V and Abraham Kok, who had attended the exhumation. This can be seen clearly when comparing a photo of Abraham at the grave site in Campbell in 1961 with one taken of Adam Kok V at the event at Wits 35 years later.

What had started the ball rolling, in February 1996, was a statement the Griqua National Conference had sent to the *Mail & Guardian*, informing the newspaper that they had sent a letter to President Nelson Mandela demanding the bones of their ancestor, Cornelius Kok II, be returned and reburied in Campbell. Tobias told the newspaper that the skeleton of Cornelius Kok II had been removed with the "full blessing" of Adam Kok IV, the chief of the Griqua community at the time and that "when the studies were completed the skeleton would, in due course, be returned to him and his people."[445] Martin Engelbrecht, one of the GNC members who had called for the return of the skeletons, said that they had been kept in a "Whites only closet".[446]

Anthony Humphreys, the son of Basil Humphreys who had worked with Adam Kok IV in the late 1950s to find the graves, was still in high school back in 1961. He had gone with his father to see the excavation process. After school he went on to study archaeology and received his PhD in 1979. He disagreed strongly with Martin Engelbrecht's statement that the skeletons had been kept in secret. Humphreys said that "The remains were openly removed to the University of the Witwatersrand Anatomy Department for study purposes and this fact could have been established by anyone who cared to enquire – they were not being held in secret waiting to be 'discovered' by some intrepid sleuth."[447]

In early 1996, Phillip Tobias was involved in a different discussion about returning remains to their place of origin. As a member of the reference committee brought together by the South African government, Tobias was in discussion with the French authorities to return the remains of Sarah Baartman. The Griqua National Congress was one of the organisations that had demanded the return of her remains and they had called on the South African government to assist. Now they were negotiating with Tobias for the return of the remains of Cornelius Kok II.

Tobias was also part of a lengthy and heated debate about whether a sample of Sarah Baartman's remains, once they were returned to South Africa, should be retained for further testing. Tobias was in support of keeping a sample in order to test Baartman's DNA, arguing that it would be a mistake to bury her remains without taking DNA samples that would allow for scientific and forensic analysis. He thought the DNA would be important to determine living relatives, or to confirm the cause of her death, or to verify that the "returned bottled organs indeed belong to the same individual as did the skeleton".[448]

Dr Yvette Abrahams, who had conducted many years of research and had written about Sarah Baartman, was also on the reference committee. She took the opposite view. Using Baartman's body for scientific research, she said, should be avoided "because it is exactly what we just spent ten years saying was wrong for the French to do".[449] Initially, Himla Soodyall, who was also involved, agreed with Tobias, but after listening to the debates, she developed a slightly different perspective. She said that it was possible that if the remains were bottled in formaldehyde, the chemical would have destroyed the soft tissue's DNA. Once the French agreed to return Baartman's remains in 2002, Soodyall believed that "healing was going to be a better antidote than the proof, evidence based science". She suggested that no DNA samples be taken, and that the soft remains be placed in the grave along with Sarah Baartman's skeleton.[450]

The demand that Tobias return the Griqua skeletons and the French return Sarah Baartman's remains prompted a strong response from Alan Morris, Tobias's PhD student who had studied the Griqua skeletons, illustrating some of the political debates under way at the time. In a letter to a colleague dated 1 February 1996, Morris wrote: "I am afraid that my opinions on this matter are not very 'politically correct'. The Griquas are really trying to build a new ethnicity through the issue of the skeletons. Saartjie Baartman herself was unrelated to the Griqua although she was probably a Khoikhoi. The Griqua are emphasizing only one of their 'roots' because it will give them an advantage when they deal with the new government over land rights." He continued, "The living Griqua people are probably split 50/50 between the urban and rural areas and NONE live a 'traditional' lifestyle. In fact, to be a Griqua one needs to be a member of the Griqua church and speak Afrikaans. I could go on about this but I won't."[451] If Morris's view were extended to Afrikaners, it would imply that urban Afrikaners were not truly Afrikaner if they no longer live their "traditional" lifestyle.

In *The Skeletons of Contact*, which Morris published in 1992, he concluded that the Griqua skeletons showed "a predominately Negro morphological pattern, but with many Khoisan features". He wrote that "a curious finding of the morphological analysis is the relatively unimportant contribution of Caucasoid features", revealing his expectation that there would have been a greater European influence.[452]

Morris claimed that his study was a significant move away from the typology of the past at the Wits anatomy department, which indeed it was. He argued that typology and discrete racial categories no longer adequately explained what existed in reality. He also explained that race and ethnicity used to be seen as one and the same but that his study approached them as "totally separate concepts". He did not want to accept the typological labels given to skeletons, he said, but instead to look at them within the archaeological and historical context in which people lived.

Morris explained that this change in physical anthropology was important because it cut across the expectation of the apartheid government that everyone could fit into one specific racial category. However, Morris continued to believe that biology trumped culture, that external cultural influences would have less meaning, and that Griqua identity should be based predominantly on biology. "... the modern Griqua continue to attract non-Griqua to the community."[453]

And, he concluded, "As the 'new South Africa' dawns, these issues of race will continue to haunt us and any study which helps us understand the relationship of biology and culture will be critically important."[454]

At about the same time, Morris made an important contribution to the changing field of physical anthropology. He produced a catalogue of skeletons that further broke from the model of race typology. The catalogue included 2 500 human skeletons in collections around South Africa and identified them according to date, location, archaeological context and form of burial. "Nowhere in the entire catalogue are any skeletons listed with specific identities of Khoekhoe or San or Negro or with any other ethno-racial tag," said Morris. In the new catalogue, he included only 365 skeletons from the Raymond Dart collection with enough contextual information. All of the skeletons in the Dart collection that were identified only according to race were excluded.

While Morris was doing research for the catalogue, he wrote a letter to Raymond Dart, asking for specific information about /Keri-/Keri. He said that Alun Hughes told him that /Keri-/Keri had died at Grootfontein, but the accession register recorded that she'd died in Oudtshoorn. "Do you perhaps remember the circumstances of /Keri-/Keri's death and how you got the body to Johannesburg?" he asked Dart. According to Morris, Dart never responded.[456]

The Griqua people make clear, as do examples from across South Africa and around the world, that no one's identity – biological or cultural – is completely fixed. It can be fluid, based on a range of factors including language, history, culture and practice. The democratic elections of 1994 brought changes to South Africa and allowed new identities to flourish. The founding of new Griqua organisations in the 1990s brought greater attention to the Griqua language, culture, history and identity. A controversial exhibit called "Miscast", curated by Pippa Skotnes, at the National Gallery of Art in Cape Town in April 1996, also brought greater public attention to some of the debates about representation, and the Bushman and Khoisan identity.

In May 1996, just months before the Wits repatriation event, Deputy President Thabo Mbeki began his speech at the adoption of South Africa's new constitution with the words, "I am an African." "I owe my being to the Khoi and the San," he said, "whose desolate souls haunt the great expanses of the beautiful Cape – they who fell victim to the most merciless genocide our native land has ever seen." Mbeki declared he was formed of the migrants from Europe, the blood of the Malay slaves who came from the East, and he was the grandchild of Hintsa, Sekhukhune, Cetshwayo and Mphephu, as well as of those who were transported from India and China. "I know what it signifies when race and colour are used to determine who is human and who, sub-human." In the wake of Desmond Tutu's reference to a rainbow nation, Mbeki's speech broadened the definition of what it means to be African. "We refuse to accept that our Africanness shall be defined by our race, colour, gender or historical origins."[457]

The Wits skeleton repatriation ceremony was held in an auditorium at the medical school. In attendance were UN representatives, Griqua community members, members of the medical community and !Khoisan X, the former Benny Alexander. Tobias told the story of the original exhumation of the grave site and the ancestral relationship between Cornelius Kok II and Adam Kok V, who was present at the event. He remarked that it was a memorable event in South African history to return these skeletons just at the time when there were moves afoot to return Saartjie Baartman's remains from Paris. Tobias recounted that he had told Adam Kok IV that when he had finished studying and publishing investigations of the skeleton he would return the bones – "and the time has come for me to honour my promise. It wasn't in writing. It was a by word-of-mouth promise from me to the captain."[458]

When Tobias sat down, Adam Kok V stood to speak. "We appreciate the positive spirit with which Wits University agreed to the return of the remains of our ancestor, but we do not appreciate the humiliating and blasphemous act of interfering with the resting place of the Royal family." He looked out over the audience. "We appreciate the need for scientific investigation, and a middle ground must be found to allow research without the humiliation of our ancestors." Many people from the Griqua community clapped and cheered. Those attending from the medical school remained silent.

Kok said, "These remains were kept in total secrecy for more than 30 years and with no report back to our family of whom it is said we gave our blessing to the exercise. We must express our great displeasure with Wits University in this regard." Kok went on to call for a return of the land around Campbell to the Griqua people. "I will not allow the continued humiliation and impoverishment of our people."[459]

After the formal speeches were over, Tobias spoke to reporters and made a surprising statement. From his examination of the skeleton of Cornelius Kok II, he unequivocally concluded that the Griqua people descended from the Khoisan and not from a mixture of races as previously thought. This was in direct opposition to what he had claimed about the Griqua in the past. Not only did the *Diamond Fields Advertiser* in Kimberley report on Tobias's statement but so did the national paper *City Press*. In their article "'Pure' Bones Tell All", the opening line was, "World renowned Wits University palaeoanthropologist Professor Phillip Tobias has finally laid to rest the assumption that the Griquas are of 'mixed race'." Tobias said, "We now have a clearer picture than we had before." Retaining the skeletons for anatomical research had been important, he said. "We did not want to make a finding on only one skeleton as racist anthropologists would have done, but needed to study the affinities of the whole population group."[460]

It is not clear why Tobias made the statement that the Griqua people were directly descended from the Khoisan only. By this time, most people who identified as Griqua were also identifying themselves as coloured as well, but they also drew on their Khoisan heritage and their African customs, such as initiation. Perhaps he wanted to support the Griqua desire to be seen as distinct from the broader category of "coloured". Perhaps Tobias wanted to support the Griqua claim for land. Martin Engelbrecht, the Griqua leader from the northern Cape, who had approached Tobias in the first place, applauded Tobias's statement because, he said, it took the Griqua demands seriously.

Tobias's statement that the Griqua people were pure Khoisan rather than of "mixed race" as previously described caused a stir amongst the scientific community. Fiona Barbour, a member of the McGregor Museum in Kimberley, asked Tobias about his statement. His response was that he had based his views entirely on Alan Morris's findings published in *Skeletons of Contact*.

Barbour was confused so she wrote to Morris directly. She began her letter by stating that the work of George Nurse and Trefor Jenkins in 1975 as well as other accounts had always referred to the Griqua as a heterogeneous group. She quoted Morris's *Skeletons of Contact* as saying that the original Griqua population "was of Baster origin but Khoi, San and Negro individuals were important additions in the early 19th century". In her letter, she said that her reading of *Skeletons of Contact* "is clearly at variance with Tobias's recent pronouncements. Can you elucidate your findings?"[461]

Alan Morris was in the US on sabbatical at the time so he did not attend the August 1996 event at Wits and he did not hear what Tobias had said on that day himself. Barbour wrote that the issue of Griqua origins had become a "hot issue" in South Africa because of contested land claims. She told Morris that the issue couldn't remain academic because it had practical and political implications. "In one sense it does not matter what the genetic composition is or was," wrote Barbour, "but the present quest for an indigenous Griqua identity looks likely to create a new myth for further conflating and confusing physical and socio-cultural evidence." She said the McGregor Museum was involved because they held a Griqua display and because they were involved in discussions about skeletal remains. Therefore, she said, we want to be "reasonably certain of our facts".[462]

Morris wrote a three-page, single-spaced letter in response. He began by saying that Barbour's interpretation of his work was much closer to the truth than was Tobias's. He denied ever saying that the Griqua were not a "mixed blood group" or that they were indisputably Khoisan in origin. Morris was supportive of the Campbell-based Kok family and their demands for the exhumed skeletons. He was also supportive of their claims to ancestral lands in the area. However, he felt very strongly that the Griqua National Congress had no authority to represent all Khoisan people. "They are using their own misunderstanding of biology to claim some sort of Khoisan purity. This is false and patently political!"[463] Morris felt so strongly about this that he went on to write a paper, which he published in 1997, entitled "The Griqua and the Khoikhoi: Biology, Ethnicity and the Construction of Identity".

Morris presented his paper at a Khoisan identities and cultural heritage conference in Cape Town in July 1997. Kate Cloete, the secretary of the Griqua National Congress, immediately opposed Morris. "There were no coloureds," she said. "All these people were descendants of the Khoi." While Morris believed that he was technically correct, many people who were previously classified as "Coloured" were asserting their own distinct identity. Also, both Martin Engelbrecht's applause of Tobias's statements and Cloete's denunciation of Morris pointed to identities that were not fixed. For the first time, they were able to engage international debates and assert that the Griqua were part of the discussion about first nations. Nationally, they could also examine how the Griqua identity be maintained in the process of forging a new national South African identity for the first time.

Just a year earlier, Morris and Tobias had entered into correspondence about race. Morris wanted to conduct a study on race typology, and to compare South Africa and the United States, and he had asked Tobias to review his proposal. Tobias responded that he thought that the project had some merit but that he thought it needed more contextual information. He thought that an unbiased study would have to look at the "worldwide anthropological preoccupation with typology and racial stereotypes which persisted to almost every corner of the world until well after World War II". Tobias wrote that "to concentrate only on South Africa (and I presume Anglophone South Africa) and the USA would raise obvious criticisms of your failure to contextualize such a study". He told Morris that he should include a study of the large Afrikaans literature on science. "It is surely a trifle naïve to think that the two, three or at most four English-speaking physical anthropologists who were working in South Africa, helped to create the climate or public opinion and had such a large impact upon S.A." Tobias also wanted Morris to look at the role of Calvinist doctrine. He argued that Afrikaner intellectuals and politicians did not "care a fig" about what Dart or Broom or other English-speaking physical anthropologists were saying. He told Morris he would have to study the role of Christian National Education under apartheid and its impact. Lastly, Tobias wrote, Morris would have to include study of Nazi philosophy that was shaped by "racial stereotypes and philosophy fed to Hitler by the German physical anthropologists. It was in light of my knowledge and study of this that I made – and still stand by – my statement that no physical anthropologist in South Africa provided Malan or Verwoerd or Strijdom with the 'scientific underpinning of their philosophy.'" Tobias closed by saying, "I honestly believe you should think this project out again."[464]

A month later, Morris responded. "The Afrikaans literature on race is nearly silent on the BIOLOGY of race and when it does indeed speak of it, it uses data drawn from the very English speaking scientists you say have such a small role." He told Tobias that he received a letter from Louis Stofberg, a member of parliament representing the Herstigte Nasionale Party (HNP), in 1985 asking for clarification of Carleton Coon's ideas to use in a parliamentary debate.

Morris wrote that "typology is a system in physical anthropology that facilitates the use of physical anthropological data to support racism" and that it prevented other views of variation to develop. "Saul Dubow's new book [*Scientific Racism in Modern South Africa,* published in 1995] has shown how important this was in the South African context. If you take umbrage with me, then Saul must give you serious heartburn." Morris closed his letter by saying that he had not enjoyed the exchange between them because of the great respect he had for Tobias.[465] As it turned out, Morris did not pursue the research project as proposed. Another letter from Tobias to Morris in late 2000 said: "I remain convinced that you should write a major work on some aspects of race and typology, avoiding some of the biases that have beset some recent publications."[466] Tobias continued to be loyal to Dart, and downplayed the racism implicit in Dart's race typology, which he himself had embraced.

After the repatriation event at Wits, there was a breakdown in communications and arrangements between the university and the Griqua community. It appears that Adam Kok V expected Wits to pay for the costs of transporting the skeletons to Campbell and for the reburial. However, Wits University responded that this was not their responsibility and that they did not have the budget to do so. Their view was that the skeletons had been returned to the Griqua people in Johannesburg and now the reburial was up to them.

An article in the *City Press* entitled "Landless People Given Bones for Reburial" explained that there was a big question about where to bury the remains. "This question confounds the Griqua paramount chief, as all Griqua ancestral lands were expropriated during the colonial and apartheid eras." Clearly disturbed, Adam Kok V returned to Kimberley immediately after the ceremony at Wits. He would have to negotiate with the current holders of the land and clarify burial arrangements. Only then would he be able to bring his relatives' remains back to Campbell.[467]

Tobias had a special coffin built for Cornelius Kok II so that his skeleton could remain separate from the rest of the Raymond Dart Skeleton Collection. The coffin sat on top of the other skeleton boxes for another decade.[468] It was not until on 24 September 2007 that the skeletons were finally buried in Campbell next to the mission church. Phillip Tobias, who was 83 years old at the time, did not attend the reburial ceremony but he sent a long letter that was read out at the event. Forty-five years after the exhumation, Tobias told the gathering that Captain Adam Kok IV had asked him to find evidence that Cornelis Kok II was buried there and to shed light on the "make-up of the Griqua nation" which he said was something of a mystery.[469]

Seven skeletons were reburied on that day. It is not clear which ones and how they were chosen. It could have been the three skeletons that were originally exhumed in April 1961 plus four more. An additional 28 skeletons that were exhumed in Campbell continue to be held at Wits.

During the mid-1990s, Tobias had the opportunity to work with a new post-apartheid government and reshape his reputation. But just as he wanted to reinvent himself, he wanted to protect Raymond Dart's legacy as well. So during the discussions about the return of the Griqua skeletons and the efforts to reclaim Sarah Baartman's remains, Tobias did not speak openly about /Keri-/Keri's skeleton. Sometime during the 1980s or early 1990s, it went missing. "I'm not going to suggest that everybody who's ever been dug up should be put back into the ground," said Tobias. "Science would come to a standstill if that were the case."[470] At about the same time, /Keri-/Keri's family joined forces with a human rights attorney to begin a land claim in the southern Kalahari.

21

A New Narrative

It was in May 1997 that Ron Clarke had found the additional hominid foot bones and a tibia in the box in the department of anatomy at Wits and matched them to the Little Foot bones he had found out at Sterkfontein three years earlier. Clarke made a cast of the right tibia bone and gave it to his field assistants Stephen Motsumi and Nkwane Molefe. On 2 July 1997, he asked them to go into the cave to see if they could find the matching section of the rest of the bone that would connect them to the rest of the skeleton. "The task I set them was like looking for a needle in a haystack," said Clarke.[471]

The two men went into the cave and began to search the walls and then the ceiling and the floor of the cave. In the darkness, the only light came from their head-lamps. The walls of the cave were damp and it wasn't easy to see if there were any broken off bones amidst the mud. After the first full day of searching, they came out wondering if they were crazy to think they would find anything. Stephen Motsumi thought it was hopeless. On the second day, they went back into the cave. After the first few hours, Motsumi's eyes got tired. But he was patient. After hours of searching, they came out of the cave again. "We think we've found it," said Motsumi. "No," said Clarke, "there's no way that you could find it so quickly. You've only been in there a day and a half." Motsumi and Molefe took Clarke down into the cave to show him the place in one wall where they thought they had found the match. Motsumi carefully fit the tibia into place. He looked up at Clarke and handed him the bone. Clarke's hands were shaking, but he fit the bone perfectly into place. All three men were beaming. "We've got the skeleton," said Clarke. The bone in the wall of the cave and the bone in their hand had been blasted apart by lime workers 65 years earlier in the late 1920s or early '30s, which meant it was even more amazing that Motsumi and Molefe had found the match.[472]

Clarke knew then that, "encased in that steep slope of ancient cave infill, they would uncover something that paleoanthropologists had wanted for so long – a complete skeleton of *Australopithecus*."[473] Clarke decided to keep the find to himself. The work was painstaking and slow. Day after day, Motsumi and Molefe began to chip away at the concrete-like breccia. Every morning Motsumi woke up and was happy. "I did the job with an open heart and that's the best," said Motsumi. The more he did the digging, the more he didn't want to stop. They worked each day from about 8am until 1:30pm before they came out of the cave. Sometimes they stopped working for a couple weeks because it was raining and wet.

About eight months after Motsumi and Molefe found the match, Lee Berger drove out to Sterkfontein to tell Clarke that he was not going to renew Clarke's contract at Wits because of his lack of productivity and that the contract with the university would come to an end in December 1998. Clarke didn't argue. In fact, he was even more determined to keep the Little Foot skeleton discovery hidden from Berger. He found a job at a German university that paid him to continue his research at Sterkfontein.

After ten months of slowly, carefully chipping away at the breccia, Motsumi and Molefe had cleared enough rock away from the lower limbs to see the thigh bones of the two legs lying side by side. They had to be meticulous so that they did not chip away at the fossil itself. But Clarke and his team were frustrated. The feet, half of the legs and one piece of an arm were there, but the rest of the skeleton was missing. The three men continued to search for three months. At last, Clarke realised that there was a fault running through the cave wall. He thought that perhaps the upper part of the skeleton had fallen during a rock fall or collapse. He asked Motsumi to search the rock below the legs instead of above them. On 11 September 1998, a year and two months after the three men had smiled together in the cave, Motsumi's chisel found the end of a hominid humerus. Just above that was the skull. Clarke had been right. A rock fall had occurred, breaking the skeleton in half and shifting the skull and the top of the skeleton below the legs and feet. The glint of tooth enamel sent a shiver down Clarke's spine. It was an upper molar which, together with the lower jaw, told the team that they found the skull. The skull was largely undamaged and complete. They soon realised that the left arm of the skeleton was above its head. Clarke's initial estimate was that Little Foot was 3.3 million years old.

The team was enormously lucky to find a nearly complete skeleton. The formal ritual of humans burying their dead in graves began about 50 000 years ago. Before formal burials began, the skeletons of ancient humans decayed in the same way that any other dead animal did. The chance of finding a nearly complete ancient hominid skeleton was slim. It was the unique circumstances of Little Foot's death that resulted in its skeleton being preserved for three million years.

Back at the university, the tension between Berger, Clarke and Tobias was nearly at breaking point. In August 1998, before the team found the upper part of the skeleton, Berger received a letter from Tobias saying that Clarke was a brilliant scientist and an excellent researcher and demanding that Berger re-hire Clarke immediately. Not knowing about Little Foot, Berger was confused by the letter and took the case to the Wits vice-chancellor at the time, Colin Bundy.

In September 1998, Clarke called Berger saying that he had something that he wanted to show him. Berger had a meeting in the Sterkfontein Valley one morning to discuss the application for the area to be declared a World Heritage Site with UNESCO so he suggested to Clarke that they meet that afternoon. Clarke had invited Phillip Tobias, Beverley Kramer (the head of the Wits anatomy department) and Berger to meet him at the research hut alongside the excavation area.

They walked around the main entrance past Robert Broom's statue and followed Clarke through the back entrance and down into the Member 2 section of the cave. They rounded the rock to enter a small chamber. Already in the cave were Stephen Motsumi and Nkwane Molefe, along with the filmmaker Paul Myburgh, who had been working on a documentary on human evolution in South Africa. What the hell is going on down here, wondered Berger. The camera lights passed across them and settled on the cave wall. There in the rock was an astounding sight. The legs and skull of a skeleton emerged from the cave wall. Berger's overwhelming sense of disbelief was tempered by the thrill of seeing a hominid fossil *in situ*. Clarke told Berger that the skeleton that they were looking at was the rest of the body of Little Foot, the foot bones that he had found years earlier.[474]

A few days later, Clarke got a phone call from Phillip Tobias's secretary Heather White while he was out at Sterkfontein to tell him that he had to get to Pretoria. He and Tobias had been asked to make a presentation to cabinet. Clarke rushed home to change, put on a suit, and then drove the 60 kilometres to the Union Buildings where cabinet was gathering. In his role as director general, Roger Jardine accompanied Clarke and Tobias to the meeting. Clarke was awed by the large cabinet room with its enormous round wooden table. He brought photos of Little Foot to pass around to each of the ministers. They looked at the side view of the skull in the rock. One of the cabinet ministers joked to another, "Hey, that looks like you."[475]

As South Africa entered the post-apartheid period, with Nelson Mandela as president, a new constitution and an emphasis on democracy, the country was looking to redefine itself. This was an opportunity to look again at the fields of archaeology and palaeoanthropology. Whereas in the past, these fields of study had carried the heavy load of a colonialist project, here was the opportunity to shift the focus so as to construct a new narrative, a narrative that would recognise Africa for its contribution to humankind. Thabo Mbeki recognised this opportunity.

On 9 December 1998, Tobias and Clarke presented the Little Foot skeleton to the world. The announcement was made at Wits University and Thabo Mbeki was there. Clarke told the story of how he had found the foot bones first and the process of finding the entire skeleton. He explained that the hominid had likely fallen into an underground cave and died, and that there had been a rock slide or roof collapse which had fully encased the body with limestone. He or she had remained there face down for the next three million years.

Clarke said that the exact species was still uncertain but the skeleton was definitely part of the genus *Australopithecus*. Tobias said, "It clearly was a creature that had sprung loose from the trees." They could tell by looking at the feet that the hominid had "arboreal habits coupled with terrestrial habits". Tobias was hugely proud of the discovery. On this occasion and for many years to come, he would say "Little Foot is a rare treasure. Not only is it the oldest hominid yet found in South Africa but at 3 million years old, it is the oldest virtually complete hominid skeleton found anywhere." Tobias enjoyed the vicarious thrill. Once again, he was close at hand for someone else's fossil find.

A unique aspect of the announcement of the skeleton was that Clarke decided to credit Stephen Motsumi and Nkwane Molefe for the fossil find. They were not in the background. They attended the press conference, spoke to the media, and met Deputy President Thabo Mbeki. Clarke told the press the dramatic story of how Motsumi and Molefe took the small tibia bone and went to find its match in the cave. Motsumi was not an excitable person. He had a calm demeanor and didn't raise his voice, but he felt proud; he felt famous.

Years later, when Clarke was asked how the decision had been made to include Motsumi and Molefe in the event and to credit them for their role in the fossil find, Clarke said, "Tobias was always supportive of the African crew. He said if they find something, he'd like to record it. He was adamant that they be acknowledged, as was I. It wasn't even a question. They were essential, part of the team."[476]

Clarke pointed out that "in Broom and Brain's day, it was different; they lived in a different era. With the changes, it became easier." Saul Sithole, Daniel Mosehle and George Moenda had all played significant roles in the field, but sadly they had received little credit for the work they did. If Clarke had found and announced the full Little Foot skeleton before the end of apartheid, would it have been received differently? "If we still had the apartheid government, they wouldn't have summoned me to the Union Buildings," was his response. "They didn't accept human evolution."[477]

Stephen Phologo Motsumi no longer works at Sterkfontein. Eleven years after the Little Foot announcement, in 2010, he developed heart problems and had to have surgery to replace a heart valve. He went back to work after the operation, but in August 2013 he had to retire. Motsumi lives in Brits, about 90 kilometres north of Johannesburg. In September 2014 he had not been back to Sterkfontein for over a year and he was in the midst of constructing a new house in the township.[478]

Motsumi had just turned 60 and looked well and in good shape, but he moved quite slowly for a man of his age. Perhaps his heart trouble had slowed him down. Born in 1954 in Hekport, Motsumi went to school until Grade 8. He was the eighth child of ten children and the youngest son, so he left school to find work to support his two younger sisters. Motsumi's family was Setswana speaking. His father worked on farms in the area and his mother stayed at home raising the children. Motsumi's first job was to work at the CSIR in Hartebeeshoek in the maintenance department. In 1972 he started working at the Sterkfontein Caves. His older brother had been working there and found him a job as a tour guide. Motsumi met Alun Hughes and remembered wishing that he could work with him in the field, or with the fossils in the lab, but there wasn't an opening at the time. For the seven years that he worked as a tour guide, he longed to work more closely with the bones. "I was interested in the rock formations," said Motsumi. "And I wanted to understand the fossils. How do you handle them, clean them and identify them? And I was interested in Mrs Ples and the Taung skull and stone tools."[479]

In 1979, Motsumi left his job as a tour guide to take a better-paying job with Consol Glass. In 1986, he called Alun Hughes to ask if there was an opening and Hughes said yes. Motsumi worked with Hughes for five years and then with Ron Clarke when he took over in 1991. When asked what his family thought of his work at Sterkfontein back in the 1970s and when he started working with bones in the 1980s and '90s, Motsumi said, "My parents knew that I was a fossil digger. They never said don't do it. They never said it's against our culture … These are not new graves at Sterkfontein. Some of these bones are two million years old."

Evolution was not taught in schools when Motsumi was young. "It is not that we are from baboons," he said. "No, that's not the story. The story is that hominids are like humans," said Motsumi. "Many teachers, even today, are against it. They don't want to teach children about fossils. But that's a mistake. My plan was always to interest my children in fossils so that one of them could become a palaeoanthropologist." Motsumi's eldest daughter was studying at the University of Johannesburg and his two youngest sons were in Grade 4 and Grade 6. "My sons are going to Maropeng with their school next week," he said. "I told my one son, 'Tell your teacher what work your father did.' My son said 'Our teacher is not going to believe us.'"[480]

Looking up and off into the distance towards a group of young people walking past, Motsumi said: "We are not professors, but I do know my job. I've been to France for training to learn about making casts. It was my first time on a plane." He shifted slightly and smiled. "Palaeontology is so exciting. Not only because of the hominids, but because of everything you find. My wish is to have my job back again for another five years."

A couple of weeks later, Motsumi went to Sterkfontein. "I saw Professor Clarke and Professor Kuman," he said, "and I saw the skeleton. It was a great, great moment for me." Then he paused and said, "But there's still a lot of work to be done."[481]

The day after the announcement of Little Foot, back in December 1998, South Africa had returned to the headlines, and to centre stage in the search for human origins. "FULL AUSTRALOPITHECUS FOSSIL FOUND IN SOUTH AFRICA" read the headline in the *Washington Post*. The article celebrated that over the years many skull fragments, portions of pelvises, and other pieces of bone had been found but this was the first time that a skull and skeleton had been found at Sterkfontein from the same individual. As a result, some questions about *Australopithecus* could be answered, such as the size of the brain in relation to the body, and other body proportions, its likely diet, and how it moved around. The University of Liverpool had dated the hominid at between 3.22 and 3.58 million years old based on the surrounding breccia. That date made it slightly older than Ethiopia's Lucy, who had been found by Don Johanson and his team in 1974.

"Like the old song about the ankle bone connected to the leg bone," said Tobias "the whole thing fits together. This is the skeleton of Little Foot, which now we'll have to find a new name for." Years later Ron Clarke would joke about this. The name they had given to small foot bones was the name that stuck for the entire skeleton. "It's silly," he said. "This big, hulking, heavy individual got stuck with the name Little Foot."[482]

Before the excitement around the announcement of the skeleton had died down, the controversy of Berger having fired Clarke hit the press. The word reached the media that despite Clarke's discovery, his contract at Wits was coming to an end at the end of December and Berger was responsible. The *Mail & Guardian* broke the story. There were placards on the street with the headline "WITS FIRES FOSSIL FINDER CLARKE".

A Zapiro cartoon entitled "Bloopers of the Century" showed four panels, each depicting a scene of a historic blooper. The fourth panel – the punch-line – was dated 1998 and showed Berger with glasses and a flip-chart speaking to Clarke, saying, "Your project is going nowhere. You're fired." A big-nosed Clarke is holding a little foot in his hand with a specimen tag. Behind him is a cave marked "Sterkfontein" and there is a voice from the cave calling out "Dr Clarke! … Come quickly!!"

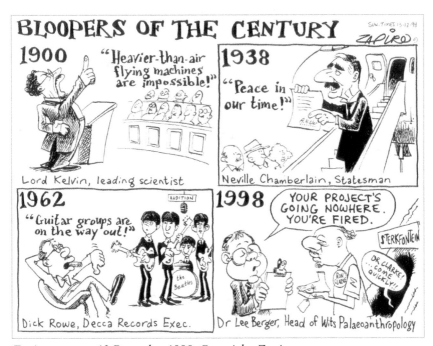

Zapiro cartoon, 13 December 1998. Copyright: Zapiro.

Wits University found itself in a difficult situation. On the one hand, there was Tobias and Clarke with the Little Foot skeleton. Neither of them was employed by the university any longer because Tobias was retired and Clarke had been fired. On the other hand, Berger was technically the head of palaeoanthropology and the chief fundraiser for paleo research. Wits formed an advisory committee to address the conflict. Tobias suggested, and the university agreed, that Sterkfontein should be run by a new separate research entity called the Sterkfontein Research Unit (SRU). This was established in March 1999. Tobias was appointed director and Clarke was deputy director. Berger left the department of anatomy, moved across town to the Wits main campus, and became the director of a new group called the Palaeoanthropology Unit for Research and Exploration (PURE). Berger remembered this as a painful time. Tobias had been his supervisor and Berger had helped to keep Sterkftontein afloat when it was under threat. They had worked together closely for a decade. Just as Dart had been a father figure for Tobias, Tobias had been like a father to Berger. "It was terrible," he said. "That was cutting the Gordian knot."[483] The resulting split – between PURE led by Berger on the main campus, and the SRU led by Tobias and Clarke in the department of anatomy at the medical school, would simmer and periodically flare; it would take another 15 years to reach resolution.

Not only was Berger in conflict on campus, but he also was in conflict with a number of his international colleagues. In 1998, Berger had published a paper in *Science* arguing that it was more likely that *Australopithecus africanus* was the true predecessor to *Homo* rather than *Australopithecus afarensis*, the species of the famous Lucy skeleton found in Afar, Ethiopia by Don Johanson and his team. Berger and his co-author Henry McHenry argued that the longer arms and shorter legs of *africanus* in South Africa, not *afarensis* in East Africa, might have been the ancestor of *Homo*. Both Donald Johanson and Tim White strongly disagreed with this hypothesis. White especially disagreed and began a fierce debate with Berger that would last for years.[484]

At the same time that Berger, Clarke and Tobias were in conflict in the late 1990s, another young scientist was entering the field in Kenya. Job Kibii admits that he hadn't planned to become a paleoanthropologist but "fate intervened". Originally, he intended to study mathematics, sociology and economics at the University of Nairobi. "The rule was," he said, "you had to register for three subjects in the first year." It was a Friday and he registered for mathematics and sociology but when he arrived at the department of economics, he was told to return on Monday. He didn't want to wait, so he followed the suggestion of a poster on the wall advertising a degree in archaeology. "That was that," he said. "By the end of the second year, I was in love with the subject." Kibii ended up dropping mathematics and majored in archaeology and sociology.

With an honour's degree in both subjects, Kibii went to the National Museums of Kenya looking for a job. There were two job vacancies but they both required a masters in archaeology. Once again, Kibii saw a pamphlet on a noticeboard, this time advertising a one-year master's in palaeoarchaeology at Wits. He applied and by mid-1999 he was on his way to Johannesburg. Kibii would remain at Wits for another fifteen years, working with Tobias, Clarke, Berger and others, and making his own unique contribution.

In December 1999, soon after Kibii arrived in Johannesburg, and one year after the Little Foot skeleton was announced to the world, the Sterkfontein Valley was declared South Africa's first World Heritage Site by UNESCO. The application said that the Sterkfontein Valley was significant because it held "an invaluable record of the stages in the emergence and evolution of humanity, over the past 3.5 million years", making it one of the world's most important sites for studying human evolution. This status protected the sites of Sterkfontein, Swartkrans, Kromdraai, Coopers, and Gladysvale from commercial development. As soon as UNESCO declared the site, plans began to develop a restaurant and gift shop with a small display museum at Sterkfontein. In addition, there were plans to build a larger visitors centre at a separate location.[485]

When the application was submitted to UNESCO in June 1998 by the Gauteng provincial government and signed off by Pallo Jordan, the minister of environmental affairs and tourism, the area was described as "The Fossil Hominid Sites of Sterkfontein, Swartkrans, Kromdraai and Environs". The application made the case by saying "The many caves in the Sterkfontein Valley have produced abundant scientific information on the evolution of modern man over the past 3.5 million years." It was only much later, after the World Heritage Site was approved by UNESCO, that it was proposed that the site be called the Cradle of Humankind.[486]

In the few years after the 1994 elections, the national government began to take an interest in palaeoanthropology, not only in terms of support for the UNESCO application. Francis Thackeray, the head of palaeontology at the Transvaal Museum, gave an "I have a dream" style speech in 1996. His dream was to have one replica of Mrs Ples in every school in South Africa within 20 years. The next year, April 1997, was the 50th anniversary of the finding of Mrs Ples by Robert Broom. There was a great deal of media coverage, and a 24-carat gold coin with Mrs Ples was produced by the South African mint. The minister of arts, culture, science and technology, Ben Ngubane, struck the first coin. Richard Dawkins struck the second one. In 1998, Thackeray got a call from Rob Adam, the deputy director general for science and technology. He said that his department wanted to support palaeontological research. He asked Thackeray to submit a proposal and to "think big". At the time, the palaeontology department at the Transvaal Museum had an annual budget of R10 000 so Thackeray applied for R500 000 to undertake research and promote public education. DACST approved the grant.[487]

In June/July 1998, PARU organised the Dual Congress with the support of PAST, particularly Standard Bank. It was the gathering of the International Association for the Study of Human Palaeontology and the International Association of Human Biologists. Over 700 delegates from around the world attended the event at Sun City and celebrated the return of South Africa to the international anthropological sciences.

Just as Deputy President Mbeki expressed interest in Little Foot and the application to UNESCO, he was putting together his thinking on African Renaissance, a concept that the continent and its people would enter a new period of cultural, economic and scientific achievement. "But Mr Deputy President," said Tobias respectfully, "Africa is the birthplace of humanity. Instead of African Renaissance, you should be talking about African Naissance."[488]

TOP: *Phillip Tobias with Mary Leakey in East Africa, holding a reconstruction of a* Homo habilis *skull. Courtesy of* Tobias in Conversation *published by Wits University Press. Copyright: School of Anatomical Sciences, University of the Witwatersrand.*

BOTTOM: *Alun Hughes, Phillip Tobias, Hertha De Villiers and Raymond Dart, February 1978. Courtesy of* Tobias in Conversation *published by Wits University Press. Copyright: School of Anatomical Sciences, University of the Witwatersrand.*

TOP LEFT: Elisabeth Vrba at Swartkrans in 1983. Copyright: 1983 David L Brill. (www.humanoriginsphotos.com)

TOP RIGHT: Bob Brain at Swartkrans in 1983. Copyright: 1983 David L Brill. (www.humanoriginsphotos.com)

BOTTOM: Bob Brain walking under an aerial grid plotting system at Swartkrans. Copyright: 1985 David L Brill. (www.humanoriginsphotos.com)

TOP: Newsweek *cover, January 1988. This issue included an article about mitochondrial DNA and human evolution. Copyright: PARS International.*

BOTTOM: *Himla Soodyall, human geneticist, National Health Laboratory Service and University of the Witwatersrand. Courtesy of Himla Soodyall.*

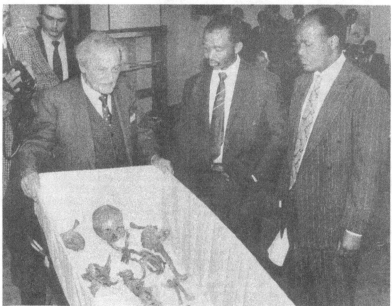

TOP: *The excavation site at Campbell in 1961. From the right, Basil Humphreys, Gerhard Fock and Phillip Tobias. The three men on the left are students named Badenhuizen, Kaufman, and Olmesdahl. Courtesy of AJB Humphreys.*

BOTTOM: *Phillip Tobias, Adam Kok V and Martin Engelbrecht stand beside the skeleton of Griqua Chief Cornelius Kok II at an event at Wits to return the bones to his descendants on 20 August 1996.* News clipping from Business Day, *21 August 1996. Courtesy of the Wits University Archive.*

TOP: *Ron Clarke, Stephen Motsumi and Nkwane Molefe at Sterkfontein soon after the announcement of Little Foot in December 1998. Courtesy of Kathy Kuman, Ron Clarke and the* South African Journal of Science.

BOTTOM: *Phillip Tobias speaks to a group of dignitaries visiting the Sterkfontein caves during the* World Conference on Sustainable Development *on 1 September 2002. Right to left are President Thabo Mbeki, UN Secretary-General Kofi Annan, Nane Annan, Jane Goodall, South African Deputy President Jacob Zuma and Wendy Luhabe Schilowa. Photographer: Dario Lopez-Mills. Copyright: AP.*

TOP: *Lee Berger and his son Matthew at the public announcement of* Australopithecus sediba *at Maropeng on 8 April 2010. Photographer: Daniel Born. Copyright: Times Media Ltd.*

BOTTOM: *Lee Berger, Deputy President Cyril Ramaphosa and other dignitaries stand by the display of the* Homo naledi *bones at the public announcement at Maropeng on 10 September 2015. Courtesy of Christa Kuljian.*

TOP LEFT: *Job Kibii, former senior researcher at Wits University's Evolutionary Studies Institute and director of excavation at Malapa, currently Head of the Palaeontology department, National Museums of Kenya. Photographer: Brett Eloff. Copyright: University of Witwatersrand.*

TOP RIGHT: *Lee Berger, Research Professor in Human Evolution and the Public Understanding of Science at Wits University's Evolutionary Studies Institute and an Explorer in Residence at the National Geographic Society. Copyright: University of the Witwatersrand.*

BOTTOM: *The six women scientists – left to right are Becca Peixotto, Alia Gurtov, Elen Feuerriegel, Marina Elliott, K Lindsay Hunter and Hannah Morris – or the "Underground Astronauts" worked to retrieve the fossils from the Dinaledi Chamber of the Rising Star Cave. Courtesy of John Hawks.*

TOP: *An artistic reconstruction by paleo-artist John Gurche of what* Homo naledi *might have looked like. Copyright: John Gurche.*

BOTTOM: *Zapiro cartoon, 13 September 2015. Copyright: Zapiro*

22

"We Are All African"

In March 1999, five years into South Africa's new democracy, Hertha de Villiers' daughter Phillippa published an article in *Elle* magazine. The introduction to the piece read: "In the heyday of apartheid, a little black girl was adopted by a white family and illegally reclassified white."

After 33 years of "half-truths and lies", Phillippa de Villiers was able to find out the truth about her biological parents. The article began with a description of a striking image – the body of a doll sitting on a rubbish heap with her hand reaching out for the head some distance away. Phillippa de Villiers said that the image helped her to make sense of her childhood. Like the doll, she had felt discarded and dismembered; she began to feel her consciousness as completely separate from her physical body. "The black Phillippa went underground," she wrote.[489]

The article told stories from Phillippa's childhood. It told the story about Phillippa's first day at school and how the bus driver refused to take her onto the bus because she was "Coloured". When Phillippa was ten, around the same time as she met palaeoanthropologist Bernard Wood, she went to the movies – *Freaky Friday* – with a friend and was told that she could not go in because she was "Coloured". The manager eventually let them in but they had to sit at the back of the cinema. When Phillippa told her mother about what happened, Hertha was furious. She went straight to the manager. "As an anthropologist, I've made a lifelong study of the physical characteristics of racial types," she told him, "and there was absolutely no way that my child is a coloured."[490]

Phillippa joined her mother for a year in Chicago when she was in Standard 3 (Grade 5). When a black teacher, Mr Harris, asked Phillippa why black people were discriminated against in South Africa, she answered, "Because they're stupid. They are too primitive to run the country." Shocked, Mr Harris spent many hours talking to her about black power. Back in school in South Africa, Phillippa remembers serious debates about whether "black people were still part monkey, trapped in some backwater of the evolutionary stream". During her apartheid childhood, adults would shout "You baboon" when speaking to a black person. She imbibed the belief that white European society represented the pinnacle of human achievement.[491]

It was in the 1980s that Phillippa began to make the connection between her head and her body. "The black girl that had been hiding in me was born in the apartheid struggle," she wrote. "All the rage I had swallowed in my life in the white world exploded." In 1985 she joined the United Democratic Front and the End Conscription Campaign and was arrested by the police after making a trip to a township outside Grahamstown where she was studying journalism at Rhodes University.

The day her father decided to tell her that she was adopted, she remembers, they were sitting on the rocks at the seaside. "You're going places where we can't go," said Bungy de Villiers. "You must be careful. You're not white. Your mother was an Australian nurse and your father is probably Aborigine." "So I'm a kaffir after all," said Phillippa, which she knew would hurt him. "You wouldn't look like one if you ironed your clothes occasionally and combed your hair," he replied. And then he said, "Your mother saved your life. How can you be so ungrateful? Now all you want is to be with your black brothers." Phillippa didn't speak to Bungy or Hertha de Villiers for five years after that. Before Bungy died in 1995, Phillippa was able to forgive her father and thank him for his courage in telling her the truth.

In early 1998, when Phillippa was 31, she decided to look for her biological parents. She went to the Child Welfare Society of Johannesburg to speak to a social worker and told her what she knew – her birth date and that her mother was Australian. She was told that it would take a while to find the file. Within two weeks, the social worker called back and said she had found it. That was when Phillippa first saw Suzie Alcock's signature on the consent form. That was when she read about Samuel Amamoo and then about John Brian North, who Alcock had said was the real father. That was when she read about the months of questioning and the visit to Phillip Tobias and his concern about her thick upper lip.

Three days after her 32nd birthday, Phillippa picked up the phone and called directory assistance in Australia. She asked for all the Alcocks in Adelaide and was given about 30 listings. She called the first three, got answering machines and then abandoned the task. She called back to directory assistance and asked for the name Amamoo. There was only one listing. It was after one in the morning in Australia. With a pounding heart, Phillippa dialled the number. "Hello," said a deep voice on the other end of the line. "Um …

I'm calling from South Africa." "Yes," said the voice. "I think you might be my father." Silence. "Hello?" "Yes, hello." "I thought you'd put the phone down on me." "No, not at all. This is a very important conversation. I've been waiting for this phone call for years." After a brief and poignant conversation, Phillippa said, "Forgive me for asking, but I've always wanted to know … Are you black?" Amamoo laughed a deep, warm laugh and said, "Very black. I'm from Ghana."

Samuel Amamoo helped Phillippa to find Sue Alcock, and they began a faltering correspondence. Amamoo told Phillippa her traditional Ghanaian name was Yaa, for a girl born on a Thursday. Phillippa decided to incorporate her new name, becoming Phillippa Yaa de Villiers.

When Phillippa showed her mother Hertha a photo that Samuel Amamoo sent of himself, Hertha repeated something that she had said over and over again in the ten years since Phillippa found out she was adopted: "There's no way you're a Negro."

In 2000, Phillippa went to Australia for the first time to meet her biological mother and father. One day Phillippa and her father made a trip to the beach. Phillippa told Sam what Hertha had said – "There's no way you're a Negro" – and Sam responded immediately. "Let's go and do a paternity test. Let's put your mind at ease. Let's put Hertha's mind at ease." They did the test and it came back saying that there was a 99.8 per cent chance of paternity.

The *Elle* magazine article resulted in another falling out between Phillippa and Hertha, who was 73 years old at the time, and long retired from Wits University. The article closed with Phillippa saying she didn't know if she would ever understand why her Australian mother brought a black child to Verwoerd's South Africa to give her up for adoption. "But I was lucky enough," said Phillippa, "to be brought up by people who loved me."[492]

Not only had Phillippa's life been shaped by the system of racial classification in South Africa, but also, given her mother's scientific career, she had grown up amidst the science of race and human evolution. As a child, she knew Raymond Dart and Phillip Tobias. She made friends with Alun Hughes and Bernard Wood and read books about the australopithecines. And she made memorable trips to Makapansgat.

In 2002, Phillippa had the opportunity to meet with Phillip Tobias. She was working with a film company that was producing a documentary called *Tobias's Bodies*. They thought they would film a discussion between Phillippa and Tobias and use it in the documentary. Thirty-six years after Tobias first examined her, they met at Phillippa's house one afternoon. Her baby son had just started to crawl, and the wooden floors creaked as Tobias and the cameraman walked in. De Villiers recalls that she was respectful of Tobias and that Tobias was formal and polite. She believed that Tobias had had powerful control over her destiny when she was a baby. The conversation between them was not, in the end, used in the film, largely because Phillippa was concerned about the anger in her family, and didn't want the issues back in public.

Tobias's Bodies was aired on the South African Broadcasting Corporation (SABC) in 2002.[493] The opening sequence shows a busy Johannesburg street. The sun has just risen and shopkeepers are opening for the morning trade. On a narrow sidewalk, women walk past the shops in shweshwe dresses and colourful head-wraps. In front of KwaMpindamshaye (a shop selling traditional medicine) a grey-bearded gentleman wearing a cream-coloured suit and tie takes off his hat. His tie carries the iconography of human evolution, depicting an upright human being followed by several stooping, bipedal apes. The man introduces himself. "Hello, my name is Phillip Tobias and I am an African. We're all African. You may think that's a peculiar statement to make. My skin is not dark. My hair is different from other African people, but we're all Africans because this was our cradle. We were all born here in this great continent."

Tobias was making the point that all humans are of one species. We are all one human race from the same origins, which is an important reminder of our common ground. He was also suggesting, however, that being African is biologically determined. In fact, that is not strictly correct. Being African is also shaped by history, culture and politics. It is a person's lived, social experience that greatly shapes their identity. The reality for people in South Africa – instead of exhibiting unity, a shared prehistory, and a shared humanity – was more in tune with Thabo Mbeki's "two nations" speech, which highlighted not a unified world of Africans, but rather a great divide between the relative prosperity of white people in South Africa and the continued socio-economic exclusion of the black majority, a reality that is mirrored globally between rich and poor. The reality of growing inequality in the world cuts against the vision of a shared heritage in Africa.[494]

More and more, people have been turning to genetic testing to gather information about their ancestral past. Some people mistakenly believe a DNA test can provide them with information about their identity, which is not always the case. Women can be tested for the mitochondrial DNA which will follow their mother's mother's mother's line. Men can be tested for both their mitochondrial DNA as well as the DNA in their Y chromosome, which only follows their father's father's father's line. That means that many relatives on both sides won't be covered at all. "No genetic test can tell you who you are,"

Himla Soodyall reminds us. We are testing ancestry, only one line or two."[495]

Genetic testing offers input on how your genes compare to other individuals from certain regions of the world. This is where the work of Phillip Tobias and Himla Soodyall overlap, and in the same television series Tobias visits Soodyall's lab to discuss the genetics of human origins research with her.

Soodyall is sitting at a counter in her laboratory, wearing a white lab coat and white plastic gloves. Tobias asks her, "What is it that turns you on about modern human origins?" With a broad smile, Soodyall answers: "The beauty of genetic work ... is that every individual is a walking example of the tree of evolution that we're trying to reconstruct. Every DNA element in the human genome has a history. Today we can extract DNA specimens from blood samples, from hair, from teeth and even from the inner lining of our cheek."

Soodyall then gives Tobias something that looks like a wooden ice-cream stick. She holds another in her hand, and they both wipe their sticks on the inside of their cheeks. "I did seven scrapes," Tobias says to the camera. "I can't wait for the results." Soodyall puts the stick into a test tube and shakes it. Tobias explains that Soodyall's work with the DNA of the San people supports the view that their DNA is closest to that of the oldest African ancestors. Soodyall adds: "Mitochondrial DNA showed quite convincingly that the root of modern human origins evolved in Africa. In fact the Khoisan people of southern Africa still harbor signatures in their genome that carry these ancient traits."[496]

Some weeks later, Tobias returns to Soodyall. "I'm back and I want to know what you've made of my DNA," he says. Soodyall tells him that his maternal DNA is consistent with European Jews. She says that the DNA from his Y chromosome, which illustrates his father's paternal line, is consistent with the DNA of both Middle Eastern Jews and Middle Eastern Arabs. While these two groups may share the same genes, they have experienced very different social and political experiences in the Middle East.

"So, Himla," says Tobias, "what you are doing is probing the past, not with the tools of the archaeologist, or the palaeoanthropologist, but you are probing the past by looking at what is recorded in the genes." "Yes," Soodyall says. "This is where genetic studies have particular importance. I could change my language in one generation. I could change my culture within the same generation. I may identify either culturally or ethnically to one of many demes [local populations]. But what's in your DNA, you cannot change. It is what has been inherited from our parents' generation, their parents before and so forth. So the DNA is a direct link to our ancestry."

But this does not mean that science always trumps language and cultural

241

influences in shaping identity. Towards the end of the episode, Tobias confirms this: "Our genes may describe our histories, but they cannot describe our culture or our mind."

At about the same time, Alan Morris had also begun to bring the study of bones and the study of DNA together to help understand human origins. He was part of the first attempt to extract ancient DNA from a human skeleton in South Africa in 2002. He and his colleague, Alec Knight from Stanford University, worked with over 20 skeletons that ranged in age from 440 to about 6 000 years old. According to Morris, despite their best efforts, "the project was a complete failure". They were unsuccessful at retrieving even one sample of DNA. Morris and Knight found this frustrating because ancient DNA had already been gathered from Neanderthal bones in Europe. The DNA they found was very difficult to work with because it had already broken into very small pieces. There were many technical challenges, but technology is advancing all the time. In another 12 years, Morris would try again.[497]

More and more often, Tobias was taking the opportunity to convey the message that the differences between human beings on the surface were not significant and that human beings had more in common that united them than the things that seemingly differentiated them. Scientists were saying that genetic data suggests that racial differences have developed very recently, possibly within the last 50 000 years. Others, including Alan Morris, said that it could be as recently as the last 10 000 years.

Over and over again, Tobias made the point that we are all human beings and that we are all members of the same species, *Homo sapiens*. The lifting of apartheid laws and the end of official racial classification provided an environment in which Tobias felt more comfortable embracing this new narrative of human origins.

23

Coming Home

In 2002 South Africa hosted the UN World Summit on Sustainable Development. During the conference, President Thabo Mbeki took Kofi Annan, the head of the United Nations, to visit the Sterkfontein Caves. They sat together on the stone steps near the Robert Broom statue. Phillip Tobias gave a presentation, explaining the history of the area and confirming that Africa was the cradle of humankind.

While there had not been many new hominid finds at Sterkfontein since Little Foot, the site was no longer struggling financially. PAST had supported Sterkfontein since 1994 with funds from the private sector – Standard Bank and Anglo-American, among others. In 2002, the department of science and technology channelled R1 million to PAST through the National Research Foundation (NRF). An additional R1 million grant was approved for 2003-04. Plans were in place to build a new building at Sterkfontein with a small museum, a restaurant and a gift shop. No longer would visitors to Sterkfontein meet at a thatch-roofed rondavel for tea as they had done for over 40 years. Soon they would visit a spacious, glass-walled viewing deck that overlooked the valley.

Just as Tobias was spreading the message that Africa was the place of origin for all of humankind, the quest for the return of Sarah Baartman's remains was finally coming to fruition. Tobias had been brought into negotiations with France in the mid-1990s, at the invitation of the department of arts, culture, science and technology He reached out to his scientific colleagues in France, but the process was not easy. He tried to convince them of the importance of the return of her remains for the dignity of the San people in South Africa, but for many years the French had said no. It was Tobias's negotiations that lay the groundwork for a political agreement. And it was the powerful poem by South African Diana Ferrus entitled, "I've come to take you home" that helped to turn the tide. The French government finally agreed, with a bill passed in parliament in 2002. At last Sarah Baartman was coming home.[498]

On 9 August 2002, Himla Soodyall stood in a school hall in Hankey in the eastern Cape participating in the Khoisan ritual to prepare Sarah Baartman's remains for burial. The burning of buku leaves and incense filled the room with aromatic smoke. Some women were singing and others wailing. The camera lights were blinding and Soodyall's eyes were burning as she broke the seal of one of the jars that contained the remains of Sarah Baartman's brain. The pungent smell of what was likely formaldehyde hit Soodyall in the face. As an asthmatic, she began to feel her lungs constrict. She couldn't breathe and had to cough. Tears ran down her face. Soodyall reached for the second jar, which held Baartman's genital tissue. She felt it starting to slip and luckily someone stepped forward to assist her and pass the contents to the person on the other side of the open coffin. Baartman's skeleton was held in place with foam padding. A member of the Khoisan community lifted the bones from the coffin and wrapped them in animal skin. Symbolically, Soodyall had returned Sarah Baartman's remains to their final resting place on the banks of the Gamtoos River. As the poet Diana Ferrus wrote, she had returned to "the lush green grass beneath the big oak trees" where "the air is cool" and "the sun does not burn." No longer were they held by science. After close to two centuries of being exhibited and then stored in a museum in Paris, Sarah Baartman's remains had completed their journey home.[499]

The crowd, the cameras and the singing moved from the school to a large outdoor marquee at the burial site. "There are many in our country who would urge constantly that we should not speak of the past," said President Thabo Mbeki at the funeral service. He called for the restoration of dignity to Baartman, to the Khoisan and to millions of Africans who had known centuries of "wretchedness". Mbeki's speech described the relationship between Sarah Baartman's individual life story and the larger realities of slavery, colonialism and white supremacy.[500]

When reflecting on the saga of Sarah Baartman, Tobias wrote: "European anatomists and anthropologists were trying to define the supposed line between human and animal. To their biased eyes, the 'Hottentot' was closest in bodily structure, behaviour and even language to non-human animals. It was seriously doubted by some whether they belonged to the same species, *Homo sapiens*, as the rest of humankind. Cuvier and his contemporaries believed that the brain and the genital anatomy (and presumed sexual behaviour) would provide convincing answers to the question: were the Khoisan human or subhuman?"[501]

But Tobias did not have to look back to the 1800s and Cuvier. The same viewpoint was still relevant in the early twentieth century as scientists tried to understand race and human evolution. It was still a question that Tobias was thinking about in the 1950s.

The return of human remains had entered the public realm, not only because of Sarah Baartman, but also because of the publication of a book entitled *Skeletons in the Cupboard* in 2000 by Ciraj Rassool and Martin Legassick. For the first time, they chronicled the skeleton trade that existed in South Africa at the turn of the twentieth century and described in detail how human remains were collected unethically and how they were brought together in numerous museum and university collections.[502]

One of the responses to the book came from Tim Maggs of the South African Museum. He was concerned that skeletons would be taken from museums and he compared such action to "burning the archive". A debate flared as a result. Sven Ouzman made the point publicly that he didn't think that any of the scientists would demand continued access to these human remains "if they had been found to belong to dead victims of the Holocaust".

Skeletons in the Cupboard was discussed at the South African Museums Association (SAMA) conference in 2001. In his address to SAMA, Mike Raath, who was the curator at Wits, expressed the need for Wits to engage in a process to address the collections they held. That process, however, did not get off the ground. The McGregor Museum hosted a launch for the book on 10 September 2001 in Kimberley as part of a workshop on "Human Remains in the Museum". While the book prompted needed discussion about human remains collections, the authors admit that the impact on museum and university management and procedures has been slow and uneven.[503]

In October 2004, four years after the publication of *Skeletons in the Cupboard* and two years after Sarah Baartman's burial, another young woman's remains were under discussion. The story of /Keri-/Keri and the story of Sarah Baartman had many similarities, joining them together across history and geography.

Ouma /Una Rooi paid a visit to Phillip Tobias at the School of Anatomical Sciences at Wits. !Gurice, also known as Abraham, who had declined to relinquish the face mask Raymond Dart had made of him at the Bain camp at Tweerivieren in 1936, was Ouma /Una's great uncle. She was /Khanako's first cousin's grandchild. Born in 1931, /Una was part of the Bain camp as a child. Her second name was Katrina, just like /Keri-/Keri. Ouma /Una made the visit to Wits with Nigel Crawhall, a linguistics professor who had worked with the ǂKhomani San on a land claim, and Annetta Bok, one of Ouma /Una's young nieces, who was also directly descended from !Gurice, /Khanako and Klein /Khanako. They walked into Tobias's cluttered office to see books piled on every surface, an articulated skeleton in the corner, paintings and photographs on the walls, and the diminutive Tobias seated behind his desk. He was 78 to Ouma /Una Rooi's 73. Tobias motioned for them to sit down and then offered them a cup of tea.[504]

Nigel Crawhall spoke about the results of the land claim that had been brought forward by the ǂKhomani San who had lived in the southern Kalahari, including Ouma /Una and her sisters, and how they had won a settlement in 1999. Ouma /Una and her family had previously lived in a hut in Swartkop, but after the land claim she had moved onto the Andriesvale farm, which was part of the settlement. It was "delicious to be back on our soil" she said. Ouma /Una was one of the few living people to speak N/u, a language that had been subsumed in order to survive under apartheid. Her children spoke Afrikaans, not N/u. Ouma /Una delighted in sharing information about plants and medicine and indigenous knowledge that had been buried for so long but was resurfacing under the new democracy. "We buried everything under the sand, and then we went back to uncover it." Annetta Bok spoke about the pride she felt in being an owner of land.[505]

The documents in the land claim had quoted Raymond Dart's 1936 Kalahari research reports. The list of the 77 people who were part of the Bain camp was presented as a genealogy in the case. On the day the land claim was settled in March 1999, President Thabo Mbeki arrived to celebrate at Askham in the Kalahari in a helicopter. There were lush clouds in the sky. Mbeki signed the award of the claim and gave the claimants access to 60 000 hectares. "This is a step towards the rebirth of a people that nearly perished because of oppression," he said. As Mbeki spoke, the clouds opened up and a light rain began to fall. When the ceremony was over and the helicopter lifted off the ground, the rain came down hard. Within two days, there were pools of water between the dunes. After 30 dry years, on that day the drought broke.[506]

Pointing to one of the black and white photographs on Tobias's office wall, Ouma /Una exclaimed, "That's !Gurice. They called him Abraham." Tobias asked her to help identify several other photographs. After a conversation about the Empire Exhibition, he asked one of his colleagues to look for /Keri-/Keri's body cast, which by that time had been taken off display. He told Rooi, Crawhall and Bok that /Keri-/Keri had died on the way to the Empire Exhibition, which was not correct, and explained that this was why her remains had ended up at Wits. They found the grim, partly broken body cast wrapped up in a corner in storage.[507]

According to Nigel Crawhall, one of the reasons for their visit to Wits was that Annetta Bok had been thinking about developing a curated exhibition at the Upington Museum, near where she lived. Tobias arranged for some plaster face masks from the Kalahari to be reproduced and given to Bok. She had been hoping that they might take /Keri-/Keri's body cast with them to Upington but they decided against it on that day for fear that it could be damaged further. Bok's intention was to develop a centre for cultural heritage in Upington to help pass on knowledge from one generation of the ǂKhomani San to the next. She knew it would be of interest to the descendants of people taken to the Bain camp, but also she thought it would attract other visitors in Upington. For a variety of reasons, the project never went ahead, and the body cast remained in the corner, behind the cupboard. As Bok and others involved in the land claim liked to say, "The land claim may have brought justice, but it could not ensure well-being."

At about the same time, a project was under way to open a museum at Wits to explain human origins and honour the San culture. It would be called The Origins Centre.

24

Getting the Message Out

As Tobias was meeting the descendants of Cornelius Kok and /Keri-/Keri, and promoting the fact that all human beings belonged to the same species, Lee Berger was continuing his search for ancient hominid fossils. Unlike Dart and Tobias, he did not work at all on the physical anthropology of living people. He only focused on ancient fossils that might shed light on how hominids developed over a million years ago. Berger's youth and American roots meant he wasn't steeped in the Imperial Age. At the dawn of the new millennium, he had the opportunity to engage in post-colonial, post-apartheid science.

For decades world attention had focused on East Africa and now Berger wanted to bring global attention back to South Africa. He thought South Africa had more to offer. One of his major areas of interest was South Africa's contribution to understanding the development of early *Homo*. In 1998 he had written a paper inspired by his PhD, that investigated the body type of *Australopithecus africanus* and compared it to the body type of *Australopithecus afarensis* and differences in the proportions of their arms and legs. Based on increased fossil evidence from Sterkfontein, together with his co-author Henry McHenry, Berger argued that *A. africanus* had "long arms and short legs," a feature that was similar to *Homo habilis*. He contended that *A. afarensis*, represented by the famous skeleton Lucy, with its shorter arms and longer legs, was more human-like. The paper closed with the suggestion that perhaps *A. afarensis* "evolved independently of the lineage leading to *Homo*".[508] The idea cut across the grain, not only in terms of the prevailing assumption that Lucy was a "mother species", but also because the scientists involved took offence.

Don Johanson, the finder of Lucy, and American archaeologist Tim White, who had worked with Lucy's skeleton extensively, and was also credited with finding one of the world's oldest hominids dating back to 4.4 million years ago, were not impressed. They critiqued Berger's science, and took personal offence. On the one hand, the debate that McHenry and Berger started with their paper was about which species led to the development of *Homo*. On the other hand it was a debate about the relative importance of South Africa and East Africa in the fossil record. The tension, especially between White and Berger, has not subsided for the past 18 years.

In 2000, Tim White wrote an article about the demise of palaeoanthropology as a result of "the careerist" making progress in the field as opposed to "the scientist". Never mentioning Berger by name, White wrote that "careerists court the media and use it to advance their careers. Scientists deal responsibly with journalists so that they can educate and accurately inform the public." He said that journalists loved words like smallest, largest, youngest, oldest, newest, first and, most important ever. Hominid fossils, he argued, were rare and precious; there were more people entering the discipline, he said, than could be sustained by the field.[509]

Not deterred by the criticism, and brushing off the conflict with Ron Clarke and the University of the Witwatersrand, Berger began a new project with the US-based *National Geographic* as the publisher of his first book. He decided that in addition to writing in scientific journals, he would write a book for a more popular audience. Historically, Dart and Tobias had spoken often on the radio to a broad audience, but writing books for a popular audience was something they had left to Robert Ardrey and Bruce Chatwin. Berger chose Brett Hilton-Barber, a well-known South African radio journalist, to work with him. In 2000, they jointly published *In the Footsteps of Eve: The Mystery of Human Origins*. The book title referred to Berger's interest in fossil footprints that he had worked with near Langebaan in the western Cape, which he described as having been made by an imaginary ancestor named Eve. The book made the case that there should be more attention given to South Africa, rather than to East Africa, in the understanding of human origins.[510] Berger argued the "South African contribution has been sidelined if not ignored" for the past 50 years.[511] The book referred to the "long arms and short legs" debate and Berger refered to Tim White as the Great White Shark, "nicknamed as such because of his aggressive intellect and inquisitor's mind".[512]

The book was not well received by the scientific community. Kevin Kuykendall, one of Berger's American colleagues at Wits, wrote a review in the *South African Journal of Science* entitled "Never Letting the Facts Get in the Way of a Good Story". Kuykendall called Berger's book "tabloid palaeoanthropology" and said that the relationship that Berger drew between his imaginary Eve that made the footprints at Langebaan and the hypothetical Mitochondrial Eve was problematic and baseless. He accused Berger of sweeping aside scientific accuracy, and concluded: "... if South African palaeoanthropology is to retain its credibility and be taken seriously, our standards must be impeccable – otherwise we will lose the faith and support of international colleagues *and* the public."[513]

Bernard Wood, the anthropologist who had worked with Tobias and De Villiers 25 years earlier, wrote a review in the *Journal of Human Evolution* entitled "Chalk and Cheese", contrasting the quality of two recently published books. The article was complimentary of Jon Kalb's book *Adventures in the Bone Trade* about the search for human origins in Ethiopia but extremely critical of Berger and Hilton-Barber's book. Wood described the book as a fairy tale in which all was going well for Berger (the young prince) until Ron Clarke (the wicked wizard) started to make trouble for him. In the fairy tale, Wood wrote that Phillip Tobias (the king) was unhappy with the wicked wizard's work at Sterkfontein. However, the king made peace with the wicked wizard leaving the young prince "looking pretty stupid". Wood said there were so many factual errors and typos in Berger's book that he went through two packets of Post-it notes by page 150, and concluded that standards had dropped so far at *National Geographic* that it was "frightening and tragic".[514]

Persevering, Berger and Hilton-Barber worked on another book together, *The Official Field Guide to the Cradle of Humankind*. Sponsored by the Gauteng provincial government, Standard Bank and PAST, the book was published by Struik, an imprint of Random House Struik. This second book received even more criticism than the first. *The South African Journal of Science* ran an editorial about it and provided space for two reviews – one by Tim White and the other by Judy Maguire from the Bernard Price Institute at Wits. Before commenting on the new book, White referred back to *In the Footsteps of Eve* as having tarnished the reputation of *National Geographic* and the University of the Witwatersrand. "Its false nationalism was a disservice both to the science and South Africa." White said the second book "epitomizes the peril that palaeoanthropology faces in the new South Africa". He critiqued Berger for presenting himself as the saviour of palaeoanthropology in South Africa with his fundraising skills. "It is true that Berger's rise to prominence signals a new era: one of smoke and mirrors, in which style triumphs over substance."[515]

One of the photographs in *The Official Field Guide*, under a section named "The Changing Face of Excavation", showed black South Africans excavating at a site called Coopers in the Sterkfontein Valley. The text suggested that technicians were being recognised at the forefront of research. "Berger has done nothing, however," wrote White, "to change the fact that South African academic paleoanthropology is still a dominantly white enterprise … where are the black South Africans with doctorates in Berger's 'new era' of South African paleoanthropology?"[516] White concluded that Berger had chosen "tabloid science in pursuit of headlines and of power gained by controlling money and resources" and asked whether this was really the model that the South African scientific community wanted to follow. "Never before in its long and distinguished history has South African palaeoanthropology stood at such a critical crossroads."

The piece written by Judy Maguire, who had worked for many years at Taung, entitled "A Second Opinion", lamented many of the same points as White. She found that the book was filled with factual inaccuracies and said: "… rarely was a guide in such a position to lead innocents astray as this one." She also found "a plethora of factual errors, serious misconceptions, contradictions, obsolete data and misstating of key concepts", all of which compromised the scientific worth of the book. In Maguire's opinion, "not to speak out against this unfortunate book is to commit the perjury of silence".[517]

Rarely in a field that had a reputation for being contentious and filled with intense rivalries had there been such impassioned conflict. Lee Berger looked back on that time and said, "Those books were easy targets because they were soft targets. They were not scientific papers. The attacks on the books originated from attacks on the science. How dare I say that *A. africanus* might have something to do with human origins? At that time, everyone thought it all happened in East Africa. That was the paradigm. It was an East Side Story. *A. afarensis* was the root."[518]

Despite the critiques and disputes, palaeoanthropology had the growing support of the new South African government and was gaining public attention. In April 2003, PAST began showing a theatre presentation called "Walking Tall", illustrating humankind's evolutionary journey, taking it to many schools. Bob Brain, who had become actively supportive of PAST as its scientific advisor, wrote that the organisation wanted "to create a proud and strong South African identity for each of its citizens as well as to bring its rich cultural heritage to the attention of the world … as part of South Africa's national drive towards unity and pride in our country and our continent".[519]

The tension between Lee Berger and Ron Clarke continued to simmer. One of the chapters in *In the Footsteps of Eve* – "Skeletons in the Closet" – was devoted to the conflict between them. In addition, the two men were in conflict about access to fossils. Tobias and Clarke held the permit for Sterkfontein and kept the work and the fossils largely to themselves. Berger thought that he and others should have greater access to the fossils.[520]

One of the debates about access to a fossil erupted between Clarke and Berger, and this time Job Kibii was involved. Kibii was working with Berger at Wits, but he was also a PhD student and Ron Clarke was his supervisor. In May 2002, Kibii had found a fossil that was a piece of a hominid pelvis. The fossil had been marked incorrectly as a bovid amongst a box of fossil fragments from Sterkfontein. Reminiscent of Clarke's finding the Little Foot bones, Kibii's discovery turned out to be a missing part of a pelvis from a partial skeleton found by Alun Hughes named Stw 431. Kibii shared his find with Clarke, who was then able to resconstruct the pieces of the *Australopithicus* pelvis. Their conclusion, which they published together in 2003, was that the pelvis flared out horizontally more so than the pelvis of a modern human. Their work helped confirm a theory about the *Australopithicus* pelvis from analysis of the Lucy skeleton 20 years earlier.

At the time, Lee Berger was also working with some of the fossil bones from the partial skeleton Stw 431. He intended to write a monograph, and asked Job Kibii to give the piece of the pelvis that he had found to him so that he could incorporate it into his study. According to Kibii, a heated debate developed but in the end the fossil remained with Clarke and Kibii. Kibii was able to maintain a good working relationship with both of the other scientists, but the episode further damaged the low level of trust between Berger and Clarke.

At the same time, another debate was underway about whether there should be more exploration at Sterkfontein. Should more resources go into exploration in the area or not? Tim White's 2000 article that assessed palaeoanthropology in the new millennium stated: "The best of the African fossil fields have probably already been found and exploited ... We do not have thousands of terrestrial fossil fields left to find and work. We don't even have dozens." As a result, White suggested that perhaps there were enough palaeoanthropologists in the field. Putting more funding into training, he said, "spreads fewer fossils among still more armchair theorists".[521] Another aspect of this issue was that many US and European universities were training palaeoanthropologists, and many of them were reliant on material in Africa. At the same time, few African students were being trained to enter the field.

Lee Berger believed that there was still more to find out in the field and he persisted in the search. But by the mid-2000s, he had a reputation for hyperbole. He had been working in South Africa for close to 20 years, from 1989 to 2008, and had continued to present himself as the leader in the field, but some argued that he had yet to make a significant contribution. Berger was enthusiastic about each and every one of his fossil finds, but they were viewed as relatively inconsequential. Many scientists continued to believe that the spotlight should shine on East Africa. Donald Johanson called Berger "the grandstander of the field". While he had helped Berger to start his career back in the late 1980s, and saw Berger as passionate about his work, he thought that Berger "often overstates the importance of what he has found".

In 2010, Berger would make an announcement about an important new fossil find in the Cradle of Humankind. The scientific sharks would circle, wanting to see if the form of Berger's announcement would outweigh the substance.

As you walk up to the entrance of Maropeng, the visitors centre at the Cradle of Humankind, you pass several stone plaques on the way. The first one says "Hominids – the ancestors of modern humans – first emerged about 7 million years ago, in Africa." Walk a few more metres and the second granite plaque reads: "The first stone tools were made and used in Africa, at least 2.6 million years ago." Keep walking through time to see the next message: "Our ancestors were able to use and control fire at least one million years ago in the Cradle of Humankind." Increasingly impressed, you keep moving along the path and forward in time, to the next marble plaque which reads: "*Homo sapiens*, the species to which we all belong, evolved in Africa approximately 200,000 years ago." The last message before you enter the building reads: "All of humanity shares an African heritage. We are one, diverse species across the globe with roots in Africa."

Maropeng was opened in December 2005. In Setswana, the word means: "returning to the place of origin." Maropeng's website explains that the name was chosen "to remind us that the ancestors of all humans, wherever they may live today, originally came from Africa. When visiting the Cradle of Humankind, people are actually 'returning to their place of origin.'" That same year, UNESCO approved two additional sites, Taung and Makapansgat, to be incorporated into the World Heritage Site that was originally named in 1999.

The building at Maropeng was designed to look like a tumulus – an ancient burial mound. The term "tumulus" is not generally used in Africa and is more common in Europe and Asia. Francis Thackeray, who was head of the Transvaal Museum in the 1990s and was then at the short-lived Institute for Human Evolution at Wits, was one of the people who served on the committee to review the design of the building. He did not support the tumulus design because it was not African-inspired. Other members of the committee believed that the design would assist in the marketing of the heritage site to an international audience, and could present the building as one large grave site.

Despite the design not being local, one major message from Maropeng is that everyone on earth has their roots in Africa. The burial mound theme gives the visitor the sense that all of our ancestors are buried here. In many African cultures, great respect for ancestors has an important spiritual role to play. Some commentators have pointed out that Maropeng houses fossils of ancient, pre-cultural hominids that did not engage in burial traditions so it makes no sense to have the building in a shape that points to the cultural behaviour and burial practices of modern humans. The topic of burial and whether it is a cultural ritual unique to humans would return to the field of palaeoanthropology in a big way ten years later, as Lee Berger announced *Homo naledi*.

In addition to the message that every human being has roots in Africa, another major message is that all human beings have a shared humanity. The exhibits take the visitor on a journey through time. First there is a boat ride that takes us down a river past displays that simulate the creation of earth. Then we walk through a vortex that simulates the big bang theory and end up before a giant globe of the earth illustrating the breaking apart of the ancient continent, Gondwana, to form the seven continents we know today. The management of Maropeng describe their work as "edu-tainment" and in fact the tour does have the feel of a theme park. Every visitor completes their tour by ending up in a large hall with multiple exhibits illustrating different aspects of human evolution. There is a display on Raymond Dart and the Taung child skull. There are questions and answers displayed about Robert Broom and Mrs Ples. There is a film on permanent display, showing Ron Clarke climbing down into the Silberg Grotto in the Sterkfontein Caves to show the Little Foot skeleton embedded in the rock.

President Thabo Mbeki formally opened Maropeng in December 2005. He reminded the audience that, "Those who appropriated history before we claimed it back, could not countenance the possibility that human life could come from so contrary a continent." He said that Maropeng was an African monument "to the human species, in all its beauty and variety, in all its tragedy and glory, in all its wonder and complexity".

This was not the first time that Thabo Mbeki had spoken about his support for the Cradle of Humankind. When Lee Berger and Brett Hilton-Barber published the *Field Guide to the Cradle of Humankind* in 2002, they asked President Thabo Mbeki to write the foreword. Mbeki wrote that research in

Africa had been important in the effort to understand human evolution. He reminded readers that hominids first became bipedal in Africa, and that it was in Africa that they first developed larger brains, first used stone tools, and first controlled the use of fire. "As Africans," he wrote, "we must overcome the debilitating effects of an unjust past that sought to inculcate the notion that black people are by nature inferior ... This book should help Africans to realise that, having given birth to humanity, we must reverse the many years of dehumanization that have characterized our recent past."[522]

Just as Jan Smuts wanted South Africa to be at the centre of palaeontology, so now did Thabo Mbeki. However, their ambitions were different. Smuts wanted to bring white South Africans together around science, while Mbeki wanted to bring together all South Africans, black and white, in a strong sense of national pride. His particular interest was to encourage black South Africans to take pride in the contributions that Africa had made. Mbeki's narrative placed South Africa at the centre of the search for human origins, but for different reasons. Rob Adam, the director general of the department of science and technology in the early 2000s thought that Mbeki's linking of his African Renaissance to the idea that Africa had been the place of origin for humans was a bit dangerous. "Scientific conclusions change and you never know what the theories will be in the future," he said.[523]

During the days of Raymond Dart and Robert Broom, Africa had been rejected as the possible place of human origins, with a preference for Asia and Europe. Mbeki and his project of African Renaissance worked to counter that view. In the narrative presented at Sterkfontein and Maropeng, Africans are no longer presented as marginalised; they are part of the overall message of what it means to be human. Many of the colonial messages have been replaced with the idea that all of humanity has a shared heritage in Africa. Mrs Ples, Little Foot and the Taung child are presented as heroes of the new narrative. Frances Thackeray had wanted a replica of Mrs Ples to be available in every school in South Africa – a dream which he subsequently updated to include "every school in the world".

Despite the progress in scientific thinking about human evolution, there are still aspects of museums that need to be changed to reflect new knowledge. For example, the logo at the Cradle of Humankind, which represents the powerful, iconic image of human evolution, itself is problematic. The image, which has been used for decades in many different forms around the world, is generally seen as the march of men, from left to right, moving from a stooped ape to an upright human. Each figure is facing forward in single file, with each one increasing in their posture, brain size, tool use and march of progress.

Generally, the images to the left are stooped over, representing ancient hominids, and the image to the far right is of a white male, striding into the future. The logo at the Cradle is a modified version of this image with four figures plodding to the right.[524]

Stephen Jay Gould was one of the first scientists to point out the prejudices embedded in this iconic image. Neither evolutionary theory nor the fossil record supports the idea of linear progress. The equation of evolution with progress reveals a social and cultural bias.[525] The images that portray progress toward lighter skinned humans are problematic. It is likely that the advances made in Africa, first with bipedalism and tool use, and then with language and symbolic thought, were all made by dark-skinned members of our genus and species. Exhibits need to reflect the fact that superficial differences such as skin tone are relatively recent.

As Maropeng was opened, the national government was developing a new research policy strategy for the palaeosciences called African Origins that recognised the growing interest in our common humanity, and promoted an awareness of the continent's rich fossil heritage. The programme identified genetics research as being an important field as well, as it helped to confirm the "Out of Africa" theory of human origins. The programme also acknowledged the growing interest not only in ancient hominid fossils but also in research that focused on the emergence of *Homo sapiens* in the past 100 000 years. One of the objectives of African Origins was to empower people who had been previously disadvantaged under apartheid to become more involved in the fields of palaeontology and archaeology.

Only a few months after Maropeng opened, Wits University opened its Origins Centre. The centre was originally conceived to promote both palaeontology and San rock art, and has made an important contribution to raising awareness in Johannesburg on both subjects. A banner outside the building held one image, that of a hominid fossil. Above the fossil was the question "Where Do You Come From?" suggesting that the Origins Centre could help answer the question. A plaque on the ground between the banner and the entrance of the museum honours Raymond Dart and describes the Taung child skull. Once inside the museum, the first room exhibits stone tools. Visitors are encouraged to open drawers with various hominid skulls inside. A large artistic wire interpretation of the continent of Africa accompanies an explanation of the Out of Africa theory. The next exhibit is a film on a loop explaining the history of modern humanity.

When a visitor turns the corner, the displays shift to focus on the topic of San culture. The rest of the museum displays are about the San people, their belief systems, art, culture and history. In fact, one of the first photographs in this section is of what looks to be /Keri-/Keri's sister, /Klein Khanako and her mother /Khanako, taken in 1937. They are not identified personally, but the photo is part of a panel about San hunter-gatherers. The juxtaposition of the hominid displays with San art and culture complicates the relationship between fossil hominids and living people, a confusion that began with Raymond Dart and Robert Broom. It is possible that visitors to the museum could mistakenly assume that the San people are in some way ancient or comparable to hominid fossils. Lara Mallen, the curator of the museum in 2015, agreed that this was a problem. She explained that the reason for the odd juxtaposition of these two parts of the museum was due to differences in fundraising for the two different components. She suggested that the fundraising for the palaeontology component of the museum did not go well. The Rock Art Research Institute (RARI) was more successful at raising money so the bulk of the museum focused on San rock art and culture.[526] The Origins Centre has been active in offering important public lectures, exhibitions and activities, which have been well received, and has played an important role in bringing recognition to the many contributions of the San people. The museum explores and celebrates both the ancient origins of humans, as well as more modern human developments in southern Africa, and will need continued funding in order to follow the latest developments in science and bring them to public attention.

But San people are not fossils. They are living people. If you are living today, you are not ancient. There is a natural inclination to create a narrative, a story that tells where humans come from. In fact, there is no clear exact point of origin. It all depends on how far back you want to go. And the narrative is getting more and more complex all the time.

25

The Spectacle of *Sediba*

In 2005, Job Kibii was the first African student to complete his PhD in palaeoarchaeology and palaeoanthropology at Wits. He spent three years working with Phillip Tobias as a post-doctoral student and in 2008, he was looking ahead to working on a new post-doctoral post with a professor who was soon to arrive at Wits, Charles Lockwood, but when Lockwood passed away, Kibii was left without a host. Kibii asked Lee Berger if he could work with him and he said yes. According to Kibii, Berger said, "I'm about to explore a new site. Why don't you come along?"

On 15 August 2008, Kibii, Berger, Berger's nine-year-old son Matthew, and their dog Tau, went out to the Cradle of Humankind. They were walking across the veldt and within a few minutes Matthew called out, "Hey Dad, I think I found a fossil!"

Matthew Berger came running up to his father with a rock in his hands. At first Berger thought that his son had found an antelope fossil, which was pretty common, but then he realised it was something much more rare. It was a hominid clavicle, something that Berger was well-trained to recognise given his PhD on the topic. When he turned the piece of rock around, he found a hominid jaw on the other side. When Kibii looked at the fossil, he also saw immediately that it was a hominid. Berger started swearing up a storm. Kibii remembers "we must have high-fived about six times" with excitement.[527]

The find wasn't as much of a coincidence as it first sounds. In early 2008, Lee Berger had begun to explore caves around the Cradle of Humankind using a new technology – Google Earth.[528] Over a few months, Berger started to map the caves and fossil sites in the area. He thought it would be a good way to explore potential new sites. Berger began to recognise what cave sites looked like from above on the satellite images so he was able to identify similar looking sites throughout the valley. The valley is filled with dolomite limestone where trees cannot easily take root. Caves in the valley are often marked with ground soil that has been disturbed and a cluster of trees. Berger visited one of these new sites that looked particularly promising. He found a small limestone cave that had been blasted out with dynamite back in the 1930s, but long since abandoned. It was two weeks later that Berger went back with Kibii and Matthew to the site. What Matthew had found turned out to belong to a young male hominid.

Immediately, Berger applied to the South African Heritage Resources Agency (SAHRA) for a permit. In less than three weeks, it was granted to both Berger and Kibii as joint permit holders, with Berger as the primary scientist. After securing the permit, Berger and Kibii went back to the site with a group of colleagues for a more thorough search. Kibii remembers that the group had been working for over three hours and was beginning to feel that they would find nothing. It was then that Berger went into a pit that they had all been in and out of throughout the day. But this time, he saw teeth on the side of the wall, which led to the team uncovering more pieces of the young male skeleton linked to the clavicle, as well as portions of a female skeleton, possibly the young male's mother.

The theory was that both skeletons were well preserved because they had died when they fell into a sink-hole and were quickly covered over with sand and earth. Berger was fortunate to find a partial skeleton from one individual, never mind two. Just as with Little Foot, it was rare to find an ancient hominid skeleton because formal burial did not begin as a practice until about 50 000 years ago.

Berger and his team could have lumped the fossils into the same category as *Australopithecus africanus* or decided that they were part of the species *Homo habilis*. However, he and his team decided that they were not yet *Homo* so they named the skeletons *Australopithecus sediba*. "Sediba" means natural spring, fountain, or wellspring in Sesotho, one of South Africa's eleven official languages. "It was deemed an appropriate name for a species that might be the point from which the genus *Homo* arises," said Berger. "I believe that this is a good candidate for being the transitional species between the southern African ape-man *Australopithecus africanus* (like the Taung child and Mrs Ples) and either *Homo habilis* or even a direct ancestor of *Homo erectus* (like Turkana Boy, Java Man or Peking Man)."[529]

This time, after nearly two decades of searching in the field, Berger had found his own special fossil (or at least his son had). While Berger did give special attention to the media, he also began to shift his reputation from showman to significant fossil finder. He waited close to two years from the time of discovery until the announcement to the world, but that was not uncommon historically, giving a scientist sufficient time to analyse the fossil and prepare. Kibii, who had been hired as the director of excavations at the Malapa cave site where the fossils had been found, remembers it as a "nerve-wracking" time because of the need to keep the analysis of the fossils quiet until the official announcement.

In addition to Google Earth, Berger used another new technology with this fossil. In February 2010, he carried the skull and parts of the skeleton to Grenoble, France, where the fossils spent two weeks creating images with a synchrotron, an advanced radiation technology. Importantly, the synchrotron was non-invasive and wouldn't destroy the fossil in any way. The synchrotron created an extremely accurate image of the skull. By studying detail of the teeth, for example, it could help provide the age of the individual at the time of death. Usually, palaeoanthropologists chip away the breccia from a skull before analysis. Now the new radiation technology could penetrate through the rock to the fossil. Berger left the skull attached to the stone.[530]

In preparation for the public announcement of *Sediba*, in early 2010 Wits established eight sub-committees to co-ordinate the launch, the marketing and the media. The fossils had been found on privately owned land within the Cradle of Humankind so the landowner Tim Nash wanted to establish a foundation that would have oversight over Malapa, but government and the management authority of the Cradle of Humankind declared that the fossils needed to be held in a public trust, and Wits University became the custodian of *Sediba*. There was a lengthy debate about the rights and responsibilities of the land owner and the university. The university, the government, the sponsors including Standard Bank and PAST, as well as the landowner, all had different perspectives on how the bones should be handled.

These tensions raised the important issue of land ownership. While the Sterkfontein Caves and the land around Swartkrans are now owned by Wits University, there are still numerous privately owned farms in the Cradle of Humankind. Current land ownership raises the question of who previously lived on the land. While the myth of the empty land was perpetuated for centuries, research in the 1970s proved this myth to be false. Wits University archaeologist Revil Mason wrote about Early Iron Age settlements that thrived in the Magaliesberg Valley, around what is now the Cradle, over 1 500 years ago. Sesotho- and Setswana-speaking people inhabited the area and left substantial architectural remains. Current historians raise the question of why this information was ignored for so long, and why Revil Mason was harassed by some and ignored by others for putting forward this evidence. Under apartheid, there was very little interest in African societies in the area because their existence countered the prevailing and convenient myth.[531]

The Wits committees had to coordinate and work out all of the tensions prior to a deadline to publish the scheduled *Science* article. The fossil was announced to the public on 8 April 2010, 20 months after the original find, a date carefully chosen to precede by one day the publication of a cover story in *Science* on the 9th.

The public announcement was held at Maropeng at the Cradle of Humankind. Deputy President Kgalema Motlanthe spoke on behalf of President Jacob Zuma. Naledi Pandor, the minister of science and technology, was there as well, as were other ministers, ambassadors, scientists and the media. Matthew Berger, who had turned eleven, was applauded for his find. With his father behind him, Matthew stood beaming next to the glass case that had been especially constructed to hold the skeleton. Phillip Tobias also attended the event and Berger deferred to him as the doyen of palaeoanthropology. At the age of 84, Tobias had spent his entire life applauding and describing other people's fossil finds. Since he had spent more time in the lab and the library than he had in the field, Tobias never had the thrill of finding his own fossil.

Whenever Tobias was asked why he had never married and had children, he always had a ready answer. "I've had 10 000 children," he would say. "They are the students who've gone through my hands at this great university, they are really my legacy, as tangibly I believe as if I had shared my personal DNA among them. Those are my offspring."[532]

"A STORY OF TWO BOYS" was the headline on a story in *The Star*. "The schoolboy fossil hunter had found a fossil almost like him. A boy who had lived almost two million years ago. A boy who will now help us to answer that most troubling question, 'Where do we come from.'"[533] Many media outlets picked up on the relationship between Matthew Berger and the ten-year-old whose fossil he found and the connection became a theme that ran through much of the international media coverage. The marketing department at Sterkfontein and Maropeng built on this theme of two boys reaching across millions of years and they even designed a poster with that message and a photograph of Matthew. Once again, as the pattern had been across history, the tendency was to personify a fossil.

Building on this interest in naming and personifying the fossils, PAST held a naming competition, sponsored by Standard Bank, to give the younger male skeleton a proper name. There were over 15 000 entries from around the country, many from children. The winning entry was from a 17-year old girl from St Mary's School in Johannesburg, Omphemetse Keepile. She submitted the winning name "Karabo," a Sesotho word meaning "answer". "The fossil has helped researchers to see much deeper into the information that they have," said Keepile. While the name Karabo has not been used as frequently as the species name *Sediba*, the naming competition brought the fossils to even greater public attention.

Berger dated the skeletons at about two million years old. In many aspects, such as the long arms and the small size of the brain, the skeletons had features similar to older hominids, hence the placing them in the genus *Australopithecus*. But the skeletons had more modern features as well, such as a human-like pelvis and a very advanced hand, which was why Berger thought that this species might be the precursor to *Homo*. Many scientists agreed that *A. sediba* was an impressive find, but not everyone agreed with Berger's interpretation of the fossils. Donald Johanson and Richard Leakey were two who did not support the idea that *Sediba* could be the direct ancestor of *Homo*. They argued that *Sediba* was not old enough and not in the right geographic location to be ancestral to *Homo*. They argued that the East African australopithecines continued to hold that honour. Predictably, Tim White joined the fray and stated the fossils were likely not sufficiently anatomically different from *A. africanus*.

Sediba was the first major hominid fossil find in South Africa since Little Foot had been announced more than ten years earlier. Government ministers were happy with the announcement and Malapa became the focus of great attention. This is what Thabo Mbeki had been hoping for in terms of building national pride in South African science.

On 10 June 2010, Archbishop Desmond Tutu stood on a massive stage to open the kick-off concert for the FIFA Soccer World Cup in Johannesburg. He was kitted out in green and yellow, the colours of South Africa's team. Tutu looked out on the enormous crowd of 110 000 people in the stadium and said, "I'm dreaming, man. What a lovely dream. We are the world." He lifted his arms and said: "We welcome you all. For Africa is the cradle of humanity. So welcome home. All of you. All of you Germans, French. Every single one of you. We are all Africans." As he danced and yelped with his signature enthusiasm, the crowd went wild, cheering and waving their South African flags. In 2010 South Africa was in a good mood.[534]

26

Meeting the Taung Child Skull

In September 2011, five more papers were published in *Science* that offered more detail on *Sediba*'s anatomy. Lee Berger and his team at Malapa had been in the limelight for more than a year and Berger was enjoying it. But before the end of 2011, some questions came up, not about the content of the papers, but about their authors. Alan Morris pointed out that of the 21 authors of the five *Science* papers, only a few were South African and only two were black South Africans. "This fact diminishes neither the importance of the discoveries, nor the accolades being given to the discoverers," wrote Morris in the *South African Journal of Science*, "but it does suggest that the transformation of palaeoanthropology still lags behind the transformation of South African society in general." The Africa Origins platform that had been introduced in 2006 was providing support to palaeoanthropology and included the goal to produce a critical mass of South Africans working in the field. Morris pointed out that the goal was not only to bring South African discoveries onto the world stage and publish papers in prestigious international journals, but also to build a team of local South African, including black South African palaeoanthropologists.[535]

Early in 2012, Berger responded in the same journal, saying he wanted to address some of the "misconceptions regarding the composition of the Malapa team". While Berger agreed that "there [were] substantial inequities in South African science," he thought it was "unfair to use my team and these authors as illustrative of this apparently widespread social issue." Berger agreed that of the 21 authors, four were South African citizens, including himself, two of whom were black South Africans; and Job Kibii was from Kenya. He explained that of the authors having no South African affiliation, six had been brought in, he said, because of their special skills and access to equipment that were not available in South Africa. Berger disagreed that palaeoanthropology lagged behind other sciences, which was a more general problem across the sciences.

His concluding point was that it was necessary for students to receive funding and to have job opportunities when they graduated. Otherwise the field would not attract people from disadvantaged backgrounds.[536]

By this time, Job Kibii had begun to feel tensions developing between him and Berger. According to Kibii, by the end of 2010, Berger had begun to tell the *Sediba* story differently, with Matthew and their dog Tau, but leaving Kibii out of the picture. A talented South African photographer did a photoshoot at Malapa in October 2010 and the photos on his website do not identify Kibii, or any of the other excavators, in the captions. One day, according to Kibii, Berger asked Kibii to drive a journalist and her cameraman out to the site, and she spoke to him as if his only role was as a driver. Kibii told her that he had his PhD and was a published scientist. When they met up with Lee and Matthew Berger at the site, Kibii felt that it became awkward.

When the time arrived to renew the permit for Malapa, Berger submitted the renewal without Kibii's name. While Berger was the senior scientist, and had the right to make decisions about the site, Kibii had expected that they might continue as joint permit holders. Kibii has since applied for a permit at Gondolin, another fossil bearing site in the Cradle of Humankind where Elisabeth Vrba once excavated in 1979. Kibii continues to do field work there.

The field has been slow in changing from its historical dominance by white men. One of the first black South African women to receive her PhD in palaeoanthropology at Wits was Mirriam Tawane. She was from Taung and was interested in public education about human evolution. Nonhlanhla Vilakazi was another. She also completed her PhD at Wits in 2013 and while she was studying palaeontology, she was initiated to become a sangoma as well. She uses bones to help patients with their healing. Vilakazi is doing a post-doctoral fellowship at the University of Pretoria.[537] In addition to appreciating support from Lee Berger, both of these women spoke of benefitting from the support of two other women who had entered the field at Wits in the late 1990s – Lucinda Backwell and Christine Steininger. Not only had Backwell made important contributions to the understanding of bone tools, she was also repeatedly voted by students as their favourite lecturer. Along with doing field work, and supporting students, Steininger became the project manager for the Evolutionary Studies Institute and the Centre of Excellence in Palaeosciences.

Winnie Dipuo Mokokwe received her master's in palaeoanthropology in 2005. While she was disappointed because she did not feel she was encouraged to continue in an academic career, she went to work for the North West province at Taung for two years. "It was exciting," she recalls. "That site has so much potential." But when she got there, she was frustrated that she didn't have a budget, or much support – "or even a pen!" In 2008, Mokokwe worked at Maropeng with the Cradle of Humankind for two years before she became the deputy director for palaeo-sciences in the national department of science and technology. Despite her frustrations, Mokokwe was excited to work on the African Origins policy and to provide support for others in the field.[538]

A symposium of the South African Archeological Society, focusing specifically on human origins, took place in August 2014 in Johannesburg. Six white male scientists and one white woman scientist were on the panel. Eric Worby, a sociologist originally from the US and a professor at Wits, presented a paper that questioned aspects of Lee Berger's presentation of *Sediba*. Worby began by asking the audience, "How many of you consider yourselves English-speaking South Africans?" Virtually everyone in the entire audience of several hundred people raised their hands. "And how many of you consider yourselves black South Africans?" asked Worby. One man at the back of the auditorium raised his hand. "What is it that fascinates you about human origins?" Worby then asked the audience. He went on to suggest that the fascination with discovery and origins had a hold on Anglophone South Africans in particular because they had the most at stake. He argued they would like to be able to draw a direct line between "pre-racialized human forbears" and themselves as "post-racial citizens" of a rainbow nation. These hominid fossils, presented in the form of a boys' adventure story, could provide white South Africans with a sense of belonging, he said. It might give them the claim they so desired on South Africa as their home. Worby reminded the audience of Donna Haraway's work that had first critiqued the search for human origins in her book *Primate Visions* in 1989. Once again, the search had been re-animated with the needs of a particular audience. Worby's point was reminiscent of the way in which Jan Smuts once gave Europeans in South Africa the "right of return" to Africa.[539]

Frank Matlala was the solitary man who had raised his hand to the question "Do you consider yourself a black South African?" He was from Limpopo and said he had attended the event because he had heard Alan Morris speak about it on the radio. "In rural Limpopo, we didn't come into contact with white people other than white policemen," he said. "Our education instilled an inferiority complex in black people and we saw white people as divine beings. Where the white people lived was beautiful. We lived in poverty."

Matlala said that Bantu Education under apartheid limited the development of black people and guided them to be labourers. "I succeeded and opened a bakery, so I could buy books," he said. "I started reading books by Malcolm

X and Martin Luther King Jr. There was this story [out there] that blacks were not human, that they had a different brain size. I'm trying to get the real story of race."

"White people used to say black people are apes and baboons. Now with the Cradle of Humankind, black people are sceptical. Now with independence, white people want to justify their presence. A lot of people take it with a pinch of salt. If we go into the new South Africa with white people thinking they are superior and black people thinking they are inferior, we have a problem."[540]

These comments served to illustrate how racist scientific practices of the past can continue to shape wrong thinking today. They also echo the comments that Phillip Bonner, Amanda Esterhuysen and Trefor Jenkins had made in their 2007 book *A Search for Origins*. They wrote: "Many white people came to believe that they had been created, but that black people had evolved, and many black people regarded activities at Sterkfontein with suspicion because they understandably believed that the motive behind the excavation was to prove that 'blacks' had evolved. Evolution thus not only touched on religious sensitivities but cut to the core of South African identity politics."[541] Clearly, post-apartheid education has not done enough to counter the false information taught to black people under Bantu education, nor the racist information that was taught to white people under apartheid.

On 24 September, Heritage Day, in 2012, Himla Soodyall met with members of the ǂKhomani San at Askham in the southern Kalahari. She was meeting with them to report back on DNA research she had done based on cheek swabs she had taken years earlier. Soodyall's journal article, entitled "Genomic Variation in Seven Khoe-San Groups Reveals Adaptation and Complex African History", was published in *Science*.[542] The findings, she said, were a "phenomenal tribute to the indigenous Khoe and San people of southern Africa," and that "we have given the peoples of Africa an opportunity to reclaim their place in the history of the world."[543] Although Ouma /Una had passed away, she had met Himla Soodyall and Trefor Jenkins before she died and had been interested in the outcomes of the research. Despite the finding of the study, and given the oppression the Khoisan people have faced historically, the day-to-day circumstances of the ǂKhomani San continue to be harsh and difficult.

The paper explains that within Africa there is greater genetic diversity than is found anywhere else in the world. The genetic diversity in Europe, Asia and the Americas is a sub-set of the variation in African variation, which shows that modern humans had their origins in Africa. The research suggests that within Africa, click-speaking Khoisan people have the oldest DNA lineages in the world, and the greatest genetic diversity on the continent. Stereotypes persist, but the scientific view that the Khoisan were somehow less than human had to be eradicated. In fact, it wasn't in Europe that language and art and symbolic art first appeared; it was in Africa.

The findings were that the origin of modern humans did not take place in a single geographic region of Africa, but that it was likely that modern humans evolved from a complex history, within the continent, and showed that there was interaction between the peoples of East Africa and southern Africa more than 100 000 years ago. Perhaps the genetics research was showing that the debate between Lee Berger and Tim White didn't have to be so ruthless.

Yet the definition of what it means to be human is not to be found in genetics, nor in the fossils of palaeoanthropology, but rather in language, and thought, and dance and music, and in the rituals related to birth, coming of age, life and death. This is what makes us human.[544]

In early 2014, after years of debate, the Phillip Tobias Hominid Fossil Laboratory at the School of Anatomical Sciences at the Wits Medical School, was to be closed down and all of its contents moved across town to the Evolutionary Studies Institute on the Wits main campus. The laboratory was the place that had, for close to 90 years, housed one of South Africa's most famous fossils, the Taung child skull. At the age of 86, in June 2012, Phillip Tobias had passed away. He was no longer able to protest the move.[545]

In an anteroom, which was too small to hold any furniture, the main feature was a larger-than-life portrait of Phillip Tobias, which hung on the wall. The walls of the laboratory were covered with filing cabinets with wide drawers to hold fossils and storage shelves with skulls. A large table, where fossils could be displayed and studied, was in the centre of the room; there were a number of old drawings on the wall and old busts on the top of the shelves.

In the left-hand corner of the room stood what looked like a metal bank vault, painted white. Bernhard Zipfel, the curator of the Wits collections, took out his master key. "Only the holy of holies are allowed in here," he said with the mild sarcasm that was characteristic of his relaxed manner. The heavy door swung open and Zipfel stepped inside. At most the vault could only hold three or four people. Three walls were lined by shelves and each shelf held boxes of fossils marked "Makapansgat" and "Sterkfontein". "This is the largest hominid fossil collection in the world," said Zipfel, "but you can fit it in your car." Zipfel had another story to tell. "Back in 2007," he said, "when I first took on the job, PVT [Tobias] said to me, 'Since when do you allow ordinary people into the vault?' 'Since I'm curator,' I said. PVT was slightly taken aback, but then he said, 'Oh well, carry on.'"

On a shelf on the right-hand side of the vault was a medium-sized, brown, wooden box. It was marked "Taung Skull". Zipfel pulled the box down from the shelf. When asked whether it was the original box – "the one that got left behind in the taxi cab in London?" – he smiled and shook his head. "No," said Zipfel, "that box has been replaced since then but this one has been around for a while." He took the box out of the vault and set it down on the table in the middle of the room and opened it slowly and with deliberate care. Inside were three fossil pieces, held in place by foam covered with red felt. Zipfel took each piece out carefully and placed it on a display mat on the table. He picked up the main part of the skull, which showed large eye sockets and a line of upper teeth. Then he picked up the second piece, the endocast, and placed it on top of the skull, showing how the brain cast fit exactly in place. Finally, Zipfel took the mandible and fit it onto the bottom of the skull. There, cupped within Zipfel's large hands, was the complete Taung child skull.

It was hard to imagine how Raymond Dart had recognised it when it was still held together by the concrete breccia. It looked so small. It looked so fragile. "PVT used to show off the Taung child as if it was a performance," said Zipfel. "He used to bring out each piece ceremoniously. As he brought out the mandible, he would say, 'Mahvelous.' As he brought out the brain case, he would say, 'Here is the cherry on the top.' Then he would open and close the jaw, making the fossil speak. For some reason, he would make it speak with an Australian accent. 'Goodday, mate.' It wasn't difficult to imagine Tobias entertaining a crowd of visiting researchers around this table. Maybe he was imitating Dart's accent? "I have no idea," said Zipfel.

When asked whether the skull had ever been damaged in the handling of it over the years, Zipfel nodded. "Yes," he said. "There was a visiting researcher who was working with my predecessor. He said, 'Should these teeth be loose like this?' PVT told the curator that he should have watched the researcher, but you can't sit with a researcher every minute. Plus the skull had been repaired before. There was an old glue break that must have come loose."

Zipfel talked about the many people who applied to see the collections. "There is a committee of ten people, including the deputy vice-chancellor, that review applications," he said. "Very rarely do we turn anyone down." In the new location at ESI the process would remain the same. "We may change the protocol a little, but it has worked well. Everything here will move over to the new facility in the next few weeks. We're just waiting for the delivery of bullet-proof glass from Germany. Lee Berger would have liked to build something the size of the United Nations building to house the fossils, but we've already spent millions on this new facility. Given the poverty that exists just a few blocks away, I feel we need to be more modest."

Over the seven years since Zipfel had been curator, he had experienced many memorable moments – from research workshops to the varied visitors he had brought there and shown around. "But I suppose the visit I remember the most was in 2009 when the government of North West province came to Wits because they wanted to take the Taung skull back to Taung," he said.

The delegation was intimidating, remembered Zipfel, as he shook his head. Darkie Africa was the provincial minister for economic development and tourism in the North West province, where Taung is. "Just his name alone scared the daylights out of me," Zipfel said. "I expected an aggressive person." But Zipfel said Darkie Africa had turned out to be a very sensible man. While some of the politicians hoped that the Taung child skull might be a money-spinner, Mr Africa understood that taking the skull back to Taung would not draw in a huge number of visitors. And once the delegation came into the room, Zipfel said, Tobias charmed them. "Soon they lost interest in politics and were more interested in photos." Turning serious for a moment, he added: "The visitors thought of the Taung child as an ancestor. It wasn't appropriate for Tobias to anthropomorphise it and speak for it. Some people say that the Taung skull looks like someone they know. But it is not human. It's a pre-human ape."

Many people in Taung felt strongly about the Taung child skull. Dr Mirriam Tawane, who was born in a small town called Majeakgoro, which is part of the Greater Taung municipality, and who had received her PhD in palaeoanthropology from Wits University in 2012, continued to work at the Evolutionary Studies Institute. When she was growing up in the 1980s, she said, many people believed that the Taung skull was a human skull and they were concerned that it had been taken away by Wits scientists. She remembered being told that the Taung skull was from a recent human ancestor. "We all believed that it needed to be laid to rest."[546]

With her growing interest in science, Tawane studied palaeoanthropology for her honours, MA and PhD degrees. She learned that the Taung skull was not a human skull at all, but the skull of a creature that lived between two and three million years ago, which had some human-like features. While this changed her thinking about the meaning of the skull, she continued to be committed to the development of her home community and to understand how people could feel very strongly about the importance of the return of the skull. In 2008, Wits was approached by a group of young people calling themselves the Taung Skull Consortium. Together with Tawane and other scientists at Wits, they developed an annual lecture to take place on Heritage Day in September that would bring Wits scientists to Taung to talk about the fossil. Professor Tobias gave the first lecture in 2008 and Francis Thackeray the second, in the following year. Unfortunately, the annual lecture fell apart in 2010 when the date clashed with another cultural festival and then the Taung Skull Consortium dissolved as some of the young people involved moved on.[547]

Tawane said that, in addition to her scientific research, she wants to dedicate her life to the support of her local community and to science education outreach. Her post-doctoral studies were focused on the possibility of developing a visitors' centre in Taung that would work with both local people and interested tourists. As part of her initial research, she arranged to speak to a wide range of stakeholders in Taung to get their views about such an idea.

Dr Mirriam Tawane speaking at Maropeng. Courtesy of Maropeng.

Taung is about 400 kilometres south-west of Johannesburg, and from there it is another 15 kilometres to Buxton, which is where Mr De Bruyn blasted the blocks of rock out of the quarry in 1924. He had given them to the mine manager, Mr Spier, who in turn gave them to Professor Robert Young, who had carried them back to Raymond Dart. For 50 years the mine continued to operate. The mine closed down completely in 1977, and with nearly 40 years of high unemployment since then, the community of Buxton has been struggling. Small cement houses sit on dusty yards. Donkey carts are often the only means of transportation for the people who live there.

Chief Lekwene of Buxton lives within sight of the quarry. The Heritage Association, that had implemented the expansion of the World Heritage Site to include Taung in 2005, told him that his house was "an eyesore", he says. The chief is a surprisingly young man, who looks in his mid-30s but is ten years older. He wears a leather jacket over a purple and orange Adidas T-shirt and jeans. During the week he works in the construction industry in Kimberley.[548]

Tawane asked him how he felt about the naming of the quarry as part of the World Heritage Site in 2005. "Many people will tell you that it took our bread and butter away," answered Lekwene. "People used to fish and swim in the area. They used to have their cattle graze there and traditional healers used to get water from the site. We used to go through the site to get to Tamasiqua on the other side. Now we are no longer allowed access."

In 2005 the North West provincial government submitted papers to UNESCO to extend the World Heritage Site to include Taung. "Did they consult you?" asked Tawane.

"I wouldn't know," was Chief Lekwene's reply. "They go to the Paramount Chief [in Taung], not to us. I don't even have a map of the site and the buffer zone." He stood up and walked into his house, then returned after a few minutes with a blue folder in his hand. It is marked Taung Regeneration Plan and dated March 2011. "We wanted sports grounds and a park, electrification, roads and a community hall but the South African Heritage Resource Association said no. I was heartbroken."

Chief Lekwene said that even with all the problems, he would support the development of an information centre at the site. "No one from Buxton is from Buxton," he said, and then explained. The mine opened in 1919 and people came from all of the surrounding villages to work there. "My father is from Tamasiqua and came here to work in the mine. My mother came from 50 kilometres away and came to Buxton as a teacher. Most of the women who moved here came from Taung" he said, "and the men are from KwaZulu-Natal, Johannesburg, the Eastern Cape, Malawi and Lesotho."

He first heard about the Taung child skull when he was growing up. "We knew that they found a skull in the mine," he said. "We thought it was big, like a human skull. We didn't know that it was so small. Was it a white child or a black child? Was it Khoisan? I didn't know."

"We have a gift for you," Tawane told Chief Lekwene, handing him a gift bag.

When he saw what was inside the bag, he looked up at her. "Is this legal?" he asks. "Is this Mrs Ples?"

"It's a cast," explained Tawane, "of the Taung child skull."

Chief Lekwene took the gift out of the bag and held it up to admire. "Ah, thank you very much," he said. "I've always wanted to hold this guy."

Tawane had arranged a meeting for the following day with Paramount Chief Mankuroane in Taung. The chief's headman spoke first, saying that Wits had not played its role in supporting development in Taung over the years, and the meeting felt tense. "Don't you make money from the skull?" he asked. "No," answered Tawane. "Well, if you don't make money from it, then we'll take it. It's high time you returned the skull." Speaking in Setswana, Tawane conveyed her intentions. "This is close to my heart," she said. "Whatever research I do and any journal article I write, we will make it available to you."

After a long discussion about the advantages and disadvantages of further development at the Buxton site and what it might mean for the people of Taung, Tawane shared with the paramount chief and his personal assistant that when she was growing up, she had believed that the Taung child was the skull of a human being. Along with Ian McKay, also from Wits, she described how the skull was that of an ape that had human characteristics, that it walked upright and that it had human-like teeth. It was the first of its kind to be found in Africa, but modern humans did not develop until about 200 000 years ago.

As Tawane was speaking, the paramount chief's PA's eyes widened. "This is information that must be widely shared," he said, his voice rising. "We need clarity. We think that humans began in Taung. We think the Taung child skull is our ancestor. That's the wrong information."

"It is the belief that the Taung skull is human that results in people wanting the skull back at Buxton," Tawane said.

When asked what people's reaction might be when they found out that the Taung skull was not human, whether they might think that it conflicted with Christianity and their traditional beliefs, the assistant said that there could be conflict but that it would be resolved. "We need to have more discussion," he said. "You can put up photographs and explain the monkey, the ape, the human."

Tawane was curious about whether people in Taung today believed in evolution. Both the paramount chief and his assistant had Zionist Christian Church (ZCC) badges on their suit jackets. Founded in the 1920s, the ZCC is an African church; on the wall above them was a portrait of the bishop of the ZCC. "The Bible says that God created man last," said the assistant. "Evolution says something different."

Tawane and Ian McKay then spoke about how it was possible to reconcile the Bible and science, and asked who could have created the process of evolution.

Everyone stood up to leave. The meeting was over. With a broad smile, Paramount Chief Mankuroane said, "Only God knows."

The local high school had been built in Buxton in 1990 – three long brick buildings running parallel to one another, surrounded by dust. In 2015 there was neither grass nor greenery in sight, let alone any sports grounds, and at 1pm the sun was harsh and high and a small group of students sought the shelter of the shade from one of the buildings. The principal explained that the school had no library and no computers and that most students struggled financially to study beyond high school. In a small room seven teenage learners, three girls and four boys, were waiting for the expected visitors.

One of Tawane's Wits colleagues, a student named Amanda Mudau, led the discussion. "What is a World Heritage Site?" she asked. "It's about memories made in the past," answered one thin girl with long legs. "It is a place to see our background and our roots." "And have you heard of the Taung Child Skull Heritage Site at Buxton Village?" The same girl, whose name was Kopano, had a shocked look on her face and said, "Of course – we grew up here." She closed her eyes and said, "The missing link was found here. It was the link that showed that humans originated from Africa."

Opening her eyes, she went on. "It's our legend. It's our skull. We're tired of seeing the fake skull. We want the original. I strongly believe the Taung skull must come back home." Kopano continued to lead the discussion. She told us that there was an old man who was over a hundred years old, her friend's great-grandfather, who had worked at the mine in Buxton when the skull was found. It was he who had taught her the history of the skull. "He told me that if he dies and the skull is not returned, he will not rest in peace," she said. The old man died in 2013. Flicking her blue Bic pen, she leaned forward and said that she would like to write to the president. "'Mr President,' I would say, 'as you know, our village is a World Heritage Site. Nkandla has everything but we lack many things.' If the government is corrupt, what can we do? It breaks my heart." Kopano shook her head and several of her classmates nodded in agreement.

One young man said, "I don't believe the scientists. I believe in Christianity"; another learner said, "I'm not sure. We are confused."

"Scientists say that humans started with apes and then step by step they became humans through evolution," said Kopano, taking control again. "People believe in God here, but this links the scientists with God. Please bring our skull home. We don't want to earn money from the skull. We want to earn respect. If the skull comes back home, maybe a university will be built here. I'm in matric and I want to go to university. I want to go to Wits and study geology. Then I could bring the skull home. Then the old man who taught me will rest in peace."

27

Rising Star and *Homo Naledi*

It was a cold, grey and rainy Sunday morning in Vancouver, Canada and Marina Elliott was up at about 9am on 6 October 2013 having a cup of coffee and looking for an excuse not to get back to work on the PhD in physical anthropology she had been toiling on for the past five years. She decided to check her email. When she opened her laptop there was an email from her supervisor, Mark Collard, which said: "This sounds like a great opportunity. Do you have caving experience?" Elliott scrolled down to see what Mark was talking about. It was an email call-out from Lee Berger that was circulating the globe. It read: "Dear Colleagues – I need the help of the whole community and for you to reach out to as many related professional groups as possible. We need perhaps three or four individuals with excellent archaeological/palaeontological and excavation skills for a short-term project that may kick off as early as November 1st 2013 and last the month if all logistics go as planned. The catch is this – the person must be skinny and preferably small. They must not be claustrophobic, they must be fit, they should have some caving experience, climbing experience would be a bonus."

November 1st was less than a month away.

Berger's email continued. "I do not think we will have much money available for pay – but we will cover flights to South Africa, accommodation (though much will be field accom., food and of course there will be guaranteed collaboration further up the road). Anyone interested please contact me directly. My deadlines on this are extremely tight so as far as anyone can spread the word, among professional groups."[549]

Berger's email reminded Elliott of the famous advertisement Ernest Shackleton had allegedly placed in the London *Times* in 1900 when he was putting together his expedition to the Antarctic. "Men Wanted," it read, "for hazardous journey, small wages, bitter cold, long months of complete darkness, constant danger, safe return doubtful, honor and recognition in case of success."[550]

By 9:35am, Elliott had written back to her supervisor saying, "Oh my f%&*%god. I have lots of caving – and HEAPS – of climbing experience (and my own gear). I am his MAN!! Can I do this?? I would do this!!!"

By 17 October Elliott received email confirmation saying that she had been accepted for the team (one of only six out of 60 candidates) and by 6 November she had landed in Johannesburg. What exactly was it, she wondered, that Lee Berger had found.

In early August 2013, Berger had hired a local caver, Pedro Boshoff, to keep a look out for new hominid fossil sites in the Cradle of Humankind. Two other cavers, Rick Hunter and Steven Tucker, had gone into the Rising Star cave system, where they had explored before. They were at the Dragon's Back ridge, deep in the cave. Hunter wanted to take photographs of the beautiful stone formations on the ceiling and Tucker was blocking his view. The only place that Tucker could move out of the way was a crack in the ground beneath him. Tucker didn't expect the opening to go anywhere, but he climbed into it to allow Hunter to continue with his photos. As he did, he noticed that the opening went further down. The deeper Tucker lowered himself down, the smaller the opening became; at one point he could barely move his head. But he kept going with the hope that he might find an unknown section of the cave. He continued climbing down until it opened up into a chamber with beautiful formations on the ceiling. Shining his head-lamp on the ceiling, Tucker called to Hunter to come down and join him in the chamber. It was only when they turned their lamps down to the floor that they noticed the bones. The bones were the same colour as the floor so they were hard to see. But they weren't encased in stone like most other fossils and several of them were lying scattered on the floor of the cave. At first Tucker and Hunter wondered what animal the bones were from. Then they found a jawbone with teeth in it and they thought it looked similar to a human jawbone. At that point, their camera battery died.[551]

It was more than a week later before Tucker and Hunter were able to go back to the cave to take photographs. They showed the photos to geologist and fellow caver Pedro Boshoff. He recognised that the bones were potentially important and decided to contact Lee Berger. They went to Berger's house together that same evening. "You're going to want to let us in," Boshoff told him when they rang the bell at the gate. When Berger looked at the pictures, his jaw dropped and he swore out loud. "He was super excited," said Tucker. "He was jumping up and down and shouting. He went completely crazy."

That night Berger couldn't sleep. At 2am he called Terry Garcia at *National Geographic*. He explained what had happened and asked for Garcia's support. "Do what you need to do," Garcia said.

Access to the chamber within the Rising Star cave system was too narrow for Berger himself to get through and so he put out a call on Facebook and via email for a crew of skinny scientists. Within a few days, he had 60 applicants. Berger chose six scientists – all women – to carry out the difficult work of retrieving the fossils from the cave. He called them the "underground astronauts". Marina Elliot was one of them.

Marina Elliott was the first scientist to enter the chamber in November 2013. After making her way through a very tight passage called Superman Crawl, she eased along the Dragon's Back ridge and down the narrow chute. Elliott looked down and wasn't sure she had made the right decision. The chute looked like a shark's mouth, with razor-sharp "teeth" protruding from either side. On that first day, she brought out a single bone, a mandible.

Berger had set up a 60-person team to provide back-up in a camp outside the Rising Star Cave. The logistics manager, Wayne Crichton, organised food, water, generators, fuel, lights, cables, cameras, toilets, trash, tents, and the overall management of the camp. The two cavers, Rick Turner and Steve Tucker, set out power cables and cameras throughout the cave. They also set up intercoms at certain points for communication. They installed safety ropes and lights up and down the narrow shoot.[552]

Outside the cave the team erected three large tents. One tent was for the cavers, one was a science tent, that held the safe and the tables on which the scientists sorted through the fossils, and the third was the camp's main command centre. This was the tent that held the computer systems and the cameras. It was also the location from which the scientists could watch what was happening underground. About 50 metres away, there were another 18 four-man tents, a medical tent and a kitchen tent. There was a lapa with a fire pit that everyone could sit around at night to eat, talk and have some down-time.

On day two, all six cavers went underground. They worked in two-hour shifts. There were so many fossils on the floor of that small chamber and they were delicate. The women worked on a one-metre square area. As they worked, other scientists flocked around the computer screen that showed the video feed from the chamber below. Elliott said the "emotional intensity" was one of the greatest challenges. One of the organisers said that bringing up so many bones was surprising. "Prof Lee's face looked like a kid getting his first bicycle. Then he was in tears." Scientists began to catalogue the bones. "We realised that we had been wrong," said Berger. "It wasn't a skeleton. It was more than one." In the end, it was the largest single find of hominid fossils on the continent, with 1 550 individual bone specimens from 15 individuals. "We are working with an unprecedented amount of evidence," said Berger. "There are more individual specimens than have been found in the last ninety years of exploration in all of Southern Africa." Every day of the expedition was documented with a daily blog by *National Geographic*, videos and tweets.[553] Never had palaeoanthropology been followed so closely as it happened.

Berger decided to bring together the senior scientists who had worked with him on *A. sediba* to assess the new fossils, and he invited them, as well as another 30 young scientists to Johannesburg for a six-week Rising Star workshop, in May and June 2014. The workshop took place in a new fossil vault at Wits in the Evolutionary Studies Institute, a window-less but brightly lit room whose walls were lined with gleaming, wooden cupboards with shelves and glass doors. One group of scientists gathered around the skulls at a large table in one corner. Other post-cranial specialists met in smaller groups divided according to parts of the skeleton – hands, feet, scapula, pelvis etc. While the atmosphere was quiet and focused, the workshop was not without tension. Since Alan Morris's critique three years earlier, diversity was still an issue. Very few members of the team were South African or from the broader continent. Job Kibii wasn't there at all. Several of the younger members said that the senior scientists maintained a hierarchy on the project.[554] When asked about the issue of diversity in the field, Berger's reply was that he invited every South African palaeoanthropologist that was qualified. When asked if he received any applications from South Africans to conduct the fieldwork in the Rising Star Cave, Berger said, "Not one." He went on to say, "No one realises how neglected palaeoanthropology is. For the last twenty years, we've been competing with Standard Bank, Absa and everywhere else. It's just a horrific situation where if we're not going to compete and pay academics at the same level that you pay successful people in other fields, then you're not going to have any."[555]

Of the material coming out of Rising Star and being examined, Berger said: "The message we're getting is of an animal right on the cusp of the transition from *Australopithecus* to *Homo*." Yet in certain aspects, he thought that the Rising Star fossils were more similar to *Homo sapiens* than *Homo erectus*. The creature's brain was very small, comparable to the size of an orange. There was a surprising mix of features. "If you found the foot by itself, you'd think some Bushman (sic) had died," said Steve Churchill, from Duke University.[556]

Berger and his team decided that the fossils definitely fell within the genus

Homo. But it didn't look like any other species in the genus. They decided to give it a new species name – *naledi*. In Sesotho the word means "star". There were no easy ways to date the fossils, but that didn't bother Berger too much. He said that if *Homo naledi* turned out to be more than two million years old, "it would represent the earliest appearance of *Homo* that is based on more than an isolated fragment". But if the fossils turn out to be less than a million years old, "it would demonstrate that several different types of ancient humans all existed at the same time in southern Africa, including an especially small-brained form like *Homo naledi*."

Another issue that the Rising Star workshop deliberated was how the hominids had gotten into the small chamber, which they decided to call Dinaledi, meaning the chamber of stars. By day three or four of the excavation, the scientists realised that they hadn't found any animal bones or other fauna. This raised the crucial question of how the hominids had found access to that isolated chamber. There were no food remains or stone tools in there with them. The workshop explored every alternative scenario, including mass death, a catastrophe, and being drawn in by other animals, and eventually they decided that intentional disposal of the bodies was the most plausible scenario.

The scientists did not argue that the hominids had made their way through Superman Crawl and down the shark mouth chute into the chamber while carrying corpses on their backs. However, they did suggest that they had deliberately disposed of their dead. "There has to be another entrance," said Richard Leakey. "Lee just hasn't found it yet."[557]

"Disposal of the dead brings closure for the living and confers respect on the departed. Such sentiments are a hallmark of humanity, but *Homo naledi* was not human." This quote from *National Geographic* was followed by Berger's observation: "It's an animal that appears to have had the cognitive ability to recognise its separation from nature." He said, "... until now we thought ritualised behaviour towards the dead was utterly unique, and perhaps identified us [as human], a phenomenon that separated us from the animal kingdom. Now we see a species with the same capacity."[558]

While burying and respecting the dead is supposed to be a distinguishing feature of being human, many scientists themselves for over a century had not respected the dead, had exhumed bodies from graves, and then placed human skeletons in scientific collections, often without giving attention to the requests of the family members. If burying the dead is a distinct cultural aspect of being human, /Keri-/Keri should have been given a decent burial, just like Sarah Baartman.

Berger and his team prepared to announce their findings to the world. They made sure that the scientific papers came out on the same day as the announcement, along with a cover story in *National Geographic,* along with the airing of a one-hour documentary called *Dawn of Humanity.* They hoped that the world would take note.

Zeblon Vilakazi, the deputy vice-chancellor for research at the University of the Witwatersrand, stood at the podium on 10 September 2015 at Maropeng in the Cradle of Humankind. Looking out at the audience of 300 people, seated and waiting expectantly in the auditorium, he began. "It's showtime, folks!" he announced with a broad smile. "It's my great pleasure to welcome you to this groundbreaking event – pun intended – to celebrate what is indeed, as the late Professor Tobias said, Africa's gift to the world – the gift of humankind." Vilakazi reminded the audience that the proceedings were being streamed live across the world.[559] He indicated the 15 TV cameras on tripods positioned at the back of the room.

Sitting on stage with the scientists was Cyril Ramaphosa, the deputy president of South Africa, and Professor Lee Berger. In front of the stage was what looked like a large table covered by a blue velvet cloth. There was an air of expectation in the room. "The legendary Professor Phillip Tobias," said Ramaphosa when it was his turn to speak, "a stalwart of the University of the Witwatersrand, is no longer with us, and we sadly miss him, but I'm quite sure that he would have reminded us that these finds again underline the fact that Africa, our continent, is the home of great scientific discoveries, the home of our humanity, the home of our collective culture as humans. These discoveries underlie the fact that despite our individual differences, in the way we appear, and the language that we speak, the beliefs that we hold on to, and our cultural practices, we are bound together by a common ancestry."

Lee Berger, looking tanned and relaxed, took a moment to settle in at the podium. Looking out at the audience, he said that this was a great moment, built on 90 years of exploration, for human origins in South Africa. He acknowledged that he and his team were standing on the shoulders of great scientists like Raymond Dart, Bob Brain and Phillip Tobias, and then he took the audience through the process of how, from the time back in 2013, with the assistance of *National Geographic* and Wits University, he had put together his team – a 60-person expedition, with six scientists selected who could fit into the small cave entrance. He took them through how they had put everything meticulously in place before fossils could be brought to the surface – about the 3.5 kilometres of communication cables and fibre optics had had to go into the cave so that they were able to stream video footage live to the world and tweet about the process. What they had found down there, Berger said, was "the largest assemblage of fossil human relatives ever discovered in the history of the continent of Africa." Allowing the words to sink in and after pausing to allow the applause from the audience to die down, Berger thanked them and said: "This is not the Lee Berger show. This is not what this discovery is about. It is anything but that." Speaking slowly and deliberately, he continued: "Today, ladies and gentlemen, I am pleased to introduce you to a new species of human ancestor." On the two giant screens at the front of the room appeared an image, an artistic representation of an ancient hominid. The audience clapped. "He is pretty. He's worth applause," said Berger, laughing. "This is a new species within our very genus. A species that we have called *Homo naledi* – naledi meaning star."

The audience reached for their cell phones and the photographers moved in closer, but the moment of the big reveal was still a little way off. Even as there was a strong sense that they were building toward a climax, the proceedings continued for well over an hour.

Berger asked the dignitaries in the audience to step forward to join him in front of the stage behind the large table covered with blue velvet. He took Cyril Ramaphosa's hand. The audience waited in silence. "Mr Deputy President, will you please show the world *Homo naledi*?" Together they pulled back the blue velvet blanket to reveal a glass display case filled with bones. The applause was thunderous. As it subsided, the loudest sound was the clicking of cameras. With near silence in the room, photographers moved forward and people strained to see. The hush was interrupted only when Lee Berger said "The hashtags are #NalediFossils and #AlmostHuman. Please join us in celebrating these fossils on social media."

There was an absolute media explosion. It was greater than anyone expected. The story trended on Twitter and Facebook in South Africa and around the world. The next day *Homo naledi* was the headline in every South African newspaper. It was the lead story in the *New York Times*, the *Washington Post*, *The Guardian*, and on BBC. Media outlets repeated Berger's words, saying that the *Homo naledi* fossils were the richest fossil find in the history of the continent of Africa, that they represented a new species, and that *Homo naledi* deliberately disposed of its dead. CNN journalist Jonathan Mann said that *Homo naledi* "could transform what we know about the narrative of human evolution". There didn't seem to be any argument that this was a momentous and important fossil find. On that scientists and journalists around the world agreed.

However, there was immediate disagreement about whether the fossils represented a new species. Tim White said it was not a new species, but rather "primitive *Homo erectus*", signalling once more the tensions that had festered between White and Berger for almost 20 years. Other scientists also questioned the findings but Berger held firm. Some scientists criticised him because the papers were not published in *Nature* as expected, but published in *E-Life*, a lesser known, newer journal. Berger applauded *E-Life* for being open access, and not behind a paywall, so that the published papers would be open to everyone around the world to read and critique.

Other scientists argued that there was no way that *Homo naledi* could have deliberately buried their dead and that there must be another access point to the cave chamber in which the fossils were found. Berger and his team maintained they had considered multiple options and presented what they thought was the most likely scenario. There was also significant concern expressed by some scientists that Berger and his team had not yet dated the fossils, leaving a big gap in the information needed to analyse them accurately.

About a week after the announcement of *Homo naledi,* Ron Clarke voiced his own concerns. While acknowledging the "importance of the discovery in its own right," he said, "I hope you are not being duped by all this. For those of us doing serious research, all of this sensationalism doesn't do us any service."[560] The tensions between Clarke and Berger had not lessened with time. While Berger was taking on more modern methods of fossil investigation and making bold statements to the media, Clarke continued to favour the older methods of physically chipping away at the breccia around Little Foot. A month later, the *Business Day* published an article where Clarke said: "Everyone is pressuring me and saying 'why is it taking so long?' There is still a way to go, but it's getting close now." In the same article, Robert Blumenschine from PAST said: "To a lay person it may seem ridiculous that someone has been working on Little Foot for almost two decades, but it's completely legitimate. It is the way a conservative scientist acts."[561]

Berger said that Tobias would have loved to see the new discovery. "He dedicated his life to the study of the origins of the genus *Homo*. To see something from South Africa, so remarkable, so primitive, that you're looking at the root of the genus, well, I think he would have been thrilled."[562]

Berger expected controversy. He expected arguments and debates. But he found himself answering to another accusation, one that he hadn't expected at all – the accusation that *Homo naledi* was racist. This was a recurring fault-line in South African life, history and politics. The issue first arose when Mathole Motshekga, a lawyer, a member of parliament for the African National Congress, and a scholar of African indigenous knowledge, spoke on a radio interview on the same day of the announcement at Maropeng, saying that the discovery was an attempt to say that African people descended from baboons. He dismissed it as "pseudo-science". The next day he spoke at greater length in a television interview, saying that he had no objections to scientists conducting research into the past but he thought the *Homo naledi* findings had affirmed the beliefs of apartheid and colonialism – that black people were sub-human and that slavery, colonialism, oppression and exploitation were justified. He said this theory was why no African was fully respected around the world today. In a third interview, Motshega defended his stance, saying, "We are still suffering from the wounds of being called baboons or apes."[563]

During the announcement, the artist's impression of what *Homo naledi* may have looked like, drawn by well-known paleo-artist John Gurche, instantaneously appeared on social media with comments comparing it to certain politicians. Seeing these images, the former general secretary of the Congress of South African Trade Unions (COSATU), Zwelinzima Vavi, tweeted: "Science is materialism – it's facts that can be proven. No one will dig old monkey bones to back up a theory that I was once a baboon – sorry." Immediately, Vavi was attacked for not understanding the theory of evolution, but he maintained he was not related to an ape, monkey or baboon and that the common comparison of black people to baboons over many generations had resulted in people questioning the validity of scientific discoveries. "It's insults like this that make some of us to question the whole thing."[564]

The president of the South African Council of Churches (SACC), Bishop Ziphozihle Siwa, entered the debate with a statement celebrating the discovery of *Homo naledi* and the fact that it was found in South Africa. But he cautioned against insulting comments and said that it was the perception of many people in the West that black people were baboons. He agreed that Vavi's comments were "spot on".[565]

Lee Berger quickly responded to Vavi on Twitter. "I agree with Vavi; we do not come from baboons." He added that science was not challenging anyone's religion or belief system but rather exploring the fossil evidence for the origins of our species. Speaking to the popular radio talk-show host Redi Thlabi, he explained that humans and baboons were both primates but that they were no more related than a cat is to a dog. He explained that *Homo naledi* was a human relative which shares unique characteristics with humans such as the hand shape and almost exactly the same foot anatomy. Thlabi said she understood scientists were saying that *Homo naledi* was not an explanation for the origins of black people or Africans, but that it was related to the origins of all humans. "This is not about the origins of black Africans," said Berger. "It is a deep common ancestor (or relative –I'll prefer that word) of all humankind."[566]

"Are we finally going to understand the missing link?" asked Thlabi. "Scientists don't like to use that term anymore," said Berger. He went on to explain that the concept of a missing link came from Victorian scientists who believed that evolution was a simple, direct progression that was linear like a chain and that there might be a clear single link in the chain between apes and humans. As a result of genetics and the fossil record, scientists now know that the concept no longer accurately describes human evolution. The metaphor of a family tree, a bush, or even a braided stream, was now more commonly used.

Berger asked listeners to put a different image in their mind. Instead of links in a chain, he suggested the image of a glacier leading to a glacial lake. As the glacier melts, it creates streams and rivulets that intertwine with one another and eventually lead to a large lake, which represents all seven billion *Homo sapiens* alive today. Before getting to the lake, some of the streams disappear off into the soil or the gravel and don't come back to the main stream. They represent extinct hominid species. Other streams represent hominid species like Neanderthals that lead off from the mainstream but then come back, resulting in inter-breeding and sharing DNA. *Homo naledi* fits somewhere into that picture of a braided stream. Redi Thlabi interjected, "So the term 'missing link' must disappear?" "It must be eradicated," said Berger.

The well-known South African cartoonist Zapiro jumped into the fray. Along with the formal announcement of *Homo naledi*, *National Geographic* magazine published an issue with the cover headline "ALMOST HUMAN". Zapiro parodied the cover by publishing a cartoon in the *Sunday Times* showing Motshega on a magazine called the *National Theocratic* saying, "Evolution is a Eurocentric plot!" Beside him was the artistic impression of *Homo naledi* saying. "I have a brain the size of an orange ... what's your excuse?" What Motshekga was saying was that because *Homo naledi* is supposed to be a pre-human being and therefore sub-human, this meant that anyone who is depicted as *Homo naledi*, especially black people, and Africans in particular, suggests that they themselves are sub-human. For many scientists around the world, the South African backlash against *Homo naledi* was perplexing.

Richard Dawkins, British evolutionary biologist and writer, said on Twitter that the debate "breathes new life into paranoia". The "whole point is we're all African apes," he said.

But Dawkins misses a point too. It is important to make clear that modern apes and modern humans last shared a common ancestor approximately seven million years ago. The genus *Homo* emerged over two million years ago and *Homo sapiens* developed symbolic thought and art in Africa less than 150 000 years ago. The whole point is that we are not African apes.

Why did the announcement of *Homo naledi* in 2015 have such a different reaction to the announcement of *Australopithecus sediba* in 2010? Part of the answer might be that South Africa was celebrating its hosting of the World Cup in 2010. It was basking in the glory of having hosted a successful world event. In 2015, by contast, the economy was failing, the political climate was confrontational, and there was growing attention to the continuing challenges of racism in the country, which had long gone unaddressed. Another possible factor was that in 2010, during the announcement and media coverage of *Australopithecus sediba*, there was no reconstruction of what *Sediba* might look like. The images that were shared with the world were of Lee Berger and his son Matthew, and the bones themselves. John Gurche did not put together a facial reconstruction until much later. For *Homo naledi* five years later, the artistic reconstruction was shared with the public at the same time as the announcement. That image of what *Homo naledi* might look like flooded social media, print media and television. The image was used alongside insults to black people so many people found it to be offensive.

While race has provided the rationale for some of the worst atrocities in human history, many biologists, physical anthropologists, and geneticists agree that there is no valid scientific justification for the concept of race as a biological fact. There is great variation amongst human beings, but many of these variations are superficial and of recent origin.

All living humans are members of the same species *Homo sapiens*. The Out of Africa theory, and the genetic evidence that helps underpin it, which is largely accepted by most scientists, shows that the ancestors of all seven billion *Homo sapiens* on earth derived from the same group of ancestors who left the African continent as recently as 100 000 years ago. This group migrated to Asia and Europe and then on to Australia and the Americas. The physical variation that we see amongst humans around the world has developed since that time and continues to evolve and change. Despite the fact that biologically, race has little meaning, we are not beyond race. Historically, culturally and politically, race continues to shape our world. And history, culture and politics have an impact on how we view, and have viewed, the search for human origins.

Epilogue

Scientific knowledge is changing at such a rate, and advances are being made in so many interrelated scientific fields, it is difficult to keep pace with new information, or even to remain current from one day to the next. The production of scientific knowledge and the spread of that knowledge is also a social process so it will always be shaped in some way by the social and political context of the time. Dynamics that are at play today will only fully be realised when we look back at them 20 or 30 years from now. The latest sophisticated technology and equipment, different ways of examining and contextualising research, and new, more diverse participants entering various fields of science and social enquiry will take us forward into discoveries not yet imagined and back to the past to re-examine truths that may no longer be true.

The ability to extract DNA from ancient bones, for example, is one new area of science that is having an impact on the field of human origins.

Alan Morris, at the University of Cape Town, has been working on the possibility of extracting ancient DNA from a human skeleton, which was dated to be about 2 300 years old, in order to find out more about what was going on with humans that long ago. The skeleton was found at St Helena Bay, about 150 kilometres north of Cape Town up the west coast along the Atlantic Ocean. He partnered with archaeologists to take bone samples from the skeleton and they were able to extract DNA from a single tooth. With a permit from the South African Heritage Resources Agency, they sent the sample off to Sydney, Australia for analysis. Over the last 12 years since Morris's last attempt, the technology has greatly improved, so they were able to analyse a mitochondrial DNA sequence from the skeleton. As a result of data they gathered, they suspected that the male individual spent significant amounts of time in the ocean near the coast looking for food and that he lived at a time before pastoral agriculture came to the region. But the scientists' findings raised more questions than they answered.[567]

The fact is that more and more testing of ancient DNA samples is under way. As additional data is collected, there will be attempts to take the research further back in time. In November 2015, scientists reported that they had found a 110 000-year-old fossil tooth containing DNA from a Denisovan sample, an extinct species of *Homo* found in Siberia, but it still is not clear if there is enough DNA available to be analysed. As Morris puts it, "The gates for ancient DNA research in southern Africa are now open ... The 'Holy Grail' in ancient DNA work," he says, "is to extract information from skeletons found in South Africa and produce a complete genome from one ancient individual." He does not think attaining this scientific goal is too far into the future.[568]

It is also possible DNA could be gathered from one or more of the *Homo naledi* bones. And the scientists who worked in the Dinaledi chamber in the Rising Star Cave say there are thousands more bones to be collected. Lee Berger insists there are many more unexplored fossil sites in the Sterkfontein Valley. It is impossible to know what information will be brought to light in the future, not only in South Africa, but in East Africa as well, and from around the globe. There are always dangers in terms of how information can be used and abused. But at each step of the way, there is also the potential to draw lessons from our past, and to develop a new vision for the future, a vision that recognises the dignity of all human beings.

Teaching human evolution can be filled with sensitivities and setbacks, and this is particularly evident in South Africa, where a large number of South Africans are not yet convinced that humans have evolved from an ape-like ancestor. In a survey conducted at Maropeng in 2013, researchers found that 37 percent of the South African visitors surveyed did not fully accept that "humans evolved from an ape-like ancestor". One teacher who was accompanying a school group stated, "There is no full evidence to convince me."[569]

For every example of ignorance, misunderstanding and intransigence, there is a case of wonder and excitement and an eagerness to learn and grow. In order to make a contribution in this area, PAST – the trust at Wits – developed a piece of theatre called Walking Tall. In the one-hour show, young performers use enormous energy and gymnastic skill to illustrate aspects of human evolution, including the fact that human evolution has occurred over the past six to seven million years, and that different shades of skin colour have developed only in the last 10 000 years. The show is interrupted at intervals for the performers to share knowledge with and ask questions of the audience. In just over a decade, Walking Tall has reached over one million students around South Africa.

In December 2015, Job Kibii left Wits to take up an opportunity back home in Kenya, to become the director for palaeontology and palaeoanthropology at the Kenya National Museums. He is driven by the questions: Where do we come from? and Why are we the way we are today? "Every bone has a story to tell," he says. "A story about an individual, a time and a place." Kibii hopes to share his experience and knowledge, and to inspire a new generation of young people in East Africa and around the world. "My first priority is to see that we have more Africans in this field. It is changing slowly but surely." Inspired by Phillip Tobias, who worked well into his 80s, Kibii says, "I hope to be as strong and active in the field well into my 90s."[570]

Winnie Dipuo Mokokwe (now Kgotleng) who previously worked at Taung and Maropeng, has now completed her PhD about the primate fossils at Sterkfontein. She is optimistic about the field and would encourage young people to consider entering archaeology and palaeoanthropology. "This field can lead you to information about our own history, our prehistory, and where we come from as a people," she says. She is thankful for the Taung skull. "It's because of that skull that there is interest in African palaeoanthropology. It is because of the Taung skull that I am working in this field today. And I hope that the special meaning of that skull will inspire many more young people the way that it inspired me."[571]

In South Africa, where issues of race and racism are ever present, many stereotypes and prejudices persist. Like field workers working with fossils cemented in breccia, it is necessary for us to chip away at long-held prejudices and beliefs, holding new truths up to the light and examining them with care and open minds. Despite the work still to be done, certainly there is hope for the future.

The era of race typology, when scientists were looking for an imaginary and pure type for each race, is behind us. Apartheid and legislated racial classification ended officially in 1994. And the field of population genetics has brought in a new way of thinking about human evolution, not without its challenges, but taking us away from the typology of the past. The science of human origins is in a vastly different place in 2016 than it was in 1925 or 1966, or even 2008. Darwin would have been delighted to know that his hunch was correct. Africa is, indisputably, the cradle of humankind.

Acknowledgements

If anyone ever tells you that writing your second book is easier than your first because you know what to expect, don't believe them. The journey seemed just as long and challenging the second time around, if not more so, and I am very thankful to all the many people who supported me along the way.

Thank you to WiSER, the Wits Institute for Social and Economic Research, for giving me positive feedback on my initial proposal and offering me a place to do my research. My time at WiSER over the past three years has been invaluable for the work space, research support and community. Thank you to Sarah Nuttall, Achille Mbembe, Pamila Gupta, Jonathan Klaaren, Najibha Deshmukh, Adila Deshmukh, Ellison Tjirera, Tim Wright, Kirk Sides, Hlonipha Mokoena, Sarah Duff, Ruth Sack, Faeeza Ballim, Christi Kruger, Kalema Masua, Khadija Patel, Neo Muyanga, Terry Kurgan and Shireen Hassim. Thank you to Joshua Walker for helping me to track down an elusive source. And many thanks to Keith Breckenridge and Catherine Burns for reading early chapters of my draft.

One of the first people I talked to about the book, back in August 2013 was Francis Thackeray at the Evolutionary Studies Institute (ESI), which conveniently happens to be in the building right next door to WiSER on the Wits campus. Francis has been supportive ever since, by loaning me books, answering countless questions, and sharing trips to Sterkfontein, Swartkrans and Kromdraai. Many people at ESI have been supportive of this project including Merrill van der Walt, Patrick Randolph-Quinney and Bernhard Zipfel. Thank you to Bruce Rubidge and Christine Steininger of the Centre of Excellence (CoE) in Paleosciences. I am very thankful to the CoE and the National Research Foundation for the grant of financial support for my research.

Thank you to ANFASA, the Academic and Non-Fiction Author's Association of South Africa for awarding me an author's grant in 2014.

Enormous thanks to Samukelisiwe Mfuphi, ANFASA's national administrator, for always being supportive and helpful.

Elizabeth Marima and Fikile Ntuli helped me often as I trawled through the Raymond Dart papers in the Wits University Archive. Thank you to Ruth Muller who worked with me for months, tracking down news clippings, photos and images for the book.

Tersia Perregril, Klaas Manamela, Stephany Potze and Lazarus Kgasi assisted me at the Ditsong Museum of Natural History in Pretoria with the Broom archives, photographs, and the archives of the Transvaal Museum. Thank you to Nadine Wubbeling at the *South African Journal of Science* for helping me find articles, even when they did not appear on their excellent online archive.

I want to thank the more than 50 people – too many to mention individually by name here – who I interviewed for this book. I valued my exchanges with each and every one of you, as you helped me to understand scientific concepts, and piece together events in history. Please take a look at the section at the end of the book under "Interviews and Personal Correspondence." I must give special thanks to Alan Morris for sharing documents with me, even before they went up online on the Human Anatomy Archive at the University of Cape Town.

Thank you to Mirriam Tawane and Ian McKay for inviting me to join them on their trip to Taung, along with Sifelani Jirah, Brian Mogaki and Amanda Mufuniwa Mudau.

I am enormously grateful to Brendon Billings and Tobias Houlton from the School of Anatomical Sciences at Wits Medical School for supporting my search for /Keri-/Keri which continues today.

Many thanks to Makhosazana Xaba, Karen Hurt, Barry Gilder, and Catherine Hunter who read several early chapters. And to Makhosazana Xaba again, along with Jackie Mondi, who both read all of Part One. Michael Titlestad, Karen Martin and Alison Lowry read early drafts of the manuscript and gave me guidance when I was still in the wilderness, and helped me to see important challenges and how to address them. I am grateful to Saul Dubow, Xolela Mangcu, Shan Ramburuth, Penny Plowman, Roger Jardine, Ian Tattersall, Brooks Spector, Keith Breckenridge and Sarah Nuttall for reading later drafts and sharing enormously helpful comments and feedback. All of these readers, at every stage, helped me to keep going and to strengthen the book. Roger deserves a special award for fortitude on the job because not only did he read the entire manuscript – twice, but he patiently agreed to read whatever I put in front of him that I had written that day, whether it was a chapter, a paragraph, an email, or a tweet.

Book publishing is a tough business, and I am grateful to Bridget Impey and Maggie Davey of Jacana Media who have supported me since I first thought about writing my first book back in 2010. I'm thankful to Nadia Goetham and Shawn Paikin for guiding me through the final production. Special thanks to Lara Jacob for enthusiastically taking on the copy edit. And thank you to Neilwe Mashigo, Janine Daniel, Shay Heydenrych, Sibongile Machika, Megan Mance, and everyone at Jacana who had a hand in bringing this book to life. Many thanks to Alan Goldsmith for creating the three maps using the ArcGIS software by ESRI. I also appreciate the help from Veronica Klipp and Hazel Cuthbertson at Wits University Press in tracking down several important photos.

Alison Lowry has been by my side, or should I say across the table at Vovo Telo, from the start. She is one reason why writing my second book has been easier than writing my first, because I knew, right from the beginning, that she would be my editor. Even though she saw the first huge, unwieldy chunk of clay, she could see where I was going as I continued to whittle and sculpt, and she keenly stepped in with her skill to smooth the many rough edges at the end.

Writing a book can be all consuming, and I am thankful to my family, for making sure that I didn't disappear into the project, but that I continued to be involved in our lives together. Thank you to Bob, Jewel and Sarah Kuljian who were with me when I first watched *The Ascent of Man* and sent me off to study with Stephen Jay Gould. Thank you to Anne Jardine who always asked, "How are things going with our book?" And finally, thank you to Nadia and Mila and Roger Jardine whose love, encouragement, humour and warmth have sustained me.

Endnotes

1 Bronowski, *The Ascent of Man*, 1973, p 30
2 Gould, *The Mismeasure of Man*, 1996, p 28
3 *The Star*, 1 August 2013 and Adams, "Archbishop Desmond Tutu at Maropeng," 1 August 2013, www.maropeng.co.za/news/entry/archbishop-desmond-tutu-at-maropeng-following-the-footsteps-of-humankind
4 Wild, "Skinny Scientists", 2013
5 Bronowski, *The Ascent of Man*, p 28
6 Phillips, "Breaking the Chain", 2014
7 Linne, *Systema Naturae*, 1758, reprinted in 1939
8 Ibid, pp 20–22
9 Gould, *The Mismeasure of Man*, pp 401–412 and Junker, "Blumenbach's Racial Geometry", 1998. Stephen Jay Gould made reference to Blumenbach and his introduction of a fifth grouping of humans in the 1996 revised edition of his book, *The Mismeasure of Man*. The book included a chapter called "Racial Geometry" and a graphic citing Blumenbach's *Anthropological Treatises* on p 409. In Junker's article he points out that the Gould graphic used was incorrect. It did not reproduce Blumenbach's original illustration, but used a more "hierarchical" triangle instead of a linear line of five skulls. Gould is given the right of reply and argues that his argument did not stand on whether Blumenbach's figure was there or not.
10 Knox, *The Races of Men*, 1850, as quoted in Weidman and Jackson, *Race, Racism, and Science*, 2004, p 53
11 Darwin, *Descent of Man*, 1871
12 Goodrum, "The History of Human Origins Research", 2009, p 344
13 Tattersall, *The Fossil Trail*, 1995, pp 32–36
14 Ibid, pp 31–40
15 Kuklick, "The British Tradition", 2008, p 55
16 Hochschild, *Bury the Chains*, 2005
17 Chidester, "Christianity and Evolution", 2002, p 101
18 Much of the information about Broom's life told in this section was gathered from the only biography of Broom, by George Findlay. Published in 1972, Findlay's book does not have footnotes, references or a bibliography.
19 Findlay, *Dr. Robert Broom*, 1972, p 13
20 Ibid, pp 15–16
21 Ibid, p 20
22 Georges Cuvier and Robert Knox were among those professing this theory. See Dubow, *Scientific Racism*,1995, pp 23–25; Abrahams, "The Great Long National Insult", 1997; and Tobias, "Saartje Baartman: her life, her remains", 2002, p 108
23 Findlay, *Dr Robert Broom*, 1972, p 20
24 Redman, *Bone Rooms*, 2016
25 Findlay, *Dr. Robert Broom*, p 26
26 Legassick and Rassool, *Skeletons in the Cupboard*, 2015, p 3
27 Ibid p 4
28 Theal as quoted in Dubow, *Scientific Racism*, 1995, pp 68–70

29 Peringuey as quoted in Legassick and Rassool, *Skeletons in the Cupboard*, 2015, p 5

30 Legassick and Rassool, *Skeletons in the Cupboard*, p 9

31 Morris, "Trophy Skulls, Meseums and the San", 1996, p 72 and in Legassick and Rassool, *Skeletons in the Cupboard*, p 31. Instead of George Lennox (Scotty Smith) shooting people when he wanted a skeleton, Morris wrote that he thought "a more likely explanation is that he excavated graves".

32 Legassick and Rassool, *Skeletons in the Cupboard*, p 47

33 Findlay, *Dr. Robert Broom*, p 27

34 Ibid

35 Whaits as quoted in Findlay, *Dr. Robert Broom*, p 30

36 Findlay, *Dr. Robert Broom*, p 32

37 Ibid, p 47

38 Strkalj, "Robert Broom's Theory of Evolution", 2003, p 38

39 Weiner, *The Piltdown Forgery*, 1955, p 2

40 Ibid

41 Ibid, p 9

42 Smith Woodward dictated his last book entitled *The Earliest Englishman* about Piltdown Man that was published in 1948, three years after he died.

43 Plaatje, *Native Life in South Africa*, 1916, p 1

44 Broom as quoted in Dubow *Scientific Racism*, p 39 and as quoted in Strkalj, "Inventing Races", 2000, p 118

45 Broom to Osborn as quoted in Findlay, *Dr. Robert Broom*, p 41

46 See South African History Online entry on the Kora (or Korana) people at www.sahistory.org.za/article/kora

47 Findlay, *Dr. Robert Broom*, p 50

48 Ibid

49 Broom is the only source for this story, and he does not provide John's surname.

50 Findlay, *Dr. Robert Broom*, p 51

51 Ibid, p 50

52 Dart 1972, "Associations with and Impressions of Sir Grafton Elliot Smith," p 171 and Dart and Craig, *Adventures with the Missing Link*, 1959, p 26

53 This experience is mentioned in Tobias's 1984 book *Dart, Taung and the Missing Link* and in Wheelhouse and Smithford's *Dart: Scientist and Man of Grit* – two biographies of Dart – but this particular language was used by Dart when he gave an interview to Jane Dugard in Johannesburg in the early 1980s.

54 Dart and Craig, *Adventures with the Missing Link*, p 31

55 Keith, *An Autobiography*, 1950, p 480

56 Dart and Craig, *Adventures with the Missing Link*, p 32

57 Dart first told this story including the role of Josephine Salmons in his 7 February 1925 article in *Nature*, and told it again and again to the press, and in Dart and Craig, *Adventures with the Missing Link*, pp 1–3

58 Hrdlička, who visited the Taung site in 1925, quoted De Bruyn (Hrdlička, "The Taungs Ape," p 384) and then Broom told the same story in Broom and Schepers, *The South African Fossil Ape-Men: The Australopithecinae*, 1946 (p 12).

59 Dart and Craig, *Adventures with the Missing Link*, p 4

60 Ibid, pp 4–6

61 This earlier rendition of the story with a larger role for Professor Young was clear in *The Star*, 5 February 1925 and the letter from Young to Dart dated 7 February 1925 found in the Dart papers in the Wits Archive, and reviewed further in Tobias, *Dart, Taung and the Missing Link*, 1984 and Tobias, "The Discovery of the Taung Skull of *Australopithecus Africanus*", 2006.

62 Tobias, *Dart, Taung and the Missing Link*, pp 26–27

63 Dart and Craig, *Adventures with the Missing Link*, p 6

64 Ibid, pp 7–8

65 Ibid, p 9

66 Ibid

67 Dart, "*Australopithecus Africanus*", 1925

68 Ibid

69 Young to Dart, 7 February 1925 in Dart papers, Wits Archive

70 Dart and Craig, *Adventures with the Missing Link*, photo section between pp 48 and 49

71 Haines, *International Women in Science*, 2001 and www.records.ancestry.com (accessed on 26

February 2014). I met with Craig and Mary Anne Elstob on 15 June 2017 in Sandton, Johannesburg. Craig Elstob is the grandson of Josephine Salmons and was able to confirm the date and location of her death.

72 Smuts as quoted in *The Star*, 6 February 1925 and in Dart and Craig, *Adventures with the Missing Link*, pp 36–37

73 Smith Woodward in *Nature*, 14 February 1925 from Tobias, *Dart, Taung and the 'Missing Link'*, 1984, p 39 and in Dart and Craig, *Adventures with the Missing Link*, p 40

74 Elliot Smith in *Nature*, 14 February 1925 from Tobias, *Dart, Taung and the 'Missing Link'*, 1984, p 38 and in Dart and Craig, *Adventures with the Missing Link*, p 39

75 British United Press, *Telegraph-Journal*, "Dashes Hopes that Skull is Missing Link's", 11 July 1925

76 Dart and Craig, *Adventures with the Missing Link*, p 41

77 Ibid, p 40

78 Ibid, p 43

79 Ibid, p 38

80 Broom, *Finding the Missing Link*, 1950, p 27

81 Broom was appalled at Hrdlička's inability to comment on the Taung skull without additional specimens. Broom, *Finding the Missing Link*, p 28

82 "The Robert J Terry Anatomical Skeletal Collection" on the website of the National Museum of Natural History, www.anthropology.si.edu/cm/terry.htm (accessed on 6 October 2015)

83 Dart to Fourie, letter dated 16 June 1923 in the Dart papers, Wits Archives; also printed in Wheelhouse, *Raymond Arthur Dart*, 1983, p 35

84 Dart to Fourie, letter dated 7 November 1924 in the Dart papers, Wits Archives; also printed in Wheelhouse, *Raymond Arthur Dart*, 2004, p 36. In Wheelhouse's book, the caption to this letter reads: "Handwritten acknowledgement of a further Bushman specimen at the end of 1924. Although extremely busy with lectures, building up the museum and working on his Taung skull, Professor Dart never neglected to thank those who helped him." It is striking that Wheelhouse does not question the contents of the letter at all, nor Dart's request for a "Bushman specimen".

85 Dart, "The South African Negro", 1929. Dart's praise of Peringuey is on pp 311–312. His quote on hybridisation is on p 314.

86 The journal *Bantu Studies* was renamed *African Studies* in 1942.

87 Redman, *Bone Rooms*, 2016

88 See Dubow, *Scientific Racism in Modern South Africa*, 1995 and Dubow, "Human Origins, Race Typology and the Other Raymond Dart", 1996, p 6. These two publications of Dubow's provided a foundation from which to explore the history of Raymond Dart's work in South Africa. For an excellent introduction, see Dubow's chapter 2, 'Physical anthropology and the quest for the 'missing link' in *Scientific Racism in Modern South Africa* (1995).

89 For more on the creation of "whiteness" in the United States, see James Baldwin's writing, especially "On Being White and Other Lies", first published in *Essence* magazine in 1984. The essay described how European immigrants had to "become American" by learning to be white. The essay was republished in the book *Black on White* edited by David R Roediger in 1998.

90 Smuts, "South Africa in Sceince", 1925, p 245

91 Ibid, p 249

92 Dart 1925, "The Present Position of Anthropology in South Africa", *South African Journal of Science*, p 75

93 Redman, *Bone Rooms*, 2016

94 Wells, "The Foot of the South African Native", 1931, p 236

95 Ibid, p 280

96 Conversation with Bernhard Zipfel, 11 April 2016

97 Dart and Craig, *Adventures with the Missing Link*, p 52, and "Medicine in China" on the "100 Years: The Rockefeller Foundation", www.rockefeller100.org/exhibits/show/education/china-medical-board

98 Ibid, p 54

99 Dart, "Historical Succession," 1925, p 426

100 Dart as quoted in the *Rand Daily Mail*, "Vexed Problem of Colour", 16 March 1929

101 Dart as quoted in *The Star*, "Coloured or Not?", undated

102 Dart and Craig, *Adventures with the Missing Link*, p 63

103 Murray, *Wits: The Early Years*, 1982, p 179

104 Smuts to Broom as quoted in Broom, *Finding the Missing Link*, 1950, p 99

105 Smuts to Broom, as quoted in Van Der Poel, *Selections from the Smuts Papers, Volume V*, 1973

106 Murray, *Wits: The Early Years*, p 126

107 Bridie, *The Anatomist*, 1931, p 8, Wits Archives

108 Ibid, p 59

109 Murray, *Wits: The Early Years*, p 181

110 Bain to Fourie in 1925 as quoted in Wanless, "The Silence of Colonial Melancholy", 2007, p 162

111 Bain to Dart, 26 May 1936, Dart papers, Wits Archives. It was a footnote in Dubow's *Scientific Racism in Modern South Africa* (1995) that inspired me to look more closely at Dart's trip to the Kalahari. Footnote 13 on page 25 reads: "A topic suitable for further research is the high-profile scientific expedition to the Kalahari mounted by Wits academics in 1936... "

112 Ibid

113 Dart and Craig, *Adventures with the Missing Link*, p 61

114 Landi and Cecchi, "Colonial Anthropology: From the Peoples of the World", 2014, pp 27–28. Also look for a forthcoming article by Tobias Houlton and Brendon Billings about the Raymond Dart face mask collection at Wits, provisionally titled "Blood, Sweat and Plaster Casts: Reviewing the Archives, History and Scientific Value of the Raymond A Dart Collection of African Life and Death Masks"

115 Bain to Dart, 26 May 1936, Dart papers, Wits Archives

116 Van Buskirk, *Rand Daily Mail*, 8 July 1936

117 Van Buskirk, *Rand Daily Mail*, 14 July 1936

118 For more information on the history of this obsession, see Abrahams, "The Great Long National Insult", 1997

119 Van Buskirk, *Rand Daily Mail*, 8 July 1936

120 Ibid

121 Ibid

122 Dart, "The Structure of the Bushman", 1937, as quoted in Dubow, "Human Origins, Race Typology and the other Raymond Dart", p 11

123 Dart, "The Physical Characteristics of the ǀ?Auni-=Khomni Bushmen", 1937, p 228

124 Ibid, p 245

125 Ibid, p 192

126 Williams, "Facial Features of Some Kung Bushmen", 1954, p 11

127 Dart, "The Structure of the Bushman", as quoted in Dubow, "Human Origins, Race Typology and the other Raymond Dart", p 32

128 Ibid

129 Dart, "The Hut Distribution", 1937, p 173

130 Dart, "Racial Origins", in *The Bantu Speaking Tribes of South Africa*, 1937, p 8

131 Ibid, p 16

132 Smuts to Drennan, 1938, Alan Morris papers

133 Dart and Craig, *Adventures with the Missing Link*, p 68

134 For a description of ǀKeri-ǀKeri at the time, see Laing to Drennan, 1938

135 Robinson, "Johannesburg's 1936 Empire Exhibition", 2003, p 763

136 For a fuller discussion of ǀKhanako, her life and her treatment in Cape Town, see Rassool and Hayes, "Science and the Spectacle", 2002

137 *The Star*, "Senator Boydell's Plea for Bushmen", May 1937

138 Dart as quoted in the *Rand Daily Mail*, "Bushmen Should be Saved", 21 July 1936

139 Dart statement on segregation, Dart papers, Wits Archives

140 Dart was widely connected throughout higher education, the medical field, and government. The point that Dart's research and writing would have influence, no matter how indirect, on Jan Smuts and the imposition of legislation on segregation has been made by others, including Dubow in his book *Scientific Racism in Modern South Africa* and in his 2008 article, "Smuts, the United Nations and the Rhetoric of Race and Rights".

141 Ouma ǀUna tells this story in *Tracks of Sand*, directed by Hugh Brody, 2012. The film as well as the linguist Nigel Crawhall provided information on the genealogy of ǀKeri-ǀKeri and ǀKhanako's family and the relationship to ǀUna Rooi.

142 Dart, "The Problem of Race", 1937, p 3, Dart papers, Wits Archives

143 Ibid, p 14

144 Bain telegram to Dart, 9 September 1939, Dart papers, Wits Archives

145 Dart telegram to Nel, 11 September 1939, Dart papers, Wits Archives

146 Dart letter to Nel, 14 September 1939, Dart papers, Wits Archives

147 Van Zyl statement attached to Dart letter to Nel, 14 September 1939

148 Dart letter to Nel, 14 September 1939

149 Personal correspondence with Robert Gordon, 7 August 2015 and 31 August 2015. Thanks to Robert Gordon who sent me information and news clippings about Mr CF MacDonald. None of these, however, made reference to /Keri-/Keri.

150 Williams letter to Nel, 13 September 1939, Dart papers, Wits Archives

151 Despite efforts to gather information from Outdshoorn, it has not been possible to find any additional information about Dr Nel's perspective on the situation. The only information available is his brief correspondence with Dart and his willingness to make arrangements with Dart in advance of /Keri-/Keri's death, suggesting that he did not make his patient's health the priority.

152 The information in this chapter came from two meetings the author had with Brendon Billings, curator at the School of Anatomical Sciences, Wits Medical School, on 15 July 2014 and 1 August 2014, and interviews and email exchanges with many people. I initially contacted Billings because of information I first found in a footnote in the chapter by Rassool and Hayes, "Science and the Spectacle" /Khanako's South Africa, 1936–1937", in *Deep hiStories: Gender and Colonialism in Southern Africa*, 2002 on p 137. Much later, I read about /Keri-/Keri in Alan Morris's chapter "Trophy Skulls, Museums, and the San," *Miscast*, 1996, p 70

153 Dr Tobias Houlton, a post-doctoral research fellow at the School of Anatomical Sciences at the Wits Medical School, took great interest in my efforts to clarify information about /Keri-/Keri. His work and expertise have been enormously helpful. After hearing about her missing skeleton on 20 April 2016, Houlton looked for and found /Keri-/Keri's face mask. Having recently begun his own research on the Raymond Dart Gallery of African Faces, Houlton had already reorganised the face masks in order of their CF – cranio-facial – numbers. This made it possible for him to locate CF401, the number indicated on /Keri-/Keri's face mask card as related to the skeleton A43. Houlton then proceeded to conduct forensic research regarding /Keri-/Keri's photograph, the photographs of her sister /Klein Khanako and her mother /Khanako and compare these to the available face masks attributed to /Keri-/Keri (CF401) and /Klein Khanako (CF397). He compared the photographs and the face masks to the one body cast in storage in the department. Houlton concluded that CF401 and the body cast are likely from the same person. He also concluded that the body cast does not match the body type of the one 1937 photograph of /Keri-/Keri published by Raymond Dart in 1937 in *Bantu Studies*. Adding to the mystery, in an accession register at the School of Anatomical Sciences, there are two photographs listed as having been taken by Mr AM Douglas-Cronin of /Keri-/Keri, marked GF32 and GF62. After repeated searches, these photographs have not yet been found.

154 Smuts, "The White Man's Task", 1917, p 3

155 Smuts as quoted in Dubow, "Smuts, the United Nations and the Rhetoric of Race and Rights", 2008, p 62

156 Smuts, "Climate and Man in Africa", 1932, p 130

157 Tobias, "The Sterkfontein Caves and the Role of the Martinaglia Family", 1983

158 Broom, *Finding the Missing Link*, 1950, p 42

159 Ibid, pp 43–45

160 Notes on the reverse side of photograph of Broom, Barlow, Sithole and Jacobus, Ditsong Museum, as quoted in Jacobs, *Birders of Africa: History of a Network*, 2016, pp 189–190

161 To read more on this theme, see Shepherd, "Disciplining Archaeology", 2002 and Shepherd, "When the Hand that Hold the Trowel is Black", 2003, as well as Ndlovu, "Transformation Challenges in South African Archaeology", 2009 and Ndlovu, "De-colonizing the Mindset", 2011.

162 I first learned about Saul Sithole from Francis Thackeray of the Evolutionary Studies Institute in a discussion on 12 February 2014. Bob Brain mentioned Saul Sithole in his 1981 book *The Hunter or the Hunted?* (p 193). The information I gathered about Saul Sithole's life came predominantly from a chapter in Nancy Jacobs' book *Birders of Africa*, 2016. Jacobs and I began a correspondence in 2014 before the book was published. I then learned about Sithole's life from a meeting with his daughter, Zondi Zitha, on 21 November 2014.

163 The sources for the information about Saul Sithole are the same as those found in note 162.

164 Broom, *Finding the Missing Link*, p 45

165 Ibid, pp 49–51

166 Ibid, pp 49–51

167 Ibid. Also, interview with Francis Thackeray on 12 February 2014

168 Ibid, p 51

169 Ibid, p 61

170 Smuts, "Preface" to *The South African Fossil Ape-Men*, 1946, pp 5–6

171 Smuts as quoted in Schlanger, "Making the Past for South Africa's Future", 2002, p 207. Smuts wrote to Gillett on 21 January 1945 and spoke to the Institute of Race Relations in 1942.

172 Broom and Schepers, *The South African Fossil Ape-Men*, 1946, p 22

173 Dart, *The Osteodontokeratic Culture of Australopithecus Prometheus*, 1957, p viii from the Preface written in June 1956

174 It is astounding to see how many recent books and articles discuss the Dart Procedures and Dart's work on posture and poise. The Dart Procedures, which Dart developed for his son Galen, have had an impact in the field of dance and bodily poise. These books include Simmons, "Teaching with the Dart Procedures", 2004; Wheelhouse, "Dart and Alexander", 2004; Nettle-Fiol and Varnier, *Dance and the Alexander Technique*, 2011; Snook, "Dance and the Alexander Technique", 2012; Foley, "Towards a Neurophysiology of the Alexander Technique", 2012; and Barsky, "Dance and the Alexander Technique", 2013.

175 Tobias, *Into the Past*, 2005, pp 211–212

176 Dart and Craig, *Adventures with the Missing Link*, p 91 and Tobias, *Into the Past*, p 37

177 Tobias, *Into the Past*, pp 50–53

178 Tobias, "Some Little Known Chapters in the Early History of the Makapansgat Fossil Hominid Site", 1997

179 Lewis, *Bones of Contention*, 1997, pp 75–77

180 Sigmon, "The Making of a Palaeoanthropologist: John T. Robinson", 2007, p 213

181 Broom, *Finding the Missing Link*, p 62

182 Ibid, pp 62–63

183 Ibid, pp 64–65

184 Ibid, p 66

185 Ibid

186 Ibid, p 65

187 Weiner, *The Piltdown Forgery*, 1955

188 Ibid

189 Ibid, and McKie, "Piltdown Man: British Archaeology's Greatest Hoax", 2012

190 Tobias, "An Appraisal of the Case against Sir Arthur Keith", 1992, p 5

191 Dart and Craig, *Adventures with the Missing Link*, p 119

192 Dart quoted in an interview with Jane Dugard in the early 1980s

193 Gould, "The Piltdown Conspiracy", 1983

194 Tobias, "An Appraisal of the Case against Sir Arthur Keith", 1992, p 1

195 Tobias, *Into the Past*, 2005, p 211

196 Ibid, p 11

197 Ibid, p 25

198 Ibid, p 65

199 Ibid, p 84

200 Ibid, pp 58–62

201 UNESCO Statement on Race 1950,1969, p 30

202 UNESCO Statement on Race 1951, 1969, p 36

203 To view the full text of the Population Registration Act of 1950, see www.disa.ukzn.ac.za/webpages/ DC/leg19500707.028.020.030/leg19500707.028.020.030.pdf

204 Tobias, "The Problem of Race Determination, 1953, p 113

205 Ibid, p 116

206 Ibid, p 117

207 Ibid, p 121

208 Ibid, pp 122–123

209 Weidenreich, "Facts and Speculations Concerning the Origin of Homo Sapiens", 1947, p 202

210 Sigmon, "The Making of an Paleoanthropologist", 2007

211 Teilhard de Chardin as quoted in Sigmon, "The Making of an Paleoanthropologist", 2007, p 339

212 For further discussion of the absence of teaching evolution in schools under apartheid, see Esterhuysen, "Evolution: 'The Forbidden Word'", 1998, and Lever, "Science, Evolution and Schooling", 2002

213 "Challenge to D.R.C. Pastors on Evolution of Man", 18 April 1952, Newspaper Clipping File at Ditsong Museum Library

214 *Sunday Times*, "Scientists to Reply to Church Over Apeman Theory", 28 September 1952

215 Sigmon, "The Making of a Palaeoanthropologist", pp 242–243

216 *Rand Daily Mail*, "And Now the Maps, Mr. De Beer", 15 September 1952

217 Tobias, with Goran Strkalj and Jane Dugard, *Tobias in Conversation*, 2008, pp 170–171

218 Tobias, *Into the Past*, 2005, p 237

219 Ibid, pp 159–162

220 Tobias, "Fifteen Years of Study on the Kalahari Bushmen or San", 1975, p 74

221 Tobias, *Into the Past*, p 164

222 Interview, Hertha de Villiers, 25 September 2014

223 De Villiers to Tobias, letters in 1955 and 1956, Tobias papers, Wits Archives

224 De Villiers 1968

225 Interviews with De Villiers, September 2014

226 Ibid

227 Singer, "The Boskop 'Race' Problem", 1958, pp 173–174

228 Singer was quoting Tobias in "Physical Anthropology and Somatic Origins of the Hottentots", 1955 and in Singer, "Presidential Address 1962", 1962, p 207

229 Singer, "The Boskop 'Race' Problem", 1958, p 177

230 Tobias, *Tobias in Conversation*, p 227

231 Simpson as quoted in Tattersall, *The Fossil Trail*, 1995, p 90

232 Tattersall, *The Fossil Trail*, p 96

233 Morris, "Biological Anthropology at the Southern Tip of Africa", 2012, pp S158-S159

234 Interview with De Villiers, September 2014 and letter from De Villiers to Tobias, 1955, Tobias papers, Wits Archives

235 Tobias to De Villiers, 6 December 1965 and De Villiers to Tobias, personal memo, 8 December 1965, Tobias papers, Wits Archives

236 Tobias, *Tobias in Conversation*, pp 117–118

237 Ibid

238 Tobias, *Into the Past*, 2005, p 81

239 Sigmon, "The Making of a Palaeoanthropologist", 2007, pp 273–274

240 Interview with De Villiers, 9 April 2015

241 Sigmon, "The Making of a Palaeoanthropologist", p 270

242 Ibid, pp 275–276

243 Ibid, p 271

244 Information about Robinson's work with Geoffrey Hodson was gathered from Sigmon's "The Making of a Palaeoanthropologist", personal communication with Becky Sigmon and Francis Thackeray, several theosophy websites as well as an online article by Soren Hauge.

245 Luthuli, *Let My People Go*, 1963

246 Tobias speaking at the Stony Brook Symposium about *Homo habilis* in 2006

247 Ibid

248 Ibid and Tobias, *Tobias in Conversation*, 2008, p 127

249 Tobias, *Tobias in Conversation* 2008, p 129 and personal communication with Bob Brain, 2015

250 Coon, *The Origin of Races*, 1962, p 656

251 Ibid, Introduction, p vii

252 Tobias, *Into the Past*, 2005, p 125

253 Coon, *The Origin of Races*

254 Tobias, "The Meaning of Race", 1961, p 1

255 Ibid

256 Ibid, p 3

257 Ibid

258 Ibid, p 8

259 Ibid, p 24

260 De Villiers, "The Tablier and Steatopygia in Kalahari Bushwomen",1961, p 223

261 Tobias, "Fifteen Years of Study on the Kalahari Bushmen or San", 1975, p 76

262 Tobias as quoted in Humphreys, "Archaeologists in District Uncover Links with Past", *Diamond Fields Advertiser*, 3 June 1961

263 The Institute of Race Relations Report of 1957–58 as quoted in Waldman, "Klaar Gesnap As Kleurling", 2006, p 177

264 Waldman, "Klaar Gesnap As Kleurling", 2006, pp 178–180

265 Tobias as quoted in Humphreys, "Science Spotlight Plays on Griqua Remains", *Diamond Fields Advertiser*, 2 June 1961

266 Fock, "Report on 1961 Campbell Excavation, submitted to Phillip Tobias", 19 June 1961, handwritten, copy in Alan Morris papers

267 Humphreys, "Archaeologists in District Uncover Links with Past", 3 June 1961

268 Fock, "Report on 1961 Campbell Excavation", 1961

269 Humphreys, "Science Spotlight plays on Grimqua Remains", 1961

270 Singer, "Presidential Address 1962", 1962, p 205

271 Ibid

272 Jackson, "'In Ways Unacademical': The Reception of Carleton S. Coon's *The Origin of Races*", 2001, p 248

273 Putnam as quoted in Jackson, "'In Ways Unacademical'", p 253

274 Putnam as quoted in Jackson, "'In Ways Unacademical', p 258

275 Ibid, p 262

276 Mead as quoted in Jackson 2001, , "'In Ways Unacademical', p 270

277 Jackson, "'In Ways Unacademical', p 272

278 Jackson, "'In Ways Unacademical', p 253 and Dubow, "Racial Irredentism, Ethnogenesis, and White Supremacy in High-Apartheid South Africa", 2015, p 7

279 Coon to Gayre in November 1962 as quoted in Jackson, "'In Ways Unacademical', p 254

280 Robinson quoted in the *Sunday Times*, 23 September 1962

281 Jackson, "'In Ways Unacademical'", 2001 and Dubow, Racial Irredentism, Ethnogenesis, and White Supremacy", 2015

282 Tobias, *Into the Past*, 2005, p 143

283 Washburn as quoted in Jackson, "'In Ways Unacademical'", 2001, p 276

284 Robinson letter to Joe Weiner dated 29 October 1962 in Box 305, JT Robinson Correspondence 1960-62, Ditsong Museum Library, and *Pretoria News*, 20 July 1963

285 Minutes of the annual meeting of the Board of Trustees, Transvaal Museum, 28 May 1963 in Box 491: Directors Monthly Reports 1960-1964

286 *Pretoria News*, "Exhibit on Evolution Must Go, Says De Klerk", 24 July 1963 and "Museum Will Oppose a Removal of Exhibits", 25 July 1963. All of the newspaper articles about the 1963 opposition to the evolution exhibit can be found in the Transvaal Museum Press Cuttings notebook entitled: Palaeontology, 15 August 1964–22 April 1967, Ditsong Museum Library, Pretoria.

287 Tobias quoted in *Pretoria News*, "Minister's Plea Criticised", 29 July 1963. Fitzsimons quoted in *Rand Daily Mail*, "State Museum Tangle over Evolution", 26 July 1963

288 Meester quoted in *Pretoria News*, "Another Expert Lost to Museum", 20 August 1963, and *Sunday Times*, "Scientist who Defied De Klerk 'ban' on Evolution Quits", 25 August 1963

289 *Pretoria News*, "New Move in Controversy", 31 August 1963

290 *Rand Daily Mail*, "De Klerk in 'Dark' about Evolution Decision", 15 November 1963 and *Pretoria News*, "Man–Ape Show Likely to Stay at Museum", 18 November 1963

291 *Pretoria News*, "Evolution Display Will Remain", 18 December 1963

292 *Pretoria News*, "Museum Exhibit Still Offends: De Beer", 24 December 1963

293 De Villiers, "The Morphology and Incidence of the Tablier in Bushman, Griqua and Negro Females", 1968, pp 2–3

294 Ibid, p 5

295 Ibid, pp 5 and 10

296 Interview with Trefor Jenkins, 28 February 2014

297 Ibid

298 Jenkins as quoted in Posel, "Race as Common Sense", 2001, p 107

299 This section of the chapter was written based on documents provided to me by Phillippa Yaa de Villiers including: Adoption consent papers signed by S Alcock, 21 February 1966; Case Report written by Marie M. Arendt, Adoption Secretary, Child Welfare Society of Johannesburg, dated 25 February–2 May 1966; Letter on University of the Witwatersrand Medical School letterhead written and signed by Dr. Hertha De Villiers stating that she had examined Tandy (sic) Alcock, dated 5 August 1966; and Case Summary written by I. Albrecht, Department of Social Welfare, 11 August 1966.

300 Wells, "One Hundred Years: Robert Broom 30 November 1866", 1966

301 Dart's Preface in Findlay, *Dr. Robert Broom*, 1972 (no page numbers)

302 Norman Broom to George Findlay, 22 September 1966, Robert Broom papers, Ditsong Museum Library

303 Broom, Norman, Robert Broom Memorial, December 1966

304 Interview with Ron Clarke, 28 May 2014

305 Ibid

306 Tobias, "The Emergence of Man in Africa", 1968, pp 3–7

307 Ardrey, Robert to Raymond Dart, letter dated 20 November 1960, Dart papers, Wits Archives

308 Dart, "The Predatory Transition from Ape to Man", 1953 and Ardrey, *African Genesis*, 1961, p 29

309 Interview with Bob Brain, Irene, Pretoria, 6 June 2014

310 Dart and Craig, *Adventures with the Missing Link*, pp 91–108 and Esterhuysen, "Excavation at Historic Cave, Makapan's Valley", 2010, p 67

311 Ardrey, "A Slight (Archaic) Case of Murder", 1955, p 34

312 Ardrey, "South Africa: A Personal Report", 1958, p 23

313 Ardrey, "A Slight (Archaic) Case of Murder", pp 34–35

314 Ibid, p 35

315 Ardrey, *African Genesis*, 1961, p 11

316 Ibid, p 12

317 Ibid, p 318

318 Erlich, "Strange Odyssey", 1976

319 This scene is from Kubrick's film *2001: A Space Oddyssey*. See YouTube video of the "Dawn of Man" scene: www.youtube.com/watch?v=Sj_Jxqi4G3s

320 Clarke, as quoted in *Making of a Myth*, 1976

321 Rice, *Kubrick's Hope: Discovering Optimism from 2001 to Eyes Wide Shut*, 2008, p 28

322 Kubrick, letter to the *New York Times*, 27 February 1972

323 Ardrey, *African Genesis*, 1961, p 324

324 Ardrey, *The Territorial Imperative*, 1967, p 316

325 Coon as quoted in a list of "Controversial Comments Received During the First Month After Publication of *The Territorial Imperative*", in Dart's papers, clearly sent to him by Ardrey, Wits Archives

326 Dart to Price, letter dated 28 May 1970, Dart papers, Wits Archives

327 Tobias as quoted in Webster, "Robert Ardrey Dies; Writer on Behaviour", 1980

328 Interview with Brain, Irene, Pretoria, 6 June 2014

329 Ibid

330 Ibid

331 In his Introduction to his 2004 book, *Swartkrans*, Brain explains the four seven-year periods that it took to excavate Swartkrans.

332 Interview with Brain, 6 June 2014

333 Brain, *Swartkrans: A Cave's Chronicle of Early Man*, 2004, p 2

334 Interview with Brain, 6 June 2014

335 Brain, *The Hunters or the Hunted?*, 1981

336 Ibid, pp 21–26

337 *Time* magazine, 7 November 1977, p 78

338 To read more about this interpretation of the Bushmen as timeless, see Gordon, *The Bushmen Myth*, 1992

339 Interview with Brain, 6 June 2014 and in Bonner, Esterhuysen and Jenkins, *In a Search for Origins*, 2007, pp 118–120

340 Interview with Brain, 6 June 2014

341 Ibid and Bonner, Esterhuysen and Jenkins, *In a Search for Origins*, pp 118–120

342 *Time* magazine, "Puzzling Out Man's Ascent", 7 November 1977, pp 64–78

343 Tattersall, *The Fossil Trail*, 1995, pp 141–145

344 *Time* magazine, "Puzzling Out Man's Ascent", 7 November 1977, p 67

345 Tattersall, *The Fossil Trail*, p 139

346 *Time* magazine, "Puzzling Out Man's Ascent", 7 November 1977, p 67

347 Ibid, p 78

348 *Time* magazine, "Loneliness is an Enemy", 7 November 1977, pp 36–37

349 Interview with Jane Dugard, 26 May 2016, interview with Job Kibii on 27 September 2016, and informal discussion with Makhosazana Xaba on 9 June 2016

350 Mancqu, *Biko*, and "Stephen Bantu Biko", South African History Online

351 Jenkins, "From Generation to Generation", 1978

352 Ibid, p 8

353 Ibid, p 5

354 Brain, "Visitor Reaction to the 'Life's Genesis' Display Sequence at the Transvaal Museum", 1979, p

248

355 Brain, "Naturalist", 1982, p 42

356 Meredith, *Born in Africa*, 2011, p 121

357 Ree, "Evolution by Jerks", 2003, p 2

358 Vrba as quoted in Meredith 2011, *Born in Africa*, p 124

359 Vrba as quoted in Chatwin, *The Songlines*, 1987, p 249

360 Ibid

361 Morris, unpublished autobiography

362 Interview with Phillippa Yaa de Villiers, Braamfontein, 20 November 2015

363 Wood, *The Evolution of Early Man*, 1976, p 112

364 Ward, *The Star*, "Research Sets Black 'Presence' Far Back in pre-History", 13 June 1979

365 Tobias as quoted in Ward, "Research Sets Black 'Presence' Far Back in pre-History"

366 Tobias, *The Bushmen: San Hunters and Herders of Southern Africa*, 1978, p 2

367 Dart's Foreword in Tobias, *The Bushmen: San Hunters and Herders of Southern Africa*, 1978 (no page numbers)

368 Tobias in Editor's Foreword of Tobias, *The Bushmen: San Hunters and Herders of Southern Africa* (no page numbers)

369 Tobias, *The Bushmen: San Hunters and Herders of Southern Africa*, p 18

370 Ibid, p 31

371 Chatwin as quoted in Shakespeare, *Bruce Chatwin*, 1999, p 2

372 Ibid, p 3

373 Brain, *Swartkrans: A Cave's Chronicle of Early Man*, 2004, p vi

374 Chatwin, *The Songlines*, 1987

375 Ibid, p 255

376 Ibid, p 251

377 Interview with Jane Duncan, 26 May 2016

378 Tobias to Gould, letter dated 24 October 1980, Folder 9, Box 890, Gould papers, Stanford University

379 Gould, "The Hottentot Venus", 1985, p 294

380 Gould Journal 1984, Box 468, Folder 2, Travel Notebook "Southern Hemisphere Book", courtesy of the Department of Special Collections, Standford University Libraries

381 Ibid

382 Ibid

383 Ibid

384 Thanks to Alan Morris for giving me a copy of the November 1985 *National Geographic* cover.

385 Tobias, "History of Physical Anthropology in Southern Africa", 1985, p 234

386 Tattersall, Van Couvering and Delson, "The 'Ancestors' Project: An Expurgated History", 1985, p 4

387 Sullivan, "Far-Flung Fossils Gathered for Exhibit", 1984

388 Tobias, *Into the Past*, 2005, p 235

389 Messinger as quoted in Freedman, "Rift Over Fossils From South Africa", 30 May 1984

390 Ibid

391 Ibid

392 Ibid

393 Freedman, articles in the *New York Times*, 1 June 1984 and 12 June 1984. Although Freedman's article refers to the decision made by the board of trustees to put up the signs, an email exchange from 5–6 May 2016, with both Ian Tattersall and Eric Delson, who were working on the "Ancestors" exhibit at the American Museum of Natural History back in 1984, concluded that they could not recall whether the signs, in the end, were put up or not. Tattersall said, "But it was a long time ago, and just because we can't recall everything that happened at a very eventful time you can't assume it didn't, either!" Also, see Tattersall, Van Couvering and Delson, "The 'Ancestors' Project: An Expurgated History", 1985. Thanks to Eric Delson for sharing this article with me.

394 Interview with Phillippa Yaa de Villiers on 4 March 2015, and "New evidence on Negro Origins", in *New Scientist*, 13 January 1983. Also, see De Villiers and Fatti, "The Antiquity of the Negro", 1982, p 332

395 De Villiers to Tobias, 14 February 1980 and Tobias to De Villiers, 14 July 1980 and 1 August 1980

396 Wood interview with Tobias, 1989, p 223

397 De Villiers as quoted in Nevill, "Skeletons 'Don't Move Me'", 1989

398 Tobias, with Goran Strkalj and Jane Dugard. *Tobias in Conversation*, 2008, p 125

399 Weidenreich (1943) as quoted in Wolpoff, WuXin and Thorne, "Modern *Homo sapiens* Origins",

1984, p 423

400 Wolpoff, WuXin and Thorne, "Modern *Homo Sapiens* Origins", 1984, p 469

401 Ibid, p 471

402 Stringer and Andrews, "Genetic and Fossil Evidence", 1988, p 1263

403 Stringer, "Palaeontology: The 100-year Mystery of Piltdown Man", 2012, p 25

404 Tobias, "Race", in *The Social Science Encyclopedia*, 1985, reprinted in Tobias, *Images of Humanity: The Selected Writing of Phillip V Tobias*, 1991, p 346

405 Ibid, p 350

406 Ibid, p 349

407 Tobias, "History of Physical Anthropology in Southern Africa", 1985, p 32

408 Morris, "Biological Anthropology at the Southern Tip of Africa", 2012, p S153

409 Interview with Himla Soodyall, 28 May 2015

410 Ibid

411 Ibid

412 Ibid

413 Cann, Stoneking and Wilson, "Mitochondrial DNA and Human Evolution", 1987

414 Ibid

415 Tierney, "The Search for Adam and Eve", *Newsweek*, 1988

416 Ibid, p 2

417 Ibid

418 Ibid, pp 2–3

419 Ibid, p 3

420 Ibid, p 5

421 Cann, "Mothers, Labels and Misogyny", 1997

422 Tierney, "The Search for Adam and Eve", p 5

423 Lecture given by Himla Soodyall at the Origins Centre, 28 February 2015

424 Tobias, with Goran Strkalj and Jane Dugard. *Tobias in Conversation*, 2008, p 245

425 Berger and Hilton-Barber, *In the Footsteps of Eve*, 2000, p 51

426 Ibid

427 Ibid, p 63

428 Tobias, with Goran Strkalj and Jane Dugard. *Tobias in Conversation*, p 245

429 Interview with Lee Berger, 2015

430 Jacobs, *Birders of Africa*, 2016, pp 190 and 203; interview with Zondi Sithole Zitha

431 Clarke, "Fossil Find Brings Our Roots Closer", 2010 and Interview with Lee Berger, 21 October 2015

432 Interview with Berger, 21 October 2015

433 Brain, "A perspective on the PAST", 2003, p 1

434 Tobias, "Ad Hominidae: The Future of South Africa's Ancient Past", 1994, p 35

435 Ibid, p 36

436 Wolpoff et al., "Multiregional Evolution", 1994

437 Tobias, "The Bearing of Fossils and Mitochondrial DNA on the Evolution of Modern Humans", 1995, p 164. Thank you to Ian Tattersall for pointing me towards this article.

438 Ibid, p 161

439 Ibid, p 163

440 This story has been recounted in numerous books and articles. I gathered details from interviews at Sterkfontein with Ron Clarke in March, May and August 2014. Meredith, *Born in Africa*, 2011 also tracks the story, beginning on p 147.

441 Clarke and Tobias, "Sterkfontein Member 2 Foot Bones of the Oldest South African Hominid", 1995

442 Interviews with Clarke in May and August 2014, and Meredith, *Born in Africa*, 2011, p 148

443 Interview with Berger, November 2014

444 Invitations, from Tobias and the Griquas of Adam Kok V dated July 1996

445 Kok V and Tobias as quoted in Koch, "Griquas Want Chief's Bones Back", *Mail & Guardian*, 2 February 1996

446 Engelbrecht, "The Connection Between Archaeological Treasures", 2002, p 243

447 Anthony JB Humphreys, personal communication, 9 June 2015

448 Tobias, "Saartje Baartman", 2002, p 110

449 Abrahams in the 2002 documentary film "The Return of Sarah Baartman", directed by Zola Maseko

450 Soodyall, interview on 28 May 2015

451 Morris to Mr M Kibunjia, National Museums of Kenya, 1 February 1996

452 Morris, *The Skeletons of Contact*, 1992, p 157
453 Ibid
454 Ibid, p 2
455 Morris, "A Master Catalogue: Holocene Human Skeletons from South Africa", 1992. The quote is from Morris, "Searching for 'real' Hottentots, 2008, p 229
456 Morris to Dart, letter dated 6 September 1984, in the Dart papers, Wits Archives. Morris told me that Dart did not respond during a discussion we had in Hout Bay on 21 April 2015.
457 Thabo Mbeki's "I am an African" speech, 8 May 1996
458 Thank you to Sunet Swanepoel of the McGregor Museum for sending me the text of both speeches of Phillip Tobias and Adam Kok V from the 1996 ceremony, McGregor Museum Library
459 Text of Adam Kok V's speech, McGregor Museum Library
460 Makele, "'Pure' Bones Tell All", *City Press*, 1996
461 Barbour to Morris, letter dated 30 October 1996, Alan Morris papers
462 Ibid
463 Morris to Barbour, letter dated 16 October 1996, Alan Morris papers
464 Tobias to Morris, letter dated 16 September 1995, Alan Morris papers
465 Morris to Tobias, letter dated 20 October 1995, Alan Morris papers
466 Tobias to Morris, letter dated 10 August 2000, Alan Morris papers
467 Makele, "'Pure' Bones Tell All", 1996
468 Mike Raath, in an interview in Kommetjie, 22 April 2015, told me about the small coffin he had constructed.
469 Tobias speech read at the burial ceremony on 23 September 2007, dated 18 September 2007, McGregor Museum Library
470 Tobias speaking in the 2002 documentary "The Life and Times of Sara Baartman", directed by Zola Maseko
471 Meredith, *Born in Africa*, 2011, p 149
472 This reconstruction of the scene was compiled by interviews with Ron Clarke in May and August 2014 and Stephen Motsumi on 8 September 2014. Also Meredith 2011, p 149
473 Clarke as quoted in Meredith, *Born in Africa*, p 149
474 Berger and Hilton-Barber, *In the Footsteps of Eve*, 2000, pp 248–249; and interview with Clarke in May and August 2014.
475 Interview with Clarke, August 2014
476 Interview with Clarke, June 2015
477 Ibid
478 The information in the proceeding pages about Motsumi's life was gathered in an interview with Motsumi, 8 September 2014
479 Ibid
480 Ibid
481 Phone conversation with Motsumi, 18 September 2014
482 Interview Ron Clarke
483 Interview with Berger, 21 October 2015
484 McHenry and Berger, "Body Proportions in *Australopithecus afarensis*", 1998
485 Gauteng Provincial Government: Department of Agriculture, Conservation and Environment, "Application for Inclusion on the World Heritage List: The Fossil Hominid Sites of Sterkfontein, Swartkrans, Kromdraai and Environs, South Africa," 16 June 1998 www. whc.unesco.org/uploads/nominations/915bis.pdf
486 Ibid
487 Discussion with Francis Thackeray in 2015, reference in Thackeray's unpublished manuscript "Fleeting Moments", 2014, as well as an interview with Rob Adam on 17 March 2016
488 Both Ron Clarke and Roger Jardine quoted Tobias as saying this to Mbeki at the cabinet meeting in Pretoria in September 1998.
489 De Villiers, "Secrets and Lies", *Elle*, March 1999, p 61
490 Ibid, pp 61–62
491 Ibid and interview with Phillippa Yaa de Villiers on 4 March 2015
492 Ibid, p 64
493 *Tobias's Bodies*, Episode Two, six-part film series, directed by Guy Spiller, produced by Harriet Gavshon, David Jammy at Curious Pictures (now Quizzical Pictures) and SABC Education, 2002
494 See Bystrom, "The DNA of the Democratic South Africa", 2009 and Erasmus, "Throwing the

Genes", 2013 for further discussion of problems with the concept of identiy as linked to biology

495 Soodyall at Lecture at Origins Centre, 28 February 2015

496 Soodyall in *Tobias's Bodies*, Episode Two, 2002

497 Morris, "Ancient DNA Comes of Age", 2015, p 1

498 Tobias, "Saartje Baartman", 2002, p 109

499 Soodyall, "History, Ancestry and Genes", 2011, p 35

500 Mbeki, "Speech at the funeral of Sarah Bartman", 2002

501 Tobias, "Saartje Baartman", 2002, p 108

502 Legassick and Rassool, *Skeletons in the Cupboard*, 2000

503 For further discussion of the impact of *Skeletons in the Cupboard*, see the postscript in the updated edition of the book published in 2015, p 106. In addition to the impact on museums in South Africa, the postscript discusses the impact on museums in Vienna, Austria and the return of the remains of Klaas and Trooi Pienaar to South Africa. The official reburial took place in Kuruman on 12 August 2012.

504 The date of Ouma /Una's visit was determined by her note on the back of Abraham's photograph in Brendon Billings' office. Nigel Crawhall confirmed the visit (personal communication, 24 August 2015).

505 Personal communication with Nigel Crawhall, August 2015; Quotes from Ouma /Una Rooi and Annetta Bok from the Hugh Brody-directed film *Tracks Across Sand*, 2012

506 Mbeki's quote and the description of the scene at Askham come from Ancer, "San Trying to Marry Old Ways with New," and the DVD *Tracks Across Sand*.

507 Crawhall personal communication, 24 August 2015

508 McHenry and Berger, "Body Proportions in *Australopithecus afarensis* and *A*. Africanus", 1998, p 20

509 White, "A View on the Science", 2000, p 288

510 Berger and Hilton-Barber, *In the Footsteps of Eve*, 2000

511 Ibid, p 306

512 Ibid, p 201

513 Kuykendall, "Never Letting the Facts Get in the Way of a Good Story", 2001, p 177

514 Wood, "Chalk and Cheese", 2002, p 503

515 White, "Robbing the Cradle", 2002, p 516

516 Ibid

517 Maguire, "A Second Opinion", 2002, p 519.

518 Interview with Lee Berger, Killarney, Johannesburg, 21 October 2015

519 Brain, "A perspective on the PAST", 2003, p 2

520 Berger with Hilton-Barber, *In the Footsteps of Eve*, 2000, p 226, as well as an interview with Ron Clarke in May 2014. Berger confirmed, in an interview on 21 October 2015, that he thought Clarke honestly believed that he (Berger) would take over Little Foot, but he denied that that was the case.

521 White, "A View on the Science", p 290

522 Mbeki, Foreword to to *Field Guide to the Cradle of Humankind*, 2002, p 5

523 Interview with Rob Adam, 17 March 2016

524 Scott, *Rethinking Evolution in the Museum*, 2007, p 49, and Pillay, "A Critical Evaluation of Representations of Hominin Evolution", 2010, pp 24–25

525 Gould, "Ladders and Cones", 1997, pp 42–49

526 Interview with Lara Mallen, 10 March 2015

527 References about the *Sediba* find include the Wits University press release dated 8 April 2010; Balter, "Candidate Human Ancestor From South Africa", 2010; Barron, "So Many Questions", 2010; Cherry, "Claim Over 'Human Ancestor' Sparks Furore", 2010; Morris, "New Hominin Fossils From Malapa", 2010; Smillie, "Boy Wonder Found, 2010; Thompson, "Sediba Rocks the Cradle", 2010; *The Australian*, 9 April 2010; *The Star*, 9 April 2010, and the scientific paper by Berger et al, 9 April 2010; as well as interview with Kibii, 27 September 2016

528 Lee Berger in a YouTube video "Google Earth and Human Evolution", 22 November 2012 and a Google Earth blog, "Google Earth Helps Discover Rare Hominid Ancestor in South Africa", 8 April 2010

529 Berger as quoted in Wits University press release, 8 April 2010

530 Graber, "Synchrotron Focuses on New Hominid Fossil", 2010, and Eureopan Synchrotron Radiation Facility (ESRF) press release 12 April 2010, as well as interview with Job Kibii, 27 September 2016

531 For a longer discussion of these issues around land, please see Bonner, "The Myth of the Vacant Land", 2007 and Mason, "Early Iron Age Settlement at Broederstroom, 1981

532 Tobias in *Tobias's Bodies*, Episode Six, 2002

533 "A Story of Two Boys", *The Star*, as quoted in Worby, "And a Little (White) Child Shall Lead Them", 2014

534 Desmond Tutu speaking on YouTube, "World Cup 2010 Ceremony Archbishop Desmond Tutu Show", www.youtube.com/watch?v=ui_paCD4rKg

535 Morris, "On Human Evolution, *Australopithecus sediba* and Nation Building", 2011, p 3

536 Berger, "Building a nation one project at a time", 2012, p 1

537 Interview with Dr Nonhlanhla Vilakazi in Johannesburg on 13 March 2015

538 Phone interview with Winnie Dipuo Mokokwe (now Kgotleng) on 26 September 2014

539 Worby's presentation at the Archaeological Society was in Linden, Johannesburg on 23 August 2014.

540 Discussion with Frank Motlala on 2 September 201.

541 Bonner, Esterhuysen and Jenkins, *A Search for Origins*, 2007, p 277

542 Schlebusch et al, "Genomic Variation in Seven Khoe-San Groups", 2012

543 Interview with Soodyall, 25 April 2016

544 To read more about the tenacity of biology in charting what it means to be human, read Erasmus, "Throwing the Genes, 2013

545 The visit to the Tobias Laboratory with Bernhard Zipfel took place on 26 March 2014

546 Interview with Dr Mirriam Tawane, 10 October 2014

547 Ibid

548 The trip to Taung and Buxton took place on 3, 4 and 5 May 2015. I travelled with Dr MirriamTawane, Ian McKay, Brian Mogaki, Sifelani Jirah and Amanda Mufuniwa Mudau of the Evolutionary Studies Institute.

549 Personal communication from Elliott on 24 February 2015

550 Schultz, "Shackleton Probably Never Took Out an Ad", 2013

551 References for the story of the Rising Star cave and expedition include daily blogs by Andrew Howley of the National Geographic Society in November 2013; Smillie, "Historic Unearthing of Hominin Fossils", 2013; Wild, "Skinny Scientists at Palaeo Rock Face", 2013; Brahic, "Small Women Take Giant Leap for Early Man", 2014; Farber, "From Obscure Caveman to Global Rock Star", 2015; Mapumulo, "A Lucky Friday 13th", 2015; Shreeve, "Mystery Man", 2015; and Yong, "Six Tiny Cavers, Fifteen Odd Skeletons", 2015, and an interview with Marina Elliot on 4 June 2015 and an interview with Wayne Crichton, 4 June 2015.

552 Interview with Wayne Crichton, 4 June 2015

553 See note 551

554 Interviews with Rising Star Workshop participants Lauren Schroeder and Rebecca Ackerman, Cape Town, 3 December 2014; and Kibii interview, Braamfontein, 27 September 2016

555 Interview with Berger, 21 October 2015

556 Shreeve, "Mystery Man", 2015, p 43

557 Ibid p 53

558 Ibid

559 The proceedings of the *Homo naledi* announcement can be found online at "Media Briefing: New Fossil Find Unveiled in Maropeng, Johannesburg", streamed live on 10 September 2015, www.youtube.com/watch?v=QiiOJ4Y9ZLo

560 Phone conversation with Ron Clarke, 19 September 2015

561 Kahn, "Big Little Foot Feat was a Labour of Love", *Business Day*, 20 October 2015

562 Bega, Small Step in SA, a Giant Step for Science", 2015

563 Germaner, "Homo Naledi was no Baboon", 2015. Other articles that covered the debate about *Homo naledi* and racism include Lindeque, "Homo Naledi Debate", 2015; Van Onselen, "Homo Naledi: Unearthing SA's Great Hour of Science, 2015; Petersen, "I am No Grandchild of Any Ape, Monkey or Baboon", 2015; and an interview with Mathole Motshekga by Tim Modise on Transformation, 24 September 2015

564 Ibid

565 Lindeque, "Homo Naledi Debate", 2015

566 Lee Berger interviewed by Redi Tlhabi on Radio 702 on 14 September 2015

567 Morris, "Ancient DNA Comes of Age", 2015

568 Ibid

569 Lelliott, "Visitors' Views of Human Origins", 2016, p 132

570 Lloyd, "Job Kibii is Close to the Bones," 2011

571 Interview with Winnie Dipuo Mokokwe Kgotleng, 13 May 2016

Bibliography and Sources

Books and Articles

Abrahams, Yvette. "The Great Long National Insult: 'Science', Sexuality and the Khoisan in the 18th and early 19th Century." *Agenda: Empowering Women for Gender Equity*, 13, no. 32 (1997): 34–48

Abrahams, Yvette. "Colonialism, Dysfunction and Disjuncture: Sarah Bartmann's Resistance (remix)." *Agenda: Empowering Women for Gender Equity*, 17, no. 58 (2003): 12–26

Adam, Rob M. "Choosing Good Science in a Developing Country," in *Fundamental and Applied Aspects of Modern Physics*, edited by S H Connell and R Tegen. Singapore and London: World Scientific Publishing, 2001

Adams, Sian. "Archbishop Desmond Tutu at Maropeng", 1 August 2013. Available at: www.maropeng.co.za/news/entry/archbishop-desmond-tutu-at-maropeng-following-the-footsteps-of-humankind (accessed on 2 June 2016)

Alfred, Mike. *Twelve + One: Some Jo'burg Poets: Their Artistic Lives and Poetry*, Interview with Phillippa Yaa de Villiers. Braamfontein: Botsotso Publishing, 2014

Ancer, Jonathan, "San Trying to Marry Old Ways with New", *Independent Online*, 2 November 2004. Available at: www.iol.co.za/news/south-africa/san-trying-to-marry-old-ways-with-new-225778 (accessed on 21 August 2016)

Ardrey, Robert. "A Slight (Archaic) Case of Murder." Views and Reviews, *The Reporter*, 5 May 1955

Ardrey, Robert. "South Africa: A Personal Report." *The Reporter*, 27 November 1958

Ardrey, Robert. *African Genesis: A Personal Investigation into the Animal Origins and Nature of Man*. London: Collins, 1961

Ardrey, Robert. *The Territorial Imperative: A Personal Inquiry into the Animal Origins of Property and Nations*. London: Collins, 1967

Baldwin, James. "On Being 'White' ... and other Lies", in *Black on White*, edited by David R Roediger. New York: Schocken Books, 1998

Balter, Michael. "Candidate Human Ancestor From South Africa Sparks Praise and Debate." *Science*, 328, no. 5975, (2010): 154–155

Balter, Michael. "Paleoanthropologist Now Rides High on a New Fossil Tide." *Science*, 333, no. 6048, (2011): 1373–1375

Barron, Chris. "So Many Questions", *Sunday Times*, 25 April 2010

Barsky, Marsha. "Dance and the Alexander Technique: Exploring the Missing Link." *Journal of Dance Education*, 13, no. 1, (2013): 30–31

Bega, Sheree. "Small Step in SA, a Giant Step for Science." *Saturday Star*, 12 September 2015

Beningfield, Jennifer. "Telling Tales: Building, Landscape and Narratives in Post-Apartheid South Africa." *Design*, 10, no. ¾, (2006): 223–234

Berger, Lee and Brett Hilton-Barber. *In the Footsteps of Eve: The Mystery of Human Origins*. Washington DC: National Geographic, 2000

Berger, Lee and Brett Hilton-Barber. *Field Guide to the Cradle of Humankind: Sterkfontein, Swartkrans, Kromdraai and Environs World Heritage Site*. Cape Town: Random House Struik, 2002

Berger, Lee R. "Building a Nation, One Project at a Time: Reply to 'On Human Evolution, *Australopithecus sediba* and Nation Building'." *South African Journal of Science*, 108, January/February, (2012): 1

Berger, Lee R and Marc Aronson. *The Skull in the Rock*. Washington DC: National Geographic, 2012

Berger, Lee R, Rodrigo Lacruz and Darryl J de Ruiter. "Brief Communication: Revised Age Estimates of *Australopithecus*-Bearing Deposits at Sterkfontein, South Africa." *American Journal of Physical Anthropology* 119, (2002): 192–197

Berger, Lee R, Darryl J de Ruiter, Steven E Churchill, Peter Schmid, Kristian J Carlson, Paul HGM Dirks and Job M Kibii. "*Australopithecus Sediba*: A New Species of *Homo*-Like Australopith from South Africa." *Science*, 328, no. 5975, (2010): 195–204

Berger, Lee R, John Hawks, Darryl J de Ruiter, Steven Churchill, Peter Schmid, Lucas K Delezene, Tracy 1 Kivell, Heather M Garvin, Scott A Williams, Jeremy M DeSilva, Matthew M Skinner, Charles M Musiba, Noel Cameron, Trenton W Holliday, William Harcourt-Smith, Rebecca R Ackermann, Markus Bastir, Barry Bogin, Debra Bolter, Juliet Brophy, Zachary D Cofran, Kimberly A Congdon, Andrew S Deane, Mana Dembo,

Michelle Drapeau, Marina C Elliott, Elen M Feuerriegel, Daniel Garcia-Martinez, David J Green, Alia Gurtov, Joel D Irish, Ashley Kruger, Myra F Laird, Damiano Marchi, Marc R Meyer, Shahed Nalla, Enquye W Negash, Caley M Orr, Davorka Radovcic, Lauren Schroeder, Jill E Scott, Zachary Throckmorton, Matthew W Tocheri, Caroline VanSickle, Christopher S Walker, Pianpian Wei and Bernhard Zipfel. "*Homo Naledi,* a New Species of the genus *Homo* from the Dinaledi Chamber, South Africa", *eLIFE*, 10 September 2015, pp. 1–35

Blundell, Geoffrey (ed.). *Origins.* Cape Town: Double Storey Books, Juta and Co, 2006

Blumenschine, Robert and Andrea Leenen. "What's New from the PAST?" *South African Journal of Science*, 112, no. 1/2, (2016): 7–9

Bonner, Philip, Amanda Esterhuysen and Trefor Jenkins (eds). *A Search for Origins: Science, History and South Africa's Cradle of Humankind.* Johannesburg: Wits University Press, 2007

Bonner, Philip. "The Myth of the Vacant Land", "The Racial Paradox: Sterkfontein, Smuts and Segregation", and "The Legacy of Gold", in *A Search for Origins*, edited by Philip Bonner, Amanda Esterhuysen and Trefor Jenkins. Johannesburg: Wits University Press, 2007

Bonner, Philip and Amanda Esterhuysen. "The Amazing Makapan", in *Origins*, edited by Geoffrey Blundell. Cape Town: Double Storey Books, Juta and Co, 2006

Boucher, Leigh and Lynette Russell. *Settler Colonial Governance in Nineteenth-Century Victoria.* For more on settler attitudes towards Aboriginal people and their place in the hierarchy of humans, see Chapter Six on the Aborigines Protection Act of 1886.

Brahic, Catherine. "Small Women Take Giant Leap for Early Man", *Sunday Times*, 14 December 2014

Brain, CK. "Visitor Reaction to the 'Life's Genesis' Display Sequence at the Transvaal Museum." *South African Museums Association Bulletin*, 13, no. 7, (1979): 246–249

Brain, CK. *The Hunters or the Hunted?* Chicago and London: The University of Chicago Press, 1981

Brain, Bob (CK). "Naturalist", in *Irene Personalities: Twelve Career Stories by Irene Residents*, edited by Nigel Helme. Pretoria: Gutenberg Book Printers, 1982

Brain, CK. "A Perspective on the PAST." *South African Journal of Science*, Issues 5&6, no. 99, (2003): 1–2

Brain, CK. "Raymond Dart and our African Origins", in *A Century of Nature:*

Twenty-One Discoveries that Changed Science and the World, edited by Laura Garwin and Tim Lincoln. Chicago: University of Chicago Press, 2003. Available at: www.press.uchicago.edu/Misc/Chicago/284158_brain. html (accessed on 15 May 2014)

Brain, CK (ed.). *Swartkrans: A Cave's Chronicle of Early Man*, Transvaal Museum Monograph, Number 8, published by the Transvaal Museum, Pretoria, 2004

Bridie, James. *The Anatomist*. London: Constable and Co, 1931. Portions of the script kept in Dart Papers, Box 71 marked Taung, Wits Archives

British United Press. "Dashes Hopes That Skull is Missing Link's", *Telegraph Journal*, 11 July 1925

Bronowski, Jacob. *The Ascent of Man*. Boston/Toronto: Little, Brown and Company, 1973

Broom, Norman. "Address on the Opening of the Robert Broom Museum." *South African Journal of Science*, 63, no. 9, September (1967): 371–377

Broom, Robert. *Finding the Missing Link*. London: CA Watts and Co, 1950

Broom, Robert and GWH Schepers. *The South African Fossil Ape-Men: The Australopithecinae*, Transvaal Museum Memoir No. 2, published by the Transvaal Museum, Pretoria, 31 January 1946

Bystrom, Kerry. "The DNA of the Democratic South Africa: Ancestral Maps, Family Trees, Genealogical Fictions." *Journal of Southern African Studies*, 35, no. 1, (2009): 223 –235

Cann, Rebecca, Mark Stoneking and Allan C Wilson. "Mitochondrial DNA and Human Evolution." *Nature*, 325, (1987): 31–36. Available at: www. nature.com/nature/ancestor/pdf/325031.pdf (accessed on 8 June 2016)

Cann, Rebecca. "Mothers, Labels and Misogyny", in *Women in Human Evolution*, edited by Lori D Hager. New York and London: Routledge, 1997

Carruthers, Jane. "Mapungubwe: An Historical and Contemporary Analysis of a World Heritage Cultural Landscape." *Koedoe*, 49/1, (2006): 1–13. Available at: www.koedoe.co.za/index.php/koedoe/article/viewFile/89/91 (accessed on 29 August 2016)

Chatwin, Bruce. *The Songlines*. London: Penguin Books, 1987

Chernick, Ilanit and Karishma Dipa. "Fossil Team Fights Back: Undated and Not a New Species, Critics Say", *The Star*, 11 September 2015

Cherry, Michael. "Claim over 'Human Ancestor' Sparks Furore." *Nature*, 8 April 2010. Available at www.nature.com/news/2010/100408/full/news.2010.171.html (accessed on 29 June 2015)

Chidester, David. "Christianity and Evolution", in *The Architect and the Scaffold: Evolution and Education in South Africa*, edited by Wilmot

James and Lynne Wilson. Pretoria: Human Sciences Research Council/ New Africa Education, 2002

Clarke, RJ. "Observations on Some Restored Hominid Specimens in the Transvaal Museum", in *Apes to Angels: Essays in Anthropology in Honor of Phillip V. Tobias.* Edited by Geoffrey H Sperber. New York: Wiley-Liss, 1990

Clarke, RJ and PV Tobias. "Sterkfontein Member 2 Foot Bones of the Oldest South African Hominid." *Science*, 269, no. 5223, (1996): 521–524

Clarke, RJ. "First Ever Discovery of a Well-Preserved Skull and Associated Skeleton of *Australopithecus.*" *South African Journal of Science*, 94, no. 10, (1998): 460–463

Clarke, RJ. "On the Unrealistic 'Revised Age Estimates' for Sterkfontein." *South African Journal of Science*, 98, no. 9&10, (2002): 415–418

Clarke, Ron. "Dr. Broom and the Skeleton in the Cavern", in *Origins*, edited by Geoffrey Blundell. Cape Town: Double Storey Books, Juta and Co, 2006

Clarke, RJ. "A Deeper Understanding of the Stratigraphy of Sterkfontein Fossil Hominid Site." *Transactions of the Royal Society of South Africa*, 61, no. 2, (2006): 111–120

Clarke, RJ. "Latest Information on Sterkfontein's *Australopithecus* Skeleton and a New Look at *Australopithecus.*" *South African Journal of Science*, 104, no. 11&12, (2008): 443–449

Clarke, Ronald J and Timothy C Partridge, with contributions by Kathleen Kuman. *Caves of the Ape-Men: South Africa's Cradle of Humankind World Heritage Site.* Johannesburg: Wits University Press, 2010

Clarke, RJ and Beverley Kramer. "Biographical Memoir: Phillip Vallentine Tobias." *Transactions of the Royal Society of South Africa*, 67, no. 3, (2012): 169–173

Clarke, James. "Fossil Find Brings Our Roots Closer: Near Human Fossils of Woman and Boy a Groundbreaking Discovery for SA", *Sunday Independent*, 11 April 2010

Cooke, Adam. "Tobias Hands Over the Remains of Cornelius Kok 2", *The Star*, 21 August 1996

Coon, Carleton S. *The Origin of Races.* New York: Random House, 1962

Crawhall, Nigel. "Alpha to Omega: Language Origins and Demise," in *Origins*, edited by Geoffrey Blundell. Cape Town: Double Storey Books, Juta and Co, 2006

D'Errico, Francesco and Linda Backwell (eds). *From Tools to Symbols: From Early Hominids to Modern Humans.* Johannesburg: Wits University Press, 2005

Dart, Raymond. "*Australopithecus Africanus:* The Man-ape of South Africa." *Nature*, 115, no. 2884, (1925). As reprinted in the *Sunday Times Heritage* Project. Available at: www.heritage.thetimes.co.za/memorials/gp/RaymondDart/article.aspx?id=585748 (accessed on 20 February 2014)

Dart, Raymond. "A Note on Makapansgat: A Site of Early Human Occupation." *South African Journal of Science*, 22, November, (1925): 454

Dart, Raymond, "The Present Position of Anthropology in South Africa." *South African Journal of Science*, 22, November, (1925): 73–80

Dart, Raymond. "The Historical Succession of Cultural Impacts upon South Africa." *Nature*, 115, no. 2890, (1925): 425–429

Dart, Raymond A. "The South African Negro." *American Journal of Physical Anthropology*, 13, no. 2, (1929): 309–317

Dart, Raymond. "The Hut Distribution Genealogy and Homogeneity of the l?Auni-=Khomni Bushmen." *Bantu Studies*, 11, no. 1, (1937): 159–174

Dart, Raymond. "The Physical Characters of the l?Auni-=Khomni Bushmen." *Bantu Studies*, 11, no. 1, (1937): 175–246

Dart, Raymond. "Racial Origins", in *The Bantu Speaking Tribes of South Africa*, edited by Isaac Schapera. Routledge and Kegan Paul, 1937

Dart, Raymond A. "Recent Discoveries Bearing on Human History in Southern Africa." *The Journal of the Royal Anthropological Institute of Great Britain and Ireland*, 70, no. 1, (1940): 13–27

Dart, Raymond. "The Postural Aspect of Malocclusion", 1946, reprinted in *Skill and Poise* by Raymond Dart. London: STAT, 1996

Dart, Raymond A. "The Predatory Transition from Ape to Man." *International Anthropological and Linguistic Review*, 1, no. 4, (1953). Available at: www.users.muohio.edu/erlichrd/vms_site/dart.html (accessed on 15 September 2014)

Dart, Raymond. *The Osteodontokeratic Culture of Australopithecus Prometheus*, Transvaal Museum Memoir No. 10, published by the Transvaal Museum, Pretoria, 1957

Dart, Raymond A and Dennis Craig. *Adventures with the Missing Link*. London: Hamish Hamilton, 1959

Dart, Raymond A. "Associations with and Impressions of Sir Grafton Elliot Smith." *Mankind*, 8, no. 3, (1972): 171–175

Dart, Raymond. "Recollections of a Reluctant Anthropologist." *Journal of Human Evolution*, 2, no. 6, (1973): 417–427

Dart, Raymond. Foreword to *The Bushmen: San Hunters and Herders of Southern Africa*, edited by Phillip Tobias. Cape Town: Human and Rousseau Publishers, 1978

Dart, Raymond A. *Skill and Poise*. London: STAT, 1996

Dart, Raymond. "Problems in Protection of Features of Anthropological, Archaeological and Ethnological Interest in Relation to Changes in Native Sociology and General Economics in the Union". Dart papers, box labelled "Research, Includes Broadcast Talks". Cited in Dubow, Saul. "Human Origins, Race Typology and the other Raymond Dart." *African Studies*, 55, no. 1, (1996): 1–30

Darwin, Charles. *On the Origin of Species*. New York: Sterling Publishing, 2011

Darwin, Charles. *Descent of Man, and Selection in Relation to Sex*, Photoreproduction of the 1871 edition, published by London: J Murray and Princeton, New Jersey: Princeton University Press, 1981. Available at: www.teoriaevolutiva.files.wordpress.com/2014/02/darwin-c-the-descent-of-man-and-selection-in-relation-to-sex.pdf (accessed on 2 June 2016)

Dayal, Manisha, Anthony Kegley, Goran Strkalj, Mubarak Bidmos and Kevin Kuykendall. "The History and Composition of the Raymond A. Dart Collection of Human Skeletons at the University of the Witwatersrand, Johannesburg, South Africa." *American Journal of Physical Anthropology*, 140 (2009): 324–335

Dechow, Paul C. "In Memoriam: Ronald Singer, 1924–2006." *The Anatomical Record*, 289b, no. 4, (2006): 114–115. Available at: www.onlinelibrary.wiley.com/doi/10.1002/ar.b.20109/pdf (accessed on 4 November 2014)

Derricourt, Robin. "The Enigma of Raymond Dart." *The International Journal of African Historical Studies*, 42, no. 2, (2009): 257–282

De Villiers, Hertha. "The Tablier and Steatopygia in Kalahari Bushwomen." *South African Journal of Science*, 57, no. 8, (1961): 223–227

De Villiers, Hertha. *The Skull of the South African Negro*. Johannesburg: Witwatersrand University Press, 1968

De Villiers, Hertha. "The First Fossil Human Skeleton From South West Africa." *Transactions of the Royal Society of South Africa*, 40, no. 3, (1972): 187–196

De Villiers, Hertha and LP Fatti. "The Antiquity of the Negro." *South African Journal of Science*, 78, (1982): 321–332

De Villiers, Phillippa. "Secrets and Lies", *Elle*, March 1999

Diamond Fields Advertiser. "Return of Skeleton Rattles Descendent", 21 August 1996

Dirks, Paul HGM, Lee R Berger, Eric M Roberts, Jan D Kramers, John Hawks,

Patrick S Randolh-Quinney, Marina Elliott, Charles M Musiba, Steven E Churchill, Darryl J de Ruiter, Peter Schmid, Lucinda R Backwell, Georgy A Belyanin, Pedro Boshoff, K Lindsay Hunter, Elen M Feuerriegel, Alia Gurtov, James du G Harrision, Rick Hunter, Ashley Kruger, Hannah Morris, Tebogo V Makhubela, Becca Peixotto, and Steven Tucker. "Geological and Taphonomic Context for the New Hominin Species *Homo Naledi* from the Dinaledi Chamber, South Africa", *eLIFE*, 10 September 2015, pp 1–37

Dip, Dailo. "Africa: Mankind's Past and Future," in *African Renaissance: The New Struggle*, edited by Malegapuru William Makgoba. Cape Town and Sandton: Tafelberg and Mafube, 1999

Dubow, Saul. *Scientific Racism in Modern South Africa*. London: Cambridge University Press, 1995

Dubow, Saul. "Human Origins, Race Typology and the Other Raymond Dart." *African Studies*, 55, no. 1, (1996): 1–30

Dubow, Saul. *A Commonwealth of Knowledge: Science, Sensibility and White South Africa 1820–2000*. Oxford: Oxford University Press, 2006

Dubow, Saul. "White South Africa and the South Africanisation of Science: Humankind or Kinds of Humans?" in *A Search for Origins*, edited by Philip Bonner, Amanda Esterhuysen and Trefor Jenkins. Johannesburg: Wits University Press, 2007

Dubow, Saul. "Smuts, the United Nations and the Rhetoric of Race and Rights." *Journal of Contemporary History*, 43, no. 1, (2008): 45–74

Dubow, Saul. "Racial Irredentism, Ethnogenesis, and White Supremacy in High-Apartheid South Africa." *Kronos*, 41, no. 1, (2015): 236–264. Available at: www.scielo.org.za/scielo.php?script=sci_arttext&pid=S0259-01902015000100010 (accessed on 29 August 2016)

Dugmore, Heather. "Being Human and a Shot of Tequila." *Wits Review*, April 2010. Available at: www.wits.ac.za/alumni/news/features/3260/being human and a shot of tequila.html (accessed on 29 May 2015)

Duke, Lynne. "Full *Australopithecus* Fossil Found in South Africa", *Washington Post*, 10 December 1998

Ebersohn, Wessel. "Communicator." *Leadership*, 6, no. 2, (1987): 76–82

Editorial, "A Grave Message from the Cradle." *South African Journal of Science*, 98, September/October, (2002): 410

Engelbrecht, Martin. "The Connection between Archaeological Treasures and the Khoisan People", in *The Dead and their Possessions: Repatriation in Principle, Policy and Practice*, edited by Fforde, C, J Hubert, and P Turnbull. London: Routledge, 2002

Erasmus, Zimitri. "Throwing the Genes: A Renewed Biologcial Imaginary of

'Race', Place and Identification." *Theoria*, 60, no. 3, (2013): 38–53

Erlich, Richard. "Strange Odyssey: From Dart to Ardrey to Kubrick and Clarke", an essay on the website *Science Fiction and Film*, 1976. Available at: www.users.miamioh,edu/erlichrd/350/odyssey.php (accessed on 15 September 2014)

Esterhuysen, Amanda and Jeanette Smith. "Evolution: 'The Forbidden Word'." *South African Archaeological Bulletin*, 53, (1998): 135–137

Esterhuysen, Amanda. "The Emerging Stone Age" and "The Earlier Stone Age", in *A Search for Origins*, edited by Philip Bonner, Amanda Esterhuysen and Trefor Jenkins. Johannesburg: Wits University Press, 2007

Esterhuysen, Amanda. *Sterkfontein: Early Hominid Site in the 'Cradle of Humankind'*. Johannesburg: Wits University Press, 2007

Esterhuysen, Amanda B. "Excavation at Historic Cave, Makapan's Valley, Limpopo: 2001–2005." *South African Archaeological Bulletin*, 65, no. 191, (2010): 67–83

European Synchrotron Radiation Facility (ESRF) Press Release, "First Studies of Fossil of New Human Ancestor Take Place at the ESRF", 12 April 2010. Available at: www.esrf.eu/news/general-old/general-2010/first-studies-of-fossil-of-new-human-ancestor-take-place-at-the-esrf;jsessionid=3E04ADB 369F049B93A9CFD629F305D49 (accessed on 22 March 2016)

Falk, Dean. *The Fossil Chronicles: How Two Controversial Discoveries Changed Our View of Human Evolution*. Los Angeles: University of California Press, 2011

Farber, Tanya. "From Obscure Caveman to Global Rock Star", *Sunday Times*, 13 September 2015

Findlay, George. *Dr. Robert Broom: Palaeontologist and Physician: 1866–1951*. Cape Town: Balkema, 1972. [The story about the skeletons that Broom sent from Port Nolloth to the University of Edinburgh is on p 20. The quote from Broom's letter to Osborn about collecting skeletons in Douglas is on p 50. The story about the man who died in the Douglas jail is also on p 50.]

Fraser, Neil. "Tall Tales with Neil Fraser", 14 July 2008. Story about the Wits Column from the Empire Exhibition on the City of Johannesburg website at: www.joburg.org.za/index.php?option=com_ content&task=view&id=2705&Ite (accessed on 8 March 2016)

Freedman, Samuel G. "Rift over Fossils from South Africa", *New York Times*, 30 May 1984

Freedman, Samuel G. "Fossil Dispute May Bring Apartheid Denunciation", *New York Times*, 1 June 1984. Available at: www.

google.co.za/?gws_rd=ssl#q=Samuel+Freedman+fossil+dispute+may+bring+apartheid+denunciation (accessed on 19 April 2016)

Freedman, Samuel G. "Museum Board Affirms Anti-Apartheid Position", *New York Times*, 12 June 1984

Forrest, Drew. "Bones of Contention", *Mail & Guardian*, 22 April 2010

Forrest, Drew. "Cradle's Grave: Homo Naledi Strikes an All-Too-Human Chord", *Mail & Guardian*, 11–17 September 2015

Fuze, Magema M. *The Black People and Whence They Came*. Pietermaritzburg: University of Natal Press, 1979

Galloway, Alexander. "The Characteristics of the Skull of the Boskop Physical Type." *American Journal of Physical Anthropology*, 23, no. 1, (1937): 31–47

Garson, Noel. "Smuts and the Idea of Race." *South African Historical Journal*, 57, no. 1, (2001): 153–178

Germaner, Shain. "Homo Naledi was no Baboon: Fossils Prof", *Daily News*, 14 September 2015. Available at: www.iol.co.za/dailynews/news/homo-naledi-was-no-baboon-fossils-prof-1915647 (accessed on 21 November 2015)

Gibbons, Ann. "Skeletons Present an Exquisite Paleo-Puzzle." *Science*, 333, no. 6048, (2011): 1370–1372

Gitschier, Jane. "All About Mitochondrial Eve: An Interview with Rebecca Cann", in *PLOS Genetics*, 27 May 2010. Available at: www.journals.plos.org/plosgenetics/article?id=10.1371/journal.pgen.1000959 (accessed on 25 February 2014)

Goodrum, Matthew R. "The History of Human Origins Research and its Place in the History of Science: Research Problems and Historiography." *History of Science*, 47, no. 3, (2009): 337–357

Goodrum, Matthew R. "History", in *A Companion to Paleoanthropology*, first edition, edited by David R Begun. Blackwell Publishing Ltd, 2013

Gordon, Robert J. *The Bushman Myth: The Making of a Namibian Underclass*. Westview Press, Boulder, 1992

Gould, Stephen Jay. *Ever Since Darwin*. London Penguin Books, 1977

Gould, Stephen Jay. "The Piltdown Conspiracy", *Natural History Magazine*, August 1980. Later published as a chapter in *Hen's Teeth and Horses' Toes,* 1983. Available at: www.clarku.edu/~piltdown/map_prim_suspects/Teilhard_de_Chardin (accessed on 6 March 2016)

Gould, Stephen Jay. "The Hottentot Venus." *Natural History*, 10, (1982). This article was reprinted with a postscript in Gould's book *The Flamingo's Smile: Reflections in Natural History*. New York: Norton Publishing, 1985

Gould, Stephen Jay and Niles Eldredge. "Punctuated Equilibrium at the Third State." *Systematic Zoology*, 35, no. 1, (1986): 143–148

Gould, Stephen Jay. *The Mismeasure of Man*, second edition. New York: Norton, 1996

Gould, Stephen Jay. "Ladders and Cones: Constraining Evolution by Canonical Icons", in *Hidden Histories of Science*, edited by Robert B Silvers. London: Granta Books, 1997

Gqola, Pumla Dineo. *What is Slavery to Me?: Postcolonial/Slave Memory in Post-Apartheid South Africa*. Johannesburg: Wits University Press, 2010

Graber, Cynthia. "Synchrotron Focuses on New Hominid Fossil", *Scientific American*, 15 April 2010

Haines, Catherine. *International Women in Science: A Biographical Dictionary to 1950*, Santa Barbara, California: ABC Clio, 2001. Available at: www. books.google.co.za (accessed on 26 February 2014)

Haraway, Donna. *Primate Visions: Gender, Race and Nature in the World of Modern Science*. Routledge, New York: Chapman and Hall, 1989

Hechinger, Fred M. "A Liberal Fights Back", *New York Times*, 13 February 1972. Available at: www.krusch.com/kubrick/Q47.html (accessed on 28 August 2015)

Henshilwood, Christopher. "Evidence from Blombos Cave, South Africa", in *Origins*, edited by Geoffrey Blundell. Cape Town: Double Storey Books, Juta and Co, 2006

Heywood, Christopher. Introduction to *A History of South African Literature*. Cambridge: Cambridge University Press, 2010

Hochschild, Adam. *Bury the Chains*. Boston and New York: Houghton Mifflin Company, 2005

Hofmeyr, JH. "Africa and Science." *South African Journal of* Science, 26, December (1929): 1–18

Holloway, Marguerite. "Mary Leakey: Unearthing History", *Scientific American*, 16 December 1996

Houlton, Tobias MR and Brendon Billings. "Blood, Sweat and Plaster Casts: Reviewing the Archives, History and Scientific Value of the Raymond A. Dart Collection of African Life and Death Masks" (unpublished)

Hrdlička, Aleš. "The Taungs Ape." *The American Journal of Physical Anthropology*, 8 (1925): 379–392

Humphreys, Anthony JB. "The Destruction of the Griqua Mission Church at Campbell – the Ultimate Irony." *The Digging Stick*, 27, no. 2, (2010): 9–10

Humphreys, Basil. "Science Spotlight Plays on Griqua Remains", *Diamond Fields Advertiser*, 2 June 1961

Humphreys, Basil. "Archaeologists in District Uncover Links with Past", *Diamond Fields Advertiser*, 3 June 1961

Isaacson, Maureen. "A Revised Look at 'Mama Sara,' a Woman of Many Parts", *Sunday Independent*, 14 February 2010

Jackson, John P. Jr. "'In Ways Unacademical': The Reception of Carleton S. Coon's *The Origin of Races*." *Journal of the History of Biology*, 34, Summer (2001): 247–285

Jacobs, Nancy J. *Birders of Africa: History of a Network*. New Haven and London: Yale University Press, 2016

James, Wilmot and Lynne Wilson (eds). *The Architect and the Scaffold: Evolution and Education in South Africa*. Pretoria: Human Sciences Research Council, 2002. Available at: www.hsrcpress.ac.za (accessed on 9 June 2014)

Jenkins, Trefor. "Fossils and Genes: A New Anthropology of Evolution", in *A Search for Origins: Science, History and South Africa's 'Cradle of Humankind'*, edited by Philip Bonner, Amanda Esterhuysen and Trefor Jenkins. Johannesburg: Wits University Press, 2007

Jenkins, Trefor. "From Generation to Generation." Inaugural lecture delivered on 3 August 1977. Johannesburg: Wits University Press, 1978

Jenkins, Trefor. Interview with Peter Harper, Professor of Human Genetics at Cardiff University, in Cardiff, 10 October 2007. Available at: www. genmedhist.info/interviews/Jenkins (accessed on 5 November 2014)

Junker, Thomas. "Blumenbach's Racial Geometry." *Isis*, 89, (1998): 498–501

Kahn, Tamar. "New Hominin Find Raises Questions", *Business Day*, 11 September 2015

Kahn, Tamar. "Big Little Foot Feat was a Labour of Love", *Business Day*, 20 October 2015

Keith, Arthur. "The New Missing Link", *The British Medical Journal*, 14 February 1925

Keith, Arthur. *An Autobiography*. London: Watts & Co, 1950

Koch, Eddie. "Griquas Want Chief's Bones Back", *Mail & Guardian*, 2–8 February 1996

Kubrick, Stanley. "Now Kubrick Fights Back", *New York Times*, 27 February 1972. Available at: www.krusch.com/kubrick/Q47.html (accessed on 28 August 2015)

Kuklick, Henrika. "The British Tradition", in *A New History of Anthropology*, edited by Henrika Kuklick. Malden, MA, Oxford and Carlton, Victoria Australia: Blackwell Publishing, 2008

Kusimba, Sibel B. "What is a Hunter-Gatherer? Variation in the Archaeological

Record of Eastern and Southern Africa." *Journal of Archaeological Research*, 13, no. 4, (2005): 337–366

Kuykendall, KL. "Never Letting the Facts Get in the Way of a Good Story." *South African Journal of* Science, 97, nos 5&6, (2001): 177–178

Kuykendall, Kevin. "Fossil Hominids of the 'Cradle of Humankind'", in *A Search for Origins*, edited by Philip Bonner, Amanda Esterhuysen and Trefor Jenkins. Johannesburg: Wits University Press, 2007

Kuykendall, Kevin and Goran Strkalj. "A History of South African Palaeoanthropology", in *A Search for Origins: Science, History and South Africa's 'Cradle of Humankind'*, edited by Philip Bonner, Amanda Esterhuysen and Trefor Jenkins. Johannesburg: Wits University Press, 2007

Landi, Mariangele and Jacopo Moggi Cecchi. "Colonial Anthropology: From the Peoples of the World to the Fascist Man, Nello Puccioni, Lidio Cipriani", in *The Museum of Natural History of the University of Florence: The Anthropological and Ethnological Collections*, edited by J. Moggi-Cecchi and R. Stanyon. Florence: Firenze University Press, 2014

Leakey, Richard. "Zinj and the Leakeys", *Archaeology*, 3 August 2009. Available at: www.archive.archaeology.org/online/interviews/leakey/ (accessed on 1 September 2015)

Legassick, Martin and Ciraj Rassool. *Skeletons in the Cupboard: South African Museums and the Trade in Human Remains 1907–1917*, updated edition. Cape Town: Iziko Museums of South Africa, 2015

Lelliott, Anthony. "Visitors' Views of Human Origins After Visiting the Cradle of Humankind World Heritage Site." *South African Journal of Science*, 112, no. 1/2, (2016): 132–139

Lever, Jeffrey. "Science, Evolution and Schooling in South Africa", in *The Architect and the Scaffold: Evolution and Education in South Africa*, edited by Wilmot James and Lynne Wilson. Pretoria: Human Sciences Research Council and New Africa Education, 2002

Lewis, Roger. *Bones of Contention: Controversies in the Search for Human Origins*, second edition (first edition in 1987). Chicago and London: University of Chicago Press, 1997

Lindeque, Mia. "Homo Naledi Debate: 'Humans Don't Come from Baboons'", *Eye Witness News*, 14 September 2015. Available at: www. ewn.co.za/2015/09/14/Homo-naledi-Humans-dont-come-from-baboons (accessed on 21 November 2015)

Linne, Carl Von. *Systema Naturae*, tenth edition (originally published in December 1758). London: British Museum of Natural History, 1939

Luthuli, *Let My People Go*. London: Fount Paperbacks, Fontana Books, 1984

Maguire, Judy. "A Second Opinion." *South African Journal of Science*, 98, nos 9&10, (2002): 518–519

Makgoba, Malegapuru William (ed.). *African Renaissance: The New Struggle*. Cape Town and Sandton: Tafelberg and Mafube, 1999

Makele, Benison. "'Pure' Bones Tell All", *City Press*, 25 August 1996

Mapumulo, Zinhle. "A Lucky Friday 13th", *City Press*, 13 September 2015

Mason, R.J. "Early Iron Age Settlement at Broederstroom 24/73, Transvaal, South Africa." *South African Journal of Science*, 77, (1981): 401–416

Mbeki, Thabo. Prologue to *African Renaissance: The New Struggle*, edited by Malegapuru William Makgoba. Cape Town and Sandton: Tafelberg and Mafube, 1999

Mbeki, Thabo. Foreword to *Field Guide to the Cradle of Humankind*, by Lee Berger and Brett Hilton-Barber. Cape Town: Random House Struik, 2002

McHenry, Henry M and Lee R Berger. "Body Proportions in *Australopithecus Afarensis* and *A. Africanus* and the Origin of the Genus *Homo*." *Journal of Human Evolution*, 35, (1998): 1–22

McKee, Jeffrey K and Phillip V Tobias. "Taung Stratigraphy and Taphonomy: Preliminary Results based on the 1988–93 Excavations." *South African Journal of Science and Technology/Suid-Afrikaanse Tydskrif vir Wetenskap*, 90, April (1994): 233–235

McKie, Robin. "Piltdown Man: British Archaeology's Greatest Hoax", *The Guardian*, 5 February 2012. Available at: www.theguardian.com/science/2012/feb/05/piltdown-man-archaeologys-greatest-hoax (accessed on 18 February 2014)

McKie, Robin. "Scientist who Found New Human Species Accused of Playing Fast and Loose with the Truth", *The Guardian*, 25 October 2015. Available at: www.theguardian.com/science/2015/oct/25/discovery-human-species-accused-of-rushing-errors (accessed on 25 July 2016)

Meredith, Martin. *Born in Africa: The Quest for the Origins of Human Life*. Cape Town: Jonathan Ball Publishers, 2011

Morris, Alan G. *The Skeletons of Contact: A Study of Protohistoric Burials from the Lower Orange River Valley, South Africa*. Johannesburg: Witwatersrand University Press, 1992

Morris, Alan G. *A Master Catalogue: Holocene Human Skeletons From South Africa*. Johannesburg: Witwatersrand University Press, 1992

Morris, Alan G. "Trophy Skulls, Museums and the San", in *Miscast*, edited and curated by Pippa Skotnes. Cape Town: University of Cape Town Press, 1996

Morris, Alan G. "The Griqua and the Khoikhoi: Biology, Ethnicity and the

Construction of Identity." *Kronos Journal Cape History*, 24, November, (1997): 106–118

Morris, Alan G. "Searching for 'Real' Hottentots: the Khoekhoe in the History of South African Physical Anthropology." *Southern African Humanities*, 20, December, (2008): 221–233

Morris, Alan G. "New Hominin Fossils From Malapa: The Unveiling of *Australopithecus Sediba*." *South African Journal of Science*, 106, no. 3/4, (2010): 1–2

Morris, Alan. "On Human Evolution, *Australopithecus Sediba* and Nation Building." *South African Journal of Science*, 107, no. 11/12, (2011): 1–3

Morris, Alan G. "Biological Anthropology at the Southern Tip of Africa: Carrying European Baggage in an African Context." *Current Anthropology*, 53, supplement 5, (2012): S152–S160

Morris, Alan. "First Ancient Mitochondrial Human Genome From a Prepastoralist Southern African." *Genome Biology Evolution*, 6, no. 10, (2014): 2647–2653. Available at: www.ncbi.nlm.nih.gov/pmc/articles/ PMC4224329/ (accessed on 10 March 2016)

Morris, Alan. "Ancient DNA Comes of Age." *South African Journal of Science*, 111, no. 5/6, (2015): 1–2

Morrison, Toni. *Playing in the Dark: Whiteness and the Literary Imagination*. New York: Vintage Books, a Division of Random House, 1992

Murray, Bruce. *Wits: The Early Years: A History of the University of the Witwatersrand Johannesburg and its Precursors 1896–1939*. Johannesburg: Witwatersrand University Press, 1982

Murray, Bruce K. *Wits: The Open Years*. Johannesburg: Wits University Press, 1997

Murray, Charles and Richard J Herrnstein. *The Bell Curve: Intelligence and Class Structure in American Life*. The Free Press, Simon and Schuster, 1994. Available at: www.lesacreduprintemps19.files.wordpress.com/2012/11/ the-bell-curve.pdf (accessed on 2 June 2016)

Naidu, Maheshvari. "Creating an African Tourist Experience at the Cradle of Humankind World Heritage Site." *Historia*, 53, no. 2, (2008): 182–207

Natal Mercury, "Preservation of Bushmen: Senator Boydell Interviewed", 19 May 1937

Ndebele, Njabulo S, "'Iph' Indlele? Finding our Way into the Future", the First Steve Biko Memorial Lecture, *Social Dynamics*, 26, no. 1, (2000): 43–55

Ndlovu, Ndukuyakhe. "Transformation Challenges in South African Archaeology." *The South African Archaeological Bulletin*, 64, no. 189, (2009): 91–93

Ndlovu, Ndukuyakhe. "Archaeological Battles and Triumphs: A Personal Reflection.", in *Being and Becoming Indigenous Archaeologists*, edited by George Nicholas. Walnut Creek, CA: Left Coast Press, 2010

Ndlovu, Ndukuyakhe, "Decolonizing the Mind-set: South African Archaeology in a Postcolonial, Post-Apartheid Era", in *Postcolonial Archaeologies in Africa,* edited by Peter Ridgway Schmidt. Santa Fe, New Mexico: School for Advanced Research Press, 2009

Nettl-Fiol, Rebecca and Luc Varnier. *Dance and the Alexander Technique: Exploring the Missing Link*. Urbana, Chicago and Springfield: University of Illinois Press, 2011

Nevill, Glenda. "Skeletons 'Don't Move Me'", *The Citizen,* 3 February 1987

No Author. "New Evidence on Negro Origins." *New Scientist*, 13 January (1983): 90

New York Times. "Thinks New Skull Links Man and Ape", 5 February 1925. Found in Dart papers, Wits University Archive

Nieuwoudt, Stephanie. "French Museum Refused to Return Saartje Baartman", *News24 Archives*. Available at: www.news24.com/xArchive/Archive/French-museum-refuses-to-return-Saartje-Baartman-20001013 (accessed on 22 May 2015)

Nuttall, Sarah, "Subjectivities in Whiteness." *African Studies Review*, 44, no. 2, (2001): 115–140

Oppenheimer, Stephen. *Out of Africa's Eden: The Peopling of the World.* Cape Town: Jonathan Ball Publishers, 2003

Palmer, Eve. *The Plains of Camdeboo*. Johannesburg: Penguin Books, 2011.

Pearce, David. "The Origin of Complex Burial in Southern Africa", in *Origins,* edited by Geoffrey Blundell. Cape Town: Double Storey Books, Juta and Co, 2006

Petersen, Tammy, "I am No Grandchild of Any Ape, Monkey or Baboon – Vavi on Homo Naledi", *News24*. Available at: www.news24.com/SouthAfrica/News/I-am-no-grandchild-of-any-ape-monkey-or-baboon-Vavi-on-Homo-naledi-20150912 (accessed on 14 September 2015)

Phillips, Robin. "Breaking the Chain: Shakespeare's Use of the Great Chain of Being in Macbeth", *Christian Worldview Journal*, 28 April 2014. Available at: www.breakpoint.org/the-center/columns/changepoint/21632-breaking-the-chain-shakespeares-use-of-the-great-chain-of-being-in-macbeth (accessed on 2 June 2016)

Pickering, Travis, Kathy Schick and Nicholas Toth. *Breathing Life into Fossils: Taphonomic Studies in Honor of C.K. (Bob) Brain*. Gosport, Indiana: Stone Age Institute Press, 2007

Plaatje, Solomon Tshekisho. *Native Life in South Africa*, 1916. Available at: www.gutenberg.org/cache/epub/1452/pg1452.html (accessed on 18 May 2015)

Posel, Deborah. "Race as Common Sense: Racial Classification in Twentieth-Century South Africa." *African Studies Review*, 44, no. 2, (2001): 87–113

Pretoria News, "D.R.C. Opposes Linking of Animals With Man", 27 April 1963

Pretoria News, "Big Changes at City Museum", 20 July 1963

Pretoria News, "Exhibit On Evolution Must Go, Says De Klerk", 24 July 1963

Pretoria News, "Museum Will Oppose A Removal Of Exhibits", 25 July 1963

Pretoria News, "De Klerk Confirms Order to Museum", 26 July 1963

Pretoria News, "Minister's Plea Criticised", 29 July 1963

Pretoria News, "Another Expert Lost to Museum", 20 August 1963

Pretoria News, "New Move in Controversy", 31 August 1963

Pretoria News, "Decision on Evolution Row Soon", 19 September 1963

Pretoria News, "New Evolution Row Expected", 12 October 1963

Pretoria News, "New Turn in Museum Quarrel Over Evolution", 29 October 1963

Pretoria News, "Scathing Attack Made on 'Bigoted' D.R.C. Ministers", 30 October 1963

Pretoria News, "Evolution Exhibit", Editorial 1 November 1963

Pretoria News, "Man-Ape Show Likely to Stay at Museum", 18 November 1963

Pretoria News, "City Museum Exhibit Rearranged", 18 December 1963

Pretoria News, "Evolution Display Will Remain", 18 December 1963

Pretoria News, "Museum Exhibit Still Offends: De Beer", 24 December 1963

Rand Daily Mail, "Denies Monkey Ancestors: Pulpit Cold Water on Missing Link", 9 February 1925. Newspaper clipping found in Dart papers, Wits Archives

Rand Daily Mail, "Vexed Problem of Colour", 16 March 1929. Found in Raymond Dart papers, Wits University Archives

Rand Daily Mail, "Bushmen Should be Saved, Says Professor Dart: Useful Scientific Work on the Kalahari", 21 July 1936

Rand Daily Mail, "And Now the Maps, Mr. De Beer", 15 September 1952

Rand Daily Mail, "Museum Says it has not 'Capitulated' to Church", 19 September 1952

Rand Daily Mail, "Evolution, True Religion Not Incompatible", 2 October 1952

Rand Daily Mail, "State Museum Tangle Over Evolution", 26 July 1963

Rand Daily Mail, "He is Not Resigning in Protest", 4 September 1963

Rand Daily Mail, "De Klerk in 'Dark' about Evolution Decision", 15 November 1963

Rassool, Ciraj and Patricia Hayes. "Science and the Spectacle" /Khanako's South Africa, 1936–1937", in *Deep hiStories: Gender and Colonialism in Southern Africa*, edited by Wendy Woodward, Patricia Hayes and Gary Minkley. Amsterdam and New York: Rodopi, 2002

Rassool, Ciraj. "Re-storing the Skeletons of Empire: Return, Reburial and Rehumanisation in Southern Africa." *Journal of Southern African Studies*, 41, no. 3, (2015): 653–670

Rassool, Ciraj. "Human Remains, the Disciplines of the Dead, and the South African Memorial Complex", in *The Politics of Heritage in Africa*, edited by Derek Peterson, Kodzo Gavua and Ciraj Rassool. London: Cambridge University Press and the International African Institute, 2015

Redman, Samuel. *Bone Rooms: From Scientific Racism to Human Prehistory in Museums*. Cambridge: Harvard University Press, 2016

Ree, Jonathan. "Evolution by Jerks: Stephen Jay Gould's Last Work Reviewed by Jonathan Ree", *New Humanist*, 31 May 2007 (originally published in 2003). Available at: www.newhumanist.org.uk/articles/598/evolution-by-jerks (accessed on 1 March 2016)

Reed, Evelyn. "Is Man an 'Aggressive Ape?'", 1970. Available at: www.marxists.org/archive/reed-evelyn/1970/agressive-ape.htm (accessed on 28 August 2015)

Rice, Julian. *Kubrick's Hope: Discovering Optimism from 2001 to Eyes Wide Shut*. Maryland: Scarecrow Press, 2008

Robins, Steven L. "Citizens and 'Bushmen': The =khomani San, NGOs and the Making of a New Social Movement," in *From Revolution to Rights in South Africa: Social Movements, NGOs and Popular Politics After Apartheid*, edited by Steven L. Robins. Suffolk/Pietermaritzburg: James Currey/University of KwaZulu-Natal Press, 2008

Robinson, Jennifer. "Johannesburg's 1936 Empire Exhibition: Interaction, Segregation and Modernity in a South African City." *Journal of Southern African Studies*, 29, no. 3, (2003): 759–789

Robinson, John. "The Discovery of the 'Mrs. Ples' Skull Specimen." *South African Journal of Science*, 93, no. 4, (1997): 164–165

Rubidge, Bruce S. "The BPI – 50 Years of Palaeontological Activity." *Palaeontologia Africana: Annals of the Bernard Price Institute for Palaeontological Research*, 33 (1997): 1–9

Rubidge, Bruce S. "Charles Kimberlin (Bob) Brain – A Tribute." *Palaeontologia Africana: Annals of the Bernard Price Institute for Palaeontological Research*, 36, (2000): 1–9. Available at: www.wiredspace.wits.ac.za/xmlui/bitstream/handle/10539/16376/2000.V36.RUBIDGE.TRIBUTE%20TO%20BOB%20BRAIN.pdf?sequence=1&isAllowed=y (accessed on 30 August 2016)

Rubidge, Bruce. "The Roots of Early Mammals Lie in the Karoo: Robert Broom's Foundation and Subsequent Research Progress." *Transactions of the Royal Society of South Africa*, 68, no. 1, (2013): 41–52

Schlanger, Nathan. "Making the Past for South Africa's Future: The Prehistory of Field-Marshal Smuts (1920s–1940s)." *Antiquity*, 76, (2002): 200–208

Schlebusch, Carina M, Pontus Skoglund, Per Sjodin, Lucie M Gattepaille, Dena Hernandez, Flora Jay, Sen Li, Michael De Jongh, Andrew Singleton, Michael G B Blum, Himla Soodyall, and Mattias Jakobsson. "Genomic Variation in Seven Khoe-San Groups Reveals Adaptation and Complex African History." *Science*, 338, no. 6105, (2012): 374–379

Schramm, Katharina. "Claims of Descent: Race and Science in Contemporary South Africa." *Vienna Working Papers in Ethnography*, 3, (2014): 1–28

Science News Letter, "World's First Murderers?", Washington DC, 11 June 1949. Found in Dart papers, Wits Archives

Scott, Monique. *Rethinking Evolution in the Museum: Envisioning African Origins*. New York and London: Routledge, 2007. The opening pages of the book are available at: www.samples.sainsburysebooks.co.uk/9781134135912_sample_526070.pdf (accessed on 9 June 2016)

Shakespeare, Nicholas. *Bruce Chatwin*. London: The Harvill Press, 1999

Shepherd, Nick. "Disciplining Archaeology: The Invention of South African Prehistory, 1923–1953." *Kronos*, 28, (2002): 127–145

Shepherd, Nick. "'When the Hand that Holds the Trowel is Black …': Disciplinary Practices of Self-Representation and the Issue of 'Native' Labour in Archaeology." *Journal of Social Anthropology*, 3, no. 3 (2003: 334–352

Shepherd, Nick. "State of the Discipline: Science, Culture and Identity in South African Archaeology, 1870–2003." *Journal of Southern African Studies*, 29, no. 4, (2003): 823–844

Shreeve, Jamie. "Mystery Man", *National Geographic*, October 2015

Sigmon, Becky A. "The Making of a Palaeoanthropologist: John T. Robinson." *Indian Journal of Physical Anthropology and Human Genetics*, 26, no. 2, (2007): 179–341

Simmons, Robin. "Teaching with the Dart Procedures." *Direction: A Journal on the Alexander Technique*, 1, no. 3, (2004): 81–84

Singer, Ronald. "The Boskop 'Race' Problem." *Man*, 58, (1958): 315–330

Singer, Ronald. "Presidential Address 1962: The South African Archeological Society: The Future of Physical Anthropology in South Africa." *The South African Archeological Bulletin*, 17, no. 68, (1962): 205–211

Singer, Ronald and Jon K Lundy (eds). *Variation, Culture and Evolution in African Populations: Papers in Honour of Dr. Hertha de Villiers*. Johannesburg: Witwatersrand University Press, 1986

Smillie, Shaun. "Boy Wonder Found: Nine-year-old Stumbles on Significant New Species", *The Star*, 9 April 2010

Smillie, Shaun. "Hand that Rocked the Cradle: Hominin Fossil Forces us to Rethink Our Ancestry", *Cape Argus*, 9 September 2011

Smillie, Shaun, "She's the Muddle in the Middle", *The Star,* 12 April 2013

Smillie, Shaun. "Historic Unearthing of Hominin Fossils", *Independent Online*, 27 November 2013. Available at: www.iol.co.za/scitech/science/ discovery/historic-unearthing-of-hominin-fossils-1612945 (accessed on 15 October 2014)

Smith, Gail. "Fetching Saartjie", *Mail & Guardian*, 17 May 2002. Available at: www.mg.co.za/article/2002-05-17-00-fetching-saartjie (accessed on 22 May 2015)

Smuts, JC. "The White Man's Task", Speech given at the Savoy Hotel in London at a dinner in Smuts's honour, 22 May 1917. Available at: www. sahistory.org.za/archive/white-man%E2%80%99s-task-jan-smuts-22-may-1917 (accessed on 5 June 2016)

Smuts, Jan Christiaan. "Science in South Africa." *Nature*, 116, no. 2911, (1925): 245–249, from the Presidential Address to the South African Association for the Advancement of Science, delivered at Oudtshoorn, Cape Province on 6 July 1925

Smuts, JC. "South Africa in Science." *South African Journal of Science*, 22, November (1925): 1–19

Smuts, JC. "Climate and Man in Africa." *South African Journal of Science*, 29, (1932): 98–131

Smuts, Jan. Preface by Field Marshal, The Right Honourable to *The South African Fossil Ape-Men: The Australopithecinae*, written by R Broom and GWH Schepers, published by the Transvaal Museum, Pretoria, Transvaal Museum Memoir No. 2, Pretoria, 31 January 1946

Snook, Barbara. "Dance and the Alexander Technique: Exploring the Missing Link." *Research in Dance Education*, 13, no. 3, (2012): 332–335

Soodyall, Himla. "History, Ancestry and Genes", in *Life of Bone: Art Meets*

Science, edited by Joni Brenner, Elizabeth Burroughs and Karel Nel. Johannesburg: Wits University Press, 2011

Soodyall, Himla. "Genetic Evidence for our Recent African Evolution", in *Origins*, edited by Geoffrey Blundell. Cape Town: Double Storey Books, Juta and Co, 2006

Soodyall, Himla (ed.). *The Prehistory of Africa: Tracing the Lineage of Modern Man*. Johannesburg and Cape Town: Jonathan Ball Publishers, 2006

Soodyall, Himla and Trefor Jenkins. "Unravelling the History of Modern Humans in Southern Africa: The Contribution of Genetic Studies", in *A Search for Origins*, edited by Philip Bonner, Amanda Esterhuysen and Trefor Jenkins. Johannesburg: Wits University Press, 2007

Sowetan Live, "New Human Ancestor, Homo Naledi Sparks Racial Row", 17 September 2015. Available at: www.sowetanlive.co.za/news/2015/09/17/new-human-ancestor-homo-naledi-sparks-racial-row (accessed on 21 November 2015)

Spector, J Brooks. "Meet the Neighbours – Homo Naledi's Coming out Party", *Daily Maverick*, 11 September 2015. Available at: www.dailymaverick.co.za/article/2015-09-11-meet-the-neighbours-homo-naledis-coming-out-party/#.V5XSw85Pqcw (accessed on 25 July 2016)

Spector, J Brooks. "Interview: Lee Berger on Homo Naledi and Fame", *Daily Maverick*, 2 November 2015. Available at: www.dailymaverick.co.za/article/2015-11-02-interview-lee-berger-on-homo-naledi-and-fame/#.V5XTPM5Pqcw (accessed on 25 July 2016)

Stringer, Chris B and P Andrews. "Genetic and Fossil Evidence for the Origin of Modern Humans." *Science, New Series*, 239, no. 4845, (1988): 1263–1268

Stringer, Christopher B. "Out of Africa – A Personal History", in *Origins of Anatomically Modern Humans*, edited by Matthew H Nitecki and Doris V Nitecki. New York: Plenum Press, 1994

Stringer, Chris. "Modern Human Origins: Distinguishing the Models." *The African Archaeological Review*, 18, no. 2, (2001): 67–75

Stringer, Chris. "Palaeontology: The 100-year Mystery of Piltdown Man." *Nature*, 492, (2012): 177–179. Available at: www.nature.com/nature/journal/v492/n7428/full/492177a.html (accessed on 18 February 2014)

Stringer, Chris. *The Origin of Our Species*. London: Penguin Books, 2012

Stringer, Chris. "Why We are Not All Multiregionalists Now." *Trends in Ecology and Evolution*, 29, no. 5, (2014): 248–251

Stringer, Chris. "Rethinking Out of Africa", in *Edge Foundation*, 11 December 2011. Available at: www.edge.org/conversation/christopher_stringer-rethinking-out-of-africa (accessed on 3 November 2015)

Strkalj, Goran. "Inventing Races: Robert Broom's Research on the Khoisan." *Annals of the Transvaal Museum*, 37, (2000): 113–123

Strkalj, Goran. "Robert Broom's Theory of Evolution." *Transactions of the Royal Society of South Africa*, 58, no. 1, (2003): 35–39

Strkalj, Goran. "A Note on the Early History of the Taung Discovery; Debunking the 'Paperweight' Myth." *Annals of the Transvaal Museum*, 42, (2005): 97–98

Strkalj, Goran and Phillip V Tobias. "Raymond Dart as a Pioneering Primatologist." *Journal of Comparative Human Biology*, 59, (2008): 271–286

Strkalj, Goran and Katarzyna A Kaszycka. "Shedding New Light on an Old Mystery: Early Photographs of the Taung Child." *South African Journal of Science*, 108, November/December, (2012): 1–4

Sunday Times, "Scientists to Reply to Church Over Apeman Theory", 28 September 1952

Sunday Times, "Pro-Apartheid Scot's Visit Causes Stir", 23 September 1962

Sunday Times, "Scientist Who Defied De Klerk 'Ban' on Evolution Quits", 25 August 1963

Sullivan, Walter. "New York Will Get a Convention of Fossil Skulls of Man's Ancestors", *New York Times*, 26 June 1983. Available at: www.nytimes.com/1983/06/26/us/new-york-will-get-a-convention-of-fossil-skulls-of-man-s-ancestors.html (accessed 5 May 2016)

Sullivan, Walter. "Far-Flung Fossils Gathered for Exhibit", *New York Times*, 5 April 1984

Swarns, Rachel L. "Cape Town Journal: Bones in Museum Cases May Get Decent Burials", *New York Times*, 4 November 2000

Sykes, Bryan. *The Seven Daughters of Eve*. London: Random House, 2001

Tattersall, Ian, John A Van Couvering and Eric Delson. "The 'Ancestors' Project: An Expurgated History", in *Ancestors: The Hard Evidence*, proceedings of the symposium held at the American Museum of Natural History, 6–10 April 1984, edited by Eric Delson. New York: Alan R Liss, Inc, 1985

Tattersall, Ian. *The Fossil Trail*. Oxford and New York: Oxford University Press, 1995

Tattersall, Ian and Rob DeSalle. *Race?: Debunking a Scientific Myth*. Texas: A&M University Press, 2011

Tattersall, Ian. *Masters of the Planet: The Search for Our Human Origins*. New York: Palgrave Macmillan, 2012

Thackeray, J Francis. "Deceiver, Joker or Innocent? Teilhard de Chardin and Piltdown Man." *Antiquity*, 86, (2012): 228–234

The Australian, "*Australopithecus Sediba*: The Two Million Year Old Boy", 9 April 2010

The Star, "Blasted Out: How Professor Young Found the Skull", 4 February 1925

The Star, "The Ancestor of Man: Find Arouses Wide Interest", 5 February 1925

The Star, "Antiquity of Man: Knowledge Revised, The South African Revelations", 6 February 1925

The Star, "The Taungs (sic) Skull: Sir A. Keith's View, Not in the Human Line", 7 February 1925

The Star, "Coloured or Not?" Undated article. Found in Raymond Dart papers, Wits University Archives

The Star, "Attempt to Save Bushmen: Plan for Special Reserve", 17 April 1937

The Star, "Senator Boydell's Plea for Bushmen: Must be Considered Before Gemsbok in Reserve: Invitation to General Smuts", 18 May 1937

The Star, "Society of Jews and Christians: Symposium on Race Problems", 6 October 1937

The Star, "Misgivings About Nat Outlook on Science: Attempt to Deny Theory of Evolution Feared", 21 March 1952

The Star, "Big Crowds a Problem at Man in Africa Show", 29 May 1963

The Star, "No Trustees, So 'Descent of Man' Show Stays", 25 July 1963

The Star, "Interest in Man-Ape Exhibits", 18 November 1963

The Star, "Discovery of a New Hominid Species: How *Australopithecus Sediba* Fits In", 9 April 2010

The Star, "The Discovery", 12 April 2013

Thompson, Kate, "Sediba Rocks the Cradle", *Financial Mail*, 16 April 2010.

Tierney, John. "The Search for Adam and Eve", *Newsweek*, 11 January 1988. Available at: www.virginia.edu/woodson/courses/aas102%20(spring%2001)/articles/tierney.html (accessed on 22 May 2015)

Time magazine, "Loneliness is an Enemy: Embargoes May Sting, but They May Not Really Hurt", 7 November 1997.

Time magazine, "Puzzling Out Man's Ascent: A Young Leakey Carries on the Search for Human Origins", 7 November 1977

Tobias, Phillip V. "The Problem of Race Determination: Limiting Factors in the Identification of the South African Races." *Journal of Forensic Medicine*, 1, no. 2, (1953): 113–123

Tobias, Phillip V. "On a Bushman-European Hybrid Family." *Man*, 287, (1954): 1–4

Tobias, Phillip V. "Physical Anthropology and Somatic Origins of the Hottentots." *African Studies*, 14, no. 1, (1955): 1–15

Tobias, Phillip V. "Provisional Report on Nuffield-Witwatersrand University Research Expedition to Kalahari Bushmen, August-September 1958, Part One - Introduction." *South African Journal of Science*, 55, no. 1, (1959): 13–18

Tobias, Phillip V. *The Meaning of Race*. Johannesburg: South African Institute of Race Relations, 1961

Tobias, Phillip V. *The Meaning of Race*. 2nd edition, Johannesburg: South African Institute of Race Relations, 1972

Tobias, Phillip. "Remarks at the Centenary of the Birth of the Late Dr. Robert Broom, Sterkfontein, 30 November 1966." *South African Journal of Science*, 63, no. 9, (1967): 354–356

Tobias, Phillip. "The Emergence of Man in Africa." *Scientiae: CSIR Monthly Journal*, 9, no. 7, (1968): 1–16,

Tobias, Phillip V. "A New Chapter in the History of the Sterkfontein Early Hominid Site." *Journal of South Africa Biology Soc*, 14, (1973): 30–44

Tobias, Phillip V. "Fifteen Years of Study on the Kalahari Bushmen or San: A Brief History of the Kalahari Research Committee." *South African Journal of Science*, 71, no. 3, (1975): 74–77

Tobias, Phillip V. "Introduction to the Bushmen or San" and "The San: An Evolutionary Perspective", in *The Bushmen: San Hunters and Herders of Southern Africa*, edited by Phillip Tobias. Cape Town: Human and Rousseau Publishers, 1978

Tobias, PV. "The Sterkfontein Caves and the Role of the Martinaglia Family." *Adler Museum Bulletin*, special issue, (1983): 46–52

Tobias, Phillip. *Dart, Taung and the Missing Link*. Johannesburg: Witwatersrand University Press, 1984

Tobias, Phillip V. "History of Physical Anthropology in Southern Africa." *Yearbook of Physical Anthropology*, 28, supplement S6, (1985): 1–52

Tobias, Phillip V. "Thoughts on Academic Boycotts." *Minerva*, 26, no. 4, (1998): 575–579

Tobias, Phillip V. "Prehistory and Political Discrimination", in *Minerva*, 26, no. 4, (1988): 588–597

Tobias, Phillip. "Race", in *The Social Science Encyclopedia*, edited by A Kuper and J Kuper, pp 678–682. London: Routledge and Kegan Paul, 1985. As reprinted in *Images of Humanity: The Selected Writing of Phillip V Tobias*. Rivonia, South Africa: Ashanti Publishing, 1991

Tobias, Phillip. "Memories, Images and Visions: A Valedictory Address", delivered in the Great Hall at the University of the Witwatersrand on 26 March 1991. Reprinted in *Images of Humanity: The Selected Writings of Phillip V. Tobias*. Rivonia: South Africa, Ashanti Publishing, 1991

Tobias, Phillip. "An Appraisal of the Case against Sir Arthur Keith." *Current Anthropology*, 33, no. 3, (1992): 243–293. Available at: www2.clarku. edu/~piltdown/map_prim_suspects/KEITH/Keith_prosecution/apprais_ Keith.html (accessed on 7 September 2015)

Tobias, Phillip. "Ad Hominidae: The Future of South Africa's Ancient Past." *Optima*, 40, no. 1, (1994): 32–37

Tobias, Phillip V. "The Bearing of Fossils and Mitochondrial DNA on the Evolution of Modern Humans, with a Critique of the 'Mitochondrial Eve' Hypothesis." *The South African Archaeological Bulletin*, 50, no. 162, (1995): 155–167

Tobias, Phillip. "Some Little Known Chapters in the Early History of the Makapansgat Fossil Hominid Site." *Palaeontologia Africana: Annals of the Bernard Price Institute for Palaeontological Research*, 33, (1997): 67–79

Tobias, Phillip. "The South African Early Fossil Hominids and John Talbot Robinson (1923–2001)." *Journal of Human Evolution*, 43, (2002): 563–576

Tobias, Phillip. "Saartje Baartman: Her Life, Her Remains, and the Negotiations for their Repatriation from France to South Africa." *South African Journal of Science*, 98, March/April, (2002): 107–110

Tobias, Phillip. *Into the Past*. Johannesburg: Picador Africa and Wits University Press, 2005

Tobias, Phillip V. "The Discovery of the Taung Skull of *Australopithecus Africanus* Dart and the Neglected Role of Professor R. B. Young." *Transactions of the Royal Society of South Africa*, 61, no. 2, (2006): 131–138

Tobias, Phillip. "The Story of Sterkfontein Since 1895", in *A Search for Origins: Science, History and South Africa's 'Cradle of Humankind'*, edited by Philip Bonner, Amanda Esterhuysen and Trefor Jenkins. Johannesburg: Wits University Press, 2007

Tobias, Phillip V, with Goran Strkalj and Jane Dugard. *Tobias in Conversation: Genes, Fossils and Anthropology*. Johannesburg: Wits University Press, 2008.

UNESCO, "The Race Question", 18 July 1950. Available at: www.unesdoc. unesco.org/images/0012/001282/128291eo.pdf (accessed on 19 May 2016)

UNESCO, "Four Statements on the Race Question", 1950, 1951, 1964 and 1967, published in 1969. Available at: www.unesdoc.unesco.org/images/0012/001229/122962eo.pdf (accessed on 19 May 2016)

Van Buskirk, James. "Plaster Casts Made of Bushmen: Patriarch Acts as Organiser, Laboratory in the Desert", *Rand Daily Mail*, 8 July 1936

Van Buskirk, James. "Professor Psycho-Analyses Bushmen in Remote Kalahari Desert: Dwarf Singers Have Voices Recorded", *Rand Daily Mail*, 14 July 1936

Van Buskirk, James. "All Eat Hard to Grow Fat and Beautiful: Exclusive Message from Rand Expedition", *Rand Daily Mail*, 15 July 1936

Van Buskirk, James. "Scientists Watch Ancient Blood Dance: Rhythm of the Living Grass", *Rand Daily Mail*, 16 July 1936

Van Buskirk, James. "But Show Great Perception in Brain Tests: Ancient Makes First Count of His Toes", *Rand Daily Mail*, 23 July 1936

Van Der Poel, Jean. *Selections from the Smuts Papers, Volume V, September 1919–November 1934*. Cambridge: Cambridge University Press, 1973

Van Onselen, Gareth. "Homo Naledi: Unearthing SA's Great Hour of Science, and Failure of Logic", *The Star*, 20 September 2015

Waldman, Linda. "Klaar Gesnap As Kleurling: The Attempted Making and Remaking of the Griqua People." *African Studies*, 65, no. 2, (2006): 175–200

Ward, Annabelle. "Research Sets Black 'Presence' Far Back in Pre-history", *The Star*, 13 June 1979

Washburn, Sherwood. "The Origin of Races: Weidenreich's Opinion." *American Anthropologist*, 66, no. 5, (1964): 1165–1167

Watson, DMS. "Robert Broom, 1866–1951." *Obituary Notices of Fellows of the Royal Society*, 8, no. 21, published by the Royal Society, November 1952, pp 36–70. Available at: www.jstor.org/stable/768799 (accessed on 20 February 2014)

Webster, Bayard. "Robert Ardrey Dies; Writer on Behaviour", *New York Times*, 16 January 1980

Weidenreich, Franz. "Facts and Speculations Concerning the Origin of Homo Sapiens." *American Anthropologist*, 49, no. 2, (1947): 187–203

Weidman, Nadine. "Popularizing the Ancestry of Man: Robert Ardrey and the Killer Instinct." *Isis*, 102, no. 2, (2011): 269–299. Available at: www.jstor.org/stable/10.1086/660130 (accessed on 18 November 2015)

Weidman, Nadine and John P Jackson. *Race, Racism, and Science: Social Impact and Interaction*. New Brunswick, New Jersey: Rutgers University Press, 2004

Weiner, JS. *The Piltdown Forgery*. London: Oxford University Press, 1955

Wells, Lawrence. "The Foot of the South African Native." *American Journal of Physical Anthropology*, 15, no. 2, (1931): 185–289

Wells, Lawrence H. "One Hundred Years: Robert Broom 30 November 1866", delivered at University of the Witwatersrand, Johannesburg on 19 October 1966, published in the *South African Journal of Science*, 63, no. 9, (1967): 357–366

West, Mary, "Responding to Whiteness in Contemporary South African Life and Literature: An Interview with Njabulo S Ndebele." *English in Africa*, 37, no. 1, (2010): 115–140

Wheelhouse, Frances. *Raymond Arthur Dart: A Pictorial Profile, Professor Dart's Discovery of 'The Missing Link'*. Sydney: Transpareon Press, 1983

Wheelhouse, Frances and Kathaleen Smithford. *Dart: Scientist and Man of Grit*. Sydney: Transpareon Press, 2001

Wheelhouse, Frances. "Dart and Alexander." *Direction: A Journal on the Alexander Technique*, 1, no. 3, (2004): 100–105

White, Tim. "A View on the Science: Physical Anthropology at the Millennium." *American Journal of Physical Anthropology*, 113, (2000): 287–292

White, Tim. "Robbing the Cradle." *South African Journal of Science*, 98, (2002): 515–517

Wild, Sarah. "New 'Best Candidate' for Human Ancestry", *Business Day*, 9 September 2011

Wild, Sarah. "Skinny Scientists at Palaeo Rock Face", *Mail & Guardian*, 8–14 November 2013

Wilford, John Noble. "Raymond A. Dart is Dead at 95: Leader in Study of Human Origins", *New York Times*, 23 November 1988. Available at: www.nytimes.com/1988/11/23/obituaries/raymond-a-dart-is-dead-at-95-leader-in-study-of-human-origins.html (accessed on 8 August 2014)

Williams, EW. "Facial Features of Some Kung Bushmen." *South African Journal of Science*, 51, (1954): 11–17

Williams, Paige, "Digging for Glory", *The New Yorker*, 27 June 2016

Wits University, Press Statement, "Wits Scientists Reveal New Species of Hominid", 8 April 2010

Wits University, Press Statement, "New Species of Human Relative Discovered in South African Cave", 10 September 2015

Wolpoff, Milford H, WuXin Zhi and Alan G Thorne. "Modern *Homo Sapiens* Origins: A General Theory of Hominid Evolution Involving the Fossil Evidence From East Asia", in *The Origins of Modern Humans: A World Survey of the Fossil Evidence*. New York: Alan R Liss, Inc, 1984

Wolpoff, Milford H, Alan Thorne and Roger Lawn. "The Case Against Eve."
    New Scientist, 130, no. 1774, (1991): 37–41

Wolpoff, NH, AG Thorne, FH Smith, DW Frayer and GG Pope. "Multiregional
    Evolution: A World-wide Source for Modern Human Populations", in
    Origins of Anatomically Modern Humans, edited by MH Nitecki and DV
    Nitecki. New York: Plenum Press, 1994

Wolpoff, Milford H, John Hawks and Rachel Caspari. "Multiregional, Not
    Multiple Origins." *American Journal of Physical Anthropology*, 112,
    (2000): 129–136

Wolpoff, Milford H and Alan G Thorne. "The Multiregional. Evolution of
    Humans", *Scientific American*, 1 May 2002, updated from the April 1992
    issue

Wolpoff, Milford H and Rachel Caspari. "Paleoanthropology and Race", in *A
    Companion to Paleoanthropology*, first edition, edited by David R Begun.
    Blackwell Publishing Ltd, 2013

Wood, Bernard. *The Evolution of Early Man*. England: Peter Lowe Publishers,
    Eurobook Limited, 1976

Wood, Bernard. "An Interview with Phillip Tobias", reprinted for private
    circulation from *Current Anthropology*, 30, no. 2, (1989): 215–224

Wood, Bernard. "Chalk and Cheese." *Journal of Human Evolution*, 42,
    (2002): 499–504

Yong, Ed. "Six Tiny Cavers, Fifteen Odd Skeletons, and One Amazing New
    Species of Ancient Human." *The Atlantic*, 10 September 2015

Archival Sources, Policy Documents, Theses and Letters

Alan Morris papers, Human Anatomy Archive at the University of Cape Town:
    This archive is a work in progress by Alan Morris. The website reads "After
    70 years in a remote storeroom in the UCT Medical School, the records
    of the Anatomy division are now available online. The collection focuses
    on the correspondence and research of three trailblazing South African
    physical anthropologists." These three men, Matthew Drennan, Lawrence
    Wells and Alan Morris, have led the department of anatomy, now the
    department of human biology at UCT for much of the past century. While
    many of the papers that Alan Morris shared with me are not yet online,
    they will be in the near future, and this will be a very important resource
    for researchers. See the collection in progress at: www.digitalcollections.
    lib.uct.ac.za/humanitec/anatomy

Broom papers, Ditsong Museum, Pretoria

Dart papers, Wits University Archives, Johannesburg

Ditsong Museum Archives (former Transvaal Museum), Pretoria

Gould papers, courtesy of the Department of Special Collections, Stanford University Libraries

ǂKhomani San Archive at the University of Cape Town. Available at: www. digitalcollections.lib.uct.ac.za/khomani

Tobias Papers, Wits University Archives, Johannesburg: The Tobias papers include a small number of boxes of correspondence that Tobias donated to Wits University before he died. The vast bulk of his personal papers and correspondence that he donated to the university when he died in 2012 has not yet been organised and is not yet available to researchers or the public.

Ardrey, Berdine. Letter to Raymond and Marjorie Dart, 15 October 1975, Dart papers, Wits University

Ardrey, Robert. Letters to Raymond Dart, correspondence in the Dart Archive, Wits University, 21 December 1959, 29 November 1960, 19 August 1963, 5 July 1965

A Plane But Sane Woman, Letter to Raymond Dart, dated 27 February 1925, Raymond Dart papers in Box marked Taung skull, Wits Archives

Bain, Donald. Correspondence to Raymond Dart, 11 February 1936, Dart papers, Wits Archives

Bain, Donald. Letter to Raymond Dart, 26 May 1936, Dart papers, Wits Archives

Bain, Donald. Telegram to Raymond Dart, 9 September 1939, Dart papers, box marked Department of Anatomy Correspondence, Wits Archives

Barbour, Fiona. Letter to Alan Morris, 30 October 1996, Alan Morris papers. Soon to be incorporated into the Human Anatomy Archive at the University of Cape Town. See the collection in progress at: www.digitalcollections.lib. uct.ac.za/humanitec/anatomy

Broom, Norman. Letter to George Findlay, 22 September 1966, Robert Broom papers, Ditsong Museum Library

Dart, Raymond. Letter to Donald Bain, 28 May 1936, Dart papers, Wits Archives

Dart, Raymond. Letter to Donald Bain, 5 June 1936, Dart papers, Wits Archives

Dart, Raymond. "The Structure of the Bushman", Kalahari Expedition File 1936/7, Dart papers, Wits Archives

Dart, Raymond. "The Problem of Race: The Physical Anthropological Point of View", remarks prepared for Symposium hosted by Society of Jews and Christians, 19 October 1937, Dart papers, Wits Archives

Dart, Raymond. Telegram to Dr Nel, 11 September 1939, Dart papers, in box marked Department of Anatomy Correspondence, Wits Archives

Dart, Raymond. Letter to Dr Nel, 14 September 1939, Dart papers, in box marked Department of Anatomy Correspondence, Wits Archives

Dart, Raymond. Letter to Dr George Price of University College London, 28 May 1970, correspondence in Dart papers, Wits Archives

Department of Science and Technology, "African Origins: A Strategy of the Department of Science and Technology for the Palaeosciences", July 2006. Available at: www.gov.za/sites/www.gov.za/files/DST_Strategic%20 Plan%20-%20African%20Origins%20July%202007_07082006_0.pdf (accessed on 20 May 2016)

Department of Science and Technology, "The South African Strategy for the Palaeosciences: Incorporating Palaeontology, Palaeo-anthropology and Archaeology", 2012. Available at: www.gov.za/sites/www.gov.za/files/ paleostrategydstFinal.pdf (accessed on 20 May 2016)

De Villiers, Phillippa Yaa. *Original Skin*, direction and script development by Robert Colman, Home Truths Productions, Copyright Phillippa Yaa de Villiers 2009

De Villiers, Phillippa Yaa. Personal Papers shared with the author:

Adoption consent papers signed by S Alcock, 21 February 1966

Albrecht, I, Department of Social Welfare, Case Summary, 11 August 1966

Arendt, Marie M, Adoption Secretary, Child Welfare Society of Johannesburg, Case Report dated 25 February-2 May 1966

De Villiers, Hertha. Letter on University of the Witwatersrand Medical School letterhead stating that she had examined Tandy (sic) Alcock, dated 5 August 1966

De Villiers, Hertha. Letters to Phillip Tobias, dated 26 November 1955 and 24 May 1956, Tobias papers, Wits Archives

De Villiers, Hertha. Letter to Phillip Tobias, 26 November 1971, Tobias papers, Wits Archives

De Villiers, Hertha. Personal memo to Phillip Tobias, 8 December 1965. Tobias papers, Wits Archives

De Villiers, Hertha. Letter to Phillip Tobias, 14 February 1980, Tobias papers, Wits Archives

De Villiers, Hertha. "The Morphology and Incidence of the Tablier in Bushman, Griqua and Negro Females", presented in Tokyo and Kyoto, Japan, 3-10 September 1968, Dart papers, Wits Archives

De Villiers, Hertha. Letter to Phillip Tobias dated 14 February 1980, Tobias papers, Wits Archives

Fock, Dr CJ. Report on 1961 Campbell excavation, submitted to Phillip Tobias, dated 19 June 1961, handwritten, copy in Alan Morris papers, soon to be incorporated into the Human Anatomy Archive at the University of Cape Town. See the collection in progress at: www.digitalcollections.lib. uct.ac.za/humanitec/anatomy

Gauteng Provincial Government: Department of Agriculture, Conservation and Environment, "Application for Inclusion on the World Heritage List: The Fossil Hominid Sites of Sterkfontein, Swartkrans, Kromdraai and Environs, South Africa", 16 June 1998. Available at: www.whc.unesco. org/uploads/nominations/915bis.pdf (accessed on 19 March 2014)

Gould, Stephen Jay. Travel Notebook, "Southern Hemisphere Book", South Africa, July 1984. Box 468, Folder 2, Gould papers, courtesy of the Department of Special Collections, Stanford University Libraries

Hechinger, Fred M. "A Liberal Fights Back", letter to the *New York Times*, 13 February 1972

Human Tissue Act 65 of 1983. Date of Commencement 12 July 1985, as amended in 1984 and 1989. Available at: www.kznhealth.gov.za/ humantissueact.pdf (accessed on 20 May 2016)

Laing, James to Professor Matthew Drennan, 23 March 1938, original and transcript from Alan Morris papers. Author given access 21 April 2015. Soon to be incorporated into the Human Anatomy Archive at the University of Cape Town. See the collection in progress at: www.digitalcollections.lib. uct.ac.za/humanitec/anatomyLloyd, Julia.

Mbeki, Thabo. "Address at the Official Opening of the Maropeng Visitor Centre", 7 December 2005, Thabo Mbeki Foundation

Morris, Alan. Letter to Raymond Dart, 6 September 1984, Dart papers, Wits Archives

Morris, Alan. Letters to Phillip Tobias, 20 October 1995 and 19 November 1995, Alan Morris papers, soon to be incorporated into the Human Anatomy Archive at the University of Cape Town. See the collection in progress at: www.digitalcollections.lib.uct.ac.za/humanitec/anatomy

Morris, Alan. Letter to Mr M Kibunjia, National Museums of Kenya, 1 February 1996, Alan Morris papers, soon to be incorporated into the Human Anatomy Archive at the University of Cape Town. See the collection in progress at: www.digitalcollections.lib.uct.ac.za/humanitec/ anatomyMorris, Alan. Letter to Fiona Barbour, 16 October 1996, Alan Morris papers, soon to be incorporated into the Human Anatomy Archive at the University of Cape Town. See the collection in progress at: www. digitalcollections.lib.uct.ac.za/humanitec/anatomy

Morris, Alan G. "From Snakes to Skeletons: a Personal Voyage of Scientific Discovery", unpublished manuscript, draft autobiography, shared with the author in December 2014

Nel telegram to Raymond Dart, 12 September 1939, Dart papers, in box marked Department of Anatomy Correspondence, Wits Archives

Nel telegram to Raymond Dart, 16 September 1939, Dart papers, in box marked Department of Anatomy Correspondence, Wits Archives

Pillay, Maganathan. "A Critical Evaluation of Representations of Hominin Evolution in the Museums of the Cradle of Humankind World Heritage Site, South Africa", Research Report submitted to the University of the Witwatersrand Faculty of Humanities, Master of Arts in Human Geography, February 2010. Available at: wiredspace.wits.ac.za/handle/10539/8403 (accessed on 20 May 2016)

Robinson, John. Letter to Joe Weiner dated 29 October 1962, in box 305, JT Robinson Correspondence 1960–62, Ditsong Museum Library

Schenck, Marcia. "Land, Water, Truth and Love Visions of Identity and Land Access: From Bain's Bushmen to |Khomani San", an undergraduate thesis submitted in partial fulfilment for the degree of Bachelor of Arts with Honors, Mount Holyoke College, December 2008. Available at: www.ida. mtholyoke.edu/xmlui/bitstream/handle/10166/735/366.pdf? (accessed on 2 August 2014)

Smuts, Jan. Letter to Professor Matthew Drennan, 25 May 1938, original and transcript from Alan Morris archive. Author given access, 21 April 2015. Soon to be incorporated into the Human Anatomy Archive at the University of Cape Town. See the collection in progress at: www.digitalcollections.lib. uct.ac.za/humanitec/anatomy

Soodyall, Himla. "The End of Race: What Science Proves" (unpublished chapter)

Thackeray, Francis. *Fleeting Moments: An Autobiography*, unpublished document shared with the author in early 2014

Tobias, Phillip. Letter to Hertha De Villiers, 6 December 1965. Tobias papers, Wits Archives

Tobias, Phillip. Letters to Hertha De Villiers dated 8 November 1971, 22 November 1971, and 1 December 1971, Tobias papers, Wits Archives

Tobias, Phillip. Letters to Hertha De Villiers dated 14 July 1980 and 1 August 1980, Tobias papers, Wits Archives

Tobias, Phillip. Letters to Alan Morris, 16 September 1995 and 24 October 1995, Alan Morris papers, soon to be incorporated into the Human Anatomy Archive at the University of Cape Town. See the collection in progress at: www.digitalcollections.lib.uct.ac.za/humanitec/anatomy

Tobias, Phillip. Letter to Alan Morris et al regarding Sterkfontein permit, 4 August 1999, Alan Morris papers, soon to be incorporated into the Human Anatomy Archive at the University of Cape Town. See the collection in progress at: www.digitalcollections.lib.uct.ac.za/humanitec/anatomy

Tobias, Phillip. "Guide to the Raymond Dart Gallery of African Faces", Hunterian Museum, Department of Anatomy, Medical School, University of Witwatersrand, Johannesburg, third edition, 1979

Tobias, Phillip. "Anatomia Witwatersrandensis: A Brief Historical Synopsis of the Wits Anatomy Department", University of the Witwatersrand, Johannesburg. A note on the cover page of this unpublished paper states: "This brief sketch was compiled during July and August 1995 at the request of Noel Cameron, Head of the Department of Anatomy and Human Biology, University of the Witwatersrand Medical School to mark the celebration of the 75th anniversary of the foundation of the Wits Department of Anatomy."

Transvaal Museum, Minutes of the annual meeting of the Board of Trustees, 28 May 1963 in Box 491: Directors Monthly Reports, 1960–1964, Ditsong Museum Library, Pretoria

University of the Witwatersrand, "Notes on a Proposed Bushman Reserve", undated (likely 1936/37), Dart papers, Wits Archives

Wanless, Ann. "The Silence of Colonial Melancholy: The Fourie Collection of Khoisan Ethnologica", a thesis submitted to the University of the Witwatersrand, 2007. I found the letter from Donald Bain to Dr Louis Fourie from 1925 in this thesis, p 162

Worby, Eric. "And a Little (White) Child Shall Lead Them: Suturing a Pre-Racial Human Family to a Post-Racial National Fantasy through the 'Discovery' of *Australopithecus sediba*", paper presented to the workshop on The Claims of Descent: Science, Representation, Race and Redress in 21st Century South Africa, 14–16 June 2014, Halle, Germany (Thank you to Worby in this paper for asking questions about whether Berger and *Sediba* are post-racial.)

Williams, Eric. Letter to Dr Nel, 13 September 1939, Dart papers, in box marked Department of Anatomy Correspondence, Wits Archives

Williams, Eric. Telegram to Raymond Dart, 18 September 1939, Dart papers, in box marked Department of Anatomy Correspondence, Wits Archives

Young, Robert. Unpublished letter, 7 February 1925, to Raymond Dart, Dart papers, in box marked Taung Skull, Wits Archives

Interviews and Personal Correspondence with the Author

Ackerman, Rebecca. Cape Town, 3 December 2014

Adam, Rob, Former Director General, Department of Arts, Culture Science and Technology, Skype call, 17 March 2016

Berger, Lee, Evolutionary Studies Institute, Wits University, Johannesburg, 26 November 2014

Berger, Lee, Killarney, Johannesburg, 21 October 2015

Blumenschine, Robert. Remarks at All From One campaign launch, Rosebank, Johannesburg, 10 November 2015

Billings, Brendon, School of Anatomical Sciences, Wits Medical School, 15 May 2014, 15 July 2014 and 1 August 2014

Brain, Bob and Laura, Irene, Pretoria 6 June 2014

Brain, Bob, Ditsong Museum, Pretoria, 30 September 2014

Clarke, Ron, Sterkfontein, 13 March 2014 and 28 May 2014

Clarke, Ron, Sterkfontein, 11 August 2014

Clarke, Ron, Sterkfontein, 1 June 2015

Crichton, Wayne, Logistics Manager, Rising Star Cave site, 4 June 2015

De Villiers, Hertha, and Phillippa Yaa De Villiers, Parkhurst, Johannesburg, 6 September 2014

De Villiers, Hertha, Parkhurst, Johannesburg, 25 September 2014

De Villiers, Hertha, Parkhurst, Johannesburg, 2 October 2014

De Villiers, Hertha, Parkhurst, Johannesburg, 24 February 2015, and 9 April 2015

De Villiers, Phillippa Yaa, Troyeville, Johannesburg, 4 March 2015

De Villiers, Phillippa Yaa, email to the author, 28 July 2014

De Villiers, Phillippa Yaa, Braamfontein, Johannesburg, 20 November 2015

Dugard, Jane, Saxonwold, Johannesburg, 26 May 2016

Elliott, Marina, Wits University, Johannesburg, 24 February 2015

Elliott, Marina, personal correspondence, 24 February 2015

Elliott, Marina, Rising Star Cave, 4 June 2015

Faugust, Peter, telephone interview, 23 March 2015

Faugust, Peter, Norwood, Johannesburg, 27 May 2015

Gabobidiwe High School learners (including Kopano), Buxton, 5 May 2015

Gordon, Robert, personal correspondence, 7 and 31 August 2015

Humphreys, Anthony J. B., personal correspondence , 9, 10 and 11 June 2015

Jenkins, Trefor, Parkview, Johannesburg, 28 February 2014

Jenkins, Trefor, Parkview, Johannesburg, 8 August 2014

Jenkins, Trefor, telephone conversation, 31 October 2014

Jenkins, Trefor, Parkview, Johannesburg, 5 November 2014

Kibii, Job, Braamfontein, 27 September 2016

Kramer, Beverley, School of Health Sciences, 10 December 2014

Kgotleng, Winnie Dipuo Mokokwe, telephone interview, 26 September 2014 and 13 May 2016

Legodi, Phillip, School of Anatomical Sciences, 12 May 2015

Lobelo, Gavin Kaone, Mayor of Taung, at Taung, 4 May 2015

Mankuroane, Kgosikgolo, Paramount Chief, at Taung, 4 May 2015

Lekwene, Chief of Buxton, 4 May 2015

Mallen, Lara, Origins Centre, 10 March 2015

Malepe, Kagiso, traditional healer, at Taung, 4 May 2015

Matlala, Frank, telephone conversation, 2 September 2014

Metcalfe, Anthea, former Project Manager for Announcement of *Australopithecus sediba*, Wits University, Skype interview, 11 February 2014, personal correspondence, including Malapa Coordinating Committee minutes, letters and reports, September 2016

Mokokwe (now Kgotleng), Winnie Dipuo, telephone interview, 26 September 2014 and May 2016

Morris, Alan, University of Cape Town, 4 December 2014

Morris, Alan, Hout Bay, Cape Town, 21 April 2015

Morris, Alan, personal communication, unpublished autobiography, *From Snakes to Skeletons: a Personal Voyage of Scientific Discovery*, 2010

Motsumi, Stephen Phologo, Brits, 8 September 2014

Raath, Michael, former Curator of Wits Collections at Geology, BPI and Anatomical Sciences, Kommetjie, 22 April 2015

Schroeder, Lauren, Cape Town, 3 December 2014

Schroeder, Lauren, Skype, 1 October 2015

Sigmon, Becky, Department of Anthropology, University of Toronto at Mississauga, email correspondence, November 2014-February 2015

Soodyall, Himla, Department of Human Genetics at Wits, National Health Laboratory Service, Hillbrow, Johannesburg, 28 May 2015 and 25 April 2016

Spiller, Guy, Director of *Tobias's Bodies*, Mountainview, Johannesburg, 16 September 2015

Tawane, Mirriam, Evolutionary Studies Institute, Wits University, 10 October 2014

Wakashe, Themba, former Deputy Director General for Arts and Culture, 20 August 2015

Wood, Bernard, Skype, 23 November 2015

Zitha, Zondi Sithole, Mamelodi, Pretoria, 21 November 2014

Zipfel, Bernhard, Department of Anatomy, Wits Medical School, Parktown,

Johannesburg, 26 March 2014

Zipfel, Bernhard, Evolutionary Studies Institute, conversation about Lawrence Wells' paper "The Foot of the South African Native", 11 April 2016

Radio, Video, Film and Online Sources

Berger, Lee, "Google Earth and Human Evolution", Talks at Google, 22 November 2012, www.youtube.com/watch?v=IHpEmD-95CQ (accessed on 18 May 2016)

Berger, Lee, Interview with John Hawks about Malapa, *Australopithecus Sediba* and Open Science", www.youtube.com/watch?v=bhg5zi5ob74 John Hawks website, 24 June 2014

Berger, Lee, Interview with Redi Tlhabi on Radio 702. 14 September 2015. www.702.co.za/articles/5152/professor-lee-burger-responds-to-questions-surrounding-homonaledi-discovery (accessed on 20 September 2015)

"Biko, Stephen Bantu", South African History Online. Available at: www.sahistory.org.za/people/stephen-bantu-biko (accessed on 7 June 2016)

Bronowski, Jacob. *The Ascent of Man, Part One, Lower than Angels*, produced by the BBC, 1973. Available at: www.youtube.com/watch?v=CH7SJf8BnBI (accessed on 6 June 2016)

Clarke, Arthur C. Quoted from a documentary film called *Making of a Myth* about the making of *2001: Space Odyssey*, directed by Paul Joyce, Atlantic Celtic Films, Lucida Productions, 2001, www.youtube.com/watch?v=F7HGwVql_FM (accessed on 5 December 2015)

Department of Science and Technology, "African Origins: A Strategy of the Department of Science and Technology for the Palaeosciences", July 2006. Available at: www.gov.za/sites/www.gov.za/files/DST_Strategic%20Plan%20-%20African%20Origins%20July%202007_07082006_0.pdf (accessed on 22 March 2016)

European Synchrotron Radiation Facility (ESRF), Press Release, "First Studies of Fossil of new Human Ancestor Take Place at the ESRF", 12 April 2010. Available at: www.esrf.eu/news/general-old/general-2010/first-studies-of-fossil-of-new-human-ancestor-take-place-at-the-esrf (accessed on 22 March 2016)

Feuerriegel, Elen, National Geographic Society, Rising Star Expedition blog post "The View From a Caver/Scientist", 23 November 2013. Available at: www.voices.nationalgeographic.com/2013/11/23/the-view-from-a-caverscientist/ (accessed on 22 February 2015)

Foley, Gerald. "Towards a Neurophysiology of the Alexander Technique", unpublished paper dated June 2012. Available at: www.geraldfoley.co.uk/

Towards%20A%20Neurophysiology4.pdf (accessed on 8 August 2014)

Garcia, Terry, National Geographic, interviewed by John Robbie on Radio 702, 10 September 2015. Available at: www.702.co.za/articles/5087/maropeng-fossil-announcement (accessed on 20 September 2015)

Google Earth, "Google Earth Helps Discover Rare Hominid Ancestor in South Africa", 8 April 2010, www.googleblog.blogspot.co.za/2010/04/google-earth-helps-discover-rare.html (accessed on 22 March 2016)

Griqua Royal House website, "The Adam Kok Family Genealogy", www.griquaroyalhouse.com

Hauge, Soren, "Fossils Come Alive: Clairvoyant Research in Paleontological Discoveries". Available at: www.translate.google.co.za/translate?hl=en&sl=da&u=http://sorenhauge.com/artikler/fossiler-bliver-levende&prev=search (accessed on 26 February 2016)

Howley, Andrew, National Geographic Society, Rising Star Expedition blog post, "Expedition Underway to Extract Latest Fossil Find From Cradle of Humankind Cave", 6 November 2013. Available at: www.voices.nationalgeographic.com/2013/11/06/expedition-underway-to-extract-latest-fossil-find-from-cradle-of-humankind-cave/ (accessed on 22 February 2015)

Howley, Andrew, National Geographic Society, Rising Star Expedition blog post, "There Are Hominid Fossils Waiting to be Discovered", 7 November 2013. Available at: www.voices.nationalgeographic.com/2013/11/07/hominid-fossils-waiting-to-be-discovered/ (accessed on 22 February 2015)

Howley, Andrew, National Geographic Society, Rising Star Expedition blog post, "First Peek into Cave System Holding New Hominid Fossils", 8 November 2013. Available at: www.voices.nationalgeographic.com/2013/11/08/first-peek-into-cave-system-holding-new-hominid-fossils/ (accessed on 22 February 2015)

Howley, Andrew, National Geographic Society, Rising Star Expedition blog post, "Remarkably Well Preserved Hominid Skeleton Emerges", 10 November 2013. Available at: www.voices.nationalgeographic.com/2013/11/10/remarkably-well-preserved-hominid-skeleton-emerges/ (accessed on 22 February 2015)

Howley, Andrew, National Geographic Society, Rising Star Expedition blog post, "Multiple Ancient Hominids Found on Day 2 of Rising Star Expedition", 12 November 2013. Available at: www.voices.nationalgeographic.com/2013/11/12/multiple-ancient-hominids-found-on-day-2-of-rising-star-expedition/ (accessed on 22 February 2015)

Howley, Andrew, National Geographic Society, Rising Star Expedition blog post, "A 'Day Off' Still Manages to Yield Another Hominid Skeleton", 13 November 2013. Available at: www.voices.nationalgeographic. com/2013/11/13/a-day-off-still-manages-to-yield-another-hominid-skeleton/ (accessed on 22 February 2015)

Howley, Andrew, National Geographic Society, Rising Star Expedition blog post, "Know Your Hominid Skulls", 24 November 2013. Available at: www.voices.nationalgeographic.com/2013/11/24/know-your-hominid-skulls/ (accessed on 22 February 2015)

Joyce, Paul. *The Making of a Myth*, 2001. Available at: www.youtube.com/ watch?v=F7HGwVqI_FM (accessed on 5 December 2015)

ǂKhomani San Archive at the University of Cape Town at :www. digitalcollections.lib.uct.ac.za/khomani (accessed on 4 June 2016)

Kubrick, Stanley. "Now Kubrick Fights Back", letter to the *New York Times*, 27 February 1972

"Job Kibii is Close to the Bones", Maropeng website, 8 September 2011. Availabe at: www.maropeng.co.za/news/entry/job_kibii_is_close_to_the_ bones (accessed on 30 May 2014)

Maropeng website, "Reflections on the sediba fossil discovery", 9 April 2010. Available at: www.maropeng.co.za/mobile/news/entry/reflections_on_the_ sediba_fossil_discovery Quote from Professor Emeritus Phillip Tobias. (accessed on 22 March 2016)

Maropeng website, "Women Set to Change the Face of Palaeoanthropology", 29 August 2012, www.maropeng.co.za/news/entry/women_set_to_ change_the_face_of_palaeoanthropology (accessed 16 July 2014)

Maseko, Zola, director, "The Life and Times of Sara Baartman", 1998

Maseko, Zola, director, and Gail Smith, writer/narrator, "The Return of Sarah Bartman", Black Roots Productions, 2002

Mbeki, Thabo. "I am an African", speech at the adoption of the constitution of the Republic of South Africa, Cape Town, 8 May 1996. Available at: www.anc.org.za/show.php?id=4322 (accessed on 7 March 2016)

Mbeki, Thabo. "Speech at the Funeral of Sarah Bartmann, 9 August 2002",South African History Online. Available at: www.sahistory.org. za/archive/speech-funeral-sarah-bartmann-9-august-2002 (accessed on 27 September 2016)

"Medicine in China", on the "100 Years: The Rockefeller Foundation" website, www.rockefeller100.org/exhibits/show/education/china-medical-board (accessed on 9 July 2016)

"Media Briefing: New Fossil Find Unveiled in Maropeng, Johannesburg", streamed live on 10 September 2015. Available at: www.youtube.com/

watch?v=QiiOJ4Y9ZLo (accessed on 20 May 2016)

Motshekga, Mathole, interviewed by Stephen Grootes, Radio 702, 15 September 2015, www.702.co.za/index.php/articles/5182/homo-naledi-racial-debate (accessed on 20 September 2015)

Motshekga, Mathole, interviewed by Tim Modise, *Transformation*, 24 September 2015, www.biznews.com/transformation/2015/09/24/mathole-motshekga-homo-naledi-human-ancestry-link-offensive-should-be-rejected/ (accessed on 25 September 2015)

Population Registration Act of 1950. Available at: www.disa.ukzn.ac.za/webpages/DC/leg19500707.028.020.030/leg19500707.028.020.030.pdf (accessed on 23 July 2016)

Rooi, /Una, interviewed on film in DVD *Tracks Across Sand*, in section "Before the Land Claim: Evictions: 1930s", directed by Hugh Brody, 2012

Schultz, Colin, "Shackleton Probably Never Took Out an Ad Seeking Men for a Hazardous Journey", Smithsonian.com 10 September 2013, www.smithsonianmag.com/smart-news/shackleton-probably-never-took-out-an-ad-seeking-men-for-a-hazardous-journey-5552379/?no-ist (accessed on 6 June 2015)

"Stephen Jay Gould: This View of Life", a WGBH TV production, written, produced and directed by Linda Harrar, 1984

"The Robert J. Terry Anatomical Skeletal Collection", found on the website of the National Museum of Natural History at www.anthropology.si.edu/cm/terry.htm (accessed on 6 October 2015)

"The Hamann-Todd Osteological Collection", on the website of the Cleveland Museum of Natural History at www.cmnh.org/c-r/phys-anthro/collections (accessed on 6 October 2015)

Tobias, Phillip, Third Annual Human Evolution Symposium, Stony Brook University, 3 October 2006, YouTube video, www.youtube.com/watch?v=5u6LhBJCiQI (accessed on 18 May 2016)

Tobias's Bodies, six-part film series, directed by Guy Spiller, produced by Harriet Gavshon, David Jammy at Curious Pictures (now Quizzical Pictures) and SABC Education, 2002

Tracks Across Sand: The Khomani San of the Southern Kalahari, The story of the land claim, directed by Hugh Brody, 2012. Portions of this video and other documents are held at the /Khomani San Archive at the University of Cape Town at: www.digitalcollections.lib.uct.ac.za/khomani

Warren, Kerryn, Bones and Evolution Blog, "Racism and Homo naledi", 22 September 2015 at www.bonevolution.wordpress.com/2015/09/22/racism-and-homo-naledi/ (accessed on 7 October 2015)

www.records.ancestry.com (accessed on 26 February 2014)

www.clarku.edu/~PILTDOWN/map_prim_suspects/KEITH/Keith_prosecution/apprais_Keith.html (accessed on 7 September 2015)

Wits University, "Wits Scientists Reveal New Species of Hominid", 8 April 2010 www.kim.wits.ac.za/index.php?module=news&action=viewstory&id=gen11Srv0Nme53_81569_1270732348 (accessed on 29 June 2015)

"World Cup 2010 Ceremony Archbishop Desmond Tutu Show", www.youtube.com/watch?v=ui_paCD4rKg

Index

CPSIA information can be obtained
at www.ICGtesting.com
Printed in the USA
LVHW080833070723
751742LV00004B/402

9 781431 424252